Over-The-Horizon Radar

The Artech House Radar Library

Radar System Analysis by David K. Barton
Introduction to Electronic Warfare by D. Curtis Schleher
High Resolution Radar by Donald R. Wehner
Electronic Intelligence: The Analysis of Radar Signals by Richard G. Wiley
Electronic Intelligence: The Interception of Radar Signals by Richard Wiley
Pulse Train Analysis Using Personal Computers by Richard G. Wiley and Michael B. Szymanski
RGCALC: Radar Range Detection Software and User's Manual by John E. Fielding and Gary D. Reynolds
Over-The-Horizon Radar by A.A. Kolosov et al.
Principles of Secure Communications Systems by Don J. Torrieri
Principles and Applications of Millimeter-Wave Radar, Charles E. Brown and Nicholas C. Currie, eds.
Phased Array Radars by Eli Brookner
Preparing and Delivering Effective Technical Briefings by David L. Adamy
Writing and Managing Winning Technical Proposals by Timothy Whalen
Multiple-Target Tracking with Radar Applications by Samuel S. Blackman
Solid-State Radar Transmitters by Edward D. Ostroff et al.
Logarithmic Amplification by Richard Smith Hughes
Radar Propagation at Low Altitudes by M.L. Meeks
Radar Cross Section by Eugene F. Knott, John F. Shaeffer, and Michael T. Tuley
Radar Anti-Jamming Techniques by M.V. Maksimov et al.
Introduction to Synthetic Array and Imaging Radars by S.A. Hovanessian
Radar System Design and Analysis by S.A. Hovanessian
Aspects of Radar Signal Processing by Bernard Lewis, Frank Kretschmer, and Wesley Shelton
Radar Calculations Using Personal Computers by William A. Skillman
Monopulse Principles and Techniques by Samuel M. Sherman
Monopulse Radar by A.I. Leonov and K.I. Fomichev
Receiving Systems Design by Stephen J. Erst
Advanced Mathematics for Practicing Engineers by Kurt Arbenz and Alfred Wohlhauser
High Resolution Radar Imaging by Dean L. Mensa
Radar Detection by J.V. DiFranco and W.L. Rubin
Synthetic Aperture Radar, John J. Kovaly, ed.
Infrared-to-Millimeter Wavelength Detectors, Frank R. Arams, ed.
Significant Phased Array Papers, R.C. Hansen, ed.
Handbook of Radar Measurement by David K. Barton and Harold R. Ward
Statistical Theory of Extended Radar Targets by R.V. Ostrovityanov and F.A. Basalov
Radar Technology, Eli Brookner, ed.
MTI Radar, D. Curtis Schleher, ed.
The Scattering of Electromagnetic Waves from Rough Surfaces by Peter Beckmann and Andre Spizzichino
Radar Range-Performance Analysis by Lamont V. Blake
A Handbook for the Mechanical Tolerancing of Waveguide Components by W.B.W. Alison

Over-The-Horizon Radar

A.A. Kolosov, et al.

Translated by
William F. Barton

Artech House
Boston and London

Library of Congress Cataloging-in-Publication Data

Osnovy zagorizontnoi radiolokatsii. English.
Over-the-horizon radar.

Translation of: Osnovy zagorizontnoi radiolokatsii.
Bibliography: p.
1. Over-the-horizon radar. I. Kolosov, A. A.
(Andrei Aleksandrovich) II. Title.
TK6592.09408613 1987 621.3848 87-19318
ISBN 0-89006-233-1

ARTECH HOUSE, INC.
685 Canton Street
Norwood, MA 02062

International Standard Book Number: 0-89006-233-1
Library of Congress Catalog Card Number: 87-19318

Translation from the Russian of *Osnovy zagorizontnoi radiolokatsii,* copyright © 1984 by Radio i Svyaz, Moscow.

10 9 8 7 6 5 4 3 2 1

Contents

Preface ix
Chapter 1 Characteristics of Over-the-Horizon Radar 1
1.1 Introduction 1
1.2 Features of Over-the Horizon Radar 4
1.3 Characteristics of Over-the-Horizon Radar Systems 6
Chapter 2 The Radar Equation 9
2.1 Introduction 9
2.2 The Standard Radar Equation 9
2.3 The General Radar Equation 10
2.4 Another Form of the Radar Equation 16
2.5 The Signal-to-Interference Ratio 17
2.6 The Performance of a Radar System 18
2.7 Losses due to System Imperfections 22
Chapter 3 Methods for Calculating the Path Loss 27
3.1 Introduction 27
3.2 Theoretical Methods for Studying the Long-Range Propagation of High-Frequency Radio Waves 28
3.3 Calculating the Spatial Field Energy Characteristics with the Ray Optics Approximation 50
3.4 The Results of Experimental Investigations into the Spectral Characteristics of Round-the-World Signals 57
3.5 Estimating Attenuation Using a Semiphenomenological Model 68
Chapter 4 The Effective Target Cross Section 87
4.1 Introduction 87
4.2 The Characteristics of OTH Radio Wave Scattering from Objects 88
4.3 The Radar Cross Section of Magnetically Oriented Disturbances in the Ionospheric Electron Density at High Frequency 89
4.4 The Radar Cross Section of Meteor Trails 91
4.5 The Radar Cross Section of Ascending Rockets 93
4.6 The Radar Cross Section of Aircraft 98

Chapter 5 High-Frequency Radar Interference 103
5.1 Introduction 103
5.2 Internal Receiver Noises 104
5.3 Atmospheric Interference 104
5.4 Cosmic Noise 109
5.5 Industrial Interference 110
5.6 Interference from Radio Stations 112
5.7 Interference from Spurious and Out-of-Band Radio Transmissions 120
5.8 The Net (Combined) Active Interference 122
5.9 Clutter 125
5.10 Noise Resulting from Imperfections in the Receiver 130
Chapter 6 High-Frequency Anti-interference Techniques 135
6.1 Introduction 135
6.2 Selecting the Channel with the Minimum Level of Active Interference 136
6.3 Adaptive Spatial Filtering 140
6.4 Correlation Loop Adaptive Antenna Systems 144
6.5 Some Algorithms for Adaptive Spatial Filtering 153
6.6 Protection Against Narrowband Interference 164
6.7 Methods for Suppressing Narrowband Interference 168
6.8 The Influence of the Transmitted Waveform on the Anti-interference Performance and Accuracy of the Radar 171
Chapter 7 Signal Detection and Parameter Estimation 177
7.1 Introduction 177
7.2 Criteria for Optimum Radar Detection 177
7.3 Detecting Signals in a Background of Stationary Interference with Unknown Intensity 186
7.4 Detecting Signals in a Background of Nonstationary Interference from Radio Stations 195
7.5 The Detection of a Signal in Clutter 200
7.6 Signal Parameter Estimation 205
7.7 The Characteristics of Signal Detection and Parameter Estimation in Interference with Unknown Intensity 217
7.8 Track Processing 229
Chapter 8 Signal Detection and Parameter Estimation in Interference with Unknown Angular Distribution 235
8.1 Introduction 235
8.2 The Signal and Interference Model 236
8.3 Detecting a Signal in Temporally Uncorrelated Stationary Gaussian Interference with Unknown Interchannel Correlation Matrix 236

8.4 Detecting Signals in Nonstationary Active Interference 244
8.5 Detecting a Signal in Clutter with Unknown Fluctuation Spectrum 248
8.6 Estimating the Parameters of Signals in Interference with Unknown Angular Distribution 250
8.7 The Characteristics of Signal Detection and Estimation in Interference with Unknown Interchannel Correlation Matrix 254
Chapter 9 Over-the-Horizon System Design 263
9.1 Introduction 263
9.2 Over-the-Horizon Radar Design Principles 264
9.3 The WARF Over-the-Horizon Radar 273
Index 281

Preface

This book has been written in an attempt to lay out the scientific and technical fundamentals of over-the-horizon (OTH) radar. The use of the high-frequency (HF) band (3–30 MHz) and "hopped" (or "skip") propagation results in a large operating range and the ability to cover regions beyond the horizon. These radars may cover areas much larger than surveillance radars operating at much higher frequencies. These are the two driving considerations in constructing HF radars.

Existing foreign (US) OTH radars which operate at HF may be categorized as belonging to one of three groups: (1) radars designed to cover large areas over seas and oceans, or to detect moving aircraft over land or water; (2) radars designed to detect regions with distinct plasma nonuniformities, created by ascending ballistic missles and meteor tracks; and (3) radars designed to carry out ionospheric investigations.

The book consists of nine chapters. The peculiarities of HF over-the-horizon radar are examined in Ch. 1. Chapter 2 treats the radar equation. Generalized equations are derived, from which the established equations for above-the-horizon radars (that is, standard radars) follow as a special case. The potential of HF radars is considered along with the components which determine these capabilities. The propagation of HF waves is studied in Ch. 3, attention being given to methods of calculating the propagation losses over long paths. The properties and characteristics of targets which may be detected with HF radars are discussed in Ch. 4.

In Ch. 5 we consider sources of interference affecting high-frequency radar receivers. Active sources (atmospheric, cosmic, and interfering transmitters) are treated along with passive interference caused by reflections from the earth's surface and ionosphere.

Possible approaches to countering interference are considered in Ch. 6. One of the primary methods of defeating active interference in HF radars is the selection of an operating band with a minimal interference level, and the questions relating to such a selection are discussed. A substantial treatment is given to methods for automatically canceling active interference received through the side and back lobes of the receiving antenna, and also for filtering interference which is concentrated in a narrow band.

Chapters 7 and 8 are devoted to the detection of target signals and the determination of target parameters in interference. In the analysis we calculate both the properties of several forms of targets and the features of HF interference which affect radar performance. Models for atmospheric, radio, and passive interference are given. Design principles for HF radars are presented in Ch. 9.

The authors express their appreciation to A.N. Shchukin, the president of the USSR Academy of Sciences council on the complex problem of radio wave propagation, who became familiar with the manuscript and made several valuable suggestions.

The authors also express their deep gratitude for a number of useful observations made by the reviewers Yu. A. Kravtsov and N.I. Kravchenko, and to all who contributed to improving the manuscript prior to its publication. The authors thank A.B. Vinogradov, S.I. Zakharov, L.G. Pikalovoy, and D.M. Platonov for providing materials used in Secs. 5.3 to 5.5, 6.2, 7.6, 7.7, 8.6, and 8.7. L.G. Pikalov offered great help throughout all stages of the preparation of the manuscript and the book.

Chapter 1
Characteristics of Over-the-Horizon Radar

1.1 INTRODUCTION

One of the primary problems which is encountered in developing modern radar technology is that of increasing the operating range. This is true both for radars designed for use on the ground and for those designed for space research. In the latter case, it has been possible to perform radar studies of the planets. On the earth's surface, however, with operating wavelengths ranging from meters to millimeters, the maximum operating range is limited by the curvature of the earth to that afforded by a direct line of sight. There is, therefore, a great interest in high-frequency radars (3–30 MHz), which can detect targets completely obscured by the horizon.

In addition to HF radars designed for over-the-horizon operation using ionospheric (space) paths, it is also possible to build radars which use ground waves. OTH radars may also operate with either back-scattered waves, where a reflected signal is received at the transmitter location, or with forward scattering, in which the incident and scattered waves propagate in the same direction.

The first domestic investigations into detecting targets in the air with radar were carried out in 1934 with the "Rapid" apparatus [3]. Forward scattering with a ground wave was employed in this system, but at VHF (4.7 m wavelength), not HF. A further step was the construction of the forward scattering VHF radar "Reven′ " (RUS-1), which was used in 1938 in experiments for detecting aircraft, and introduced into the Soviet military the next year.

A system analogous to the Reven′ was produced outside the Soviet Union. The first radar detections of aircraft in several foreign countries were achieved by using the same method as that of the Reven′ system [3].

Systems based on forward scattering, which operate at HF rather than VHF, achieve very long operating ranges [1, 4, 8, 13]. In these systems the transmitter and receiver are separated by thousands of kilometers. The transmitting antenna illuminates the target region, and the scattered signals are received at the reception point. If there is a target between the two sites in the region illuminated by the transmitted signals, then the character of the received signals changes, allowing the target to be detected. A benefit of this system is the energy advantage afforded by the use of signals scattered directly forward, and not scattered backward. It should be noted, however, that the forward scattering principle is also subject to several shortcomings.

One of these is the low information content of the received signals: a bistatic system in which the transmitter, target region, and receiver lie in a line can only detect a target, and is not capable of measuring its range or other coordinates. It is possible to overcome this difficulty by using several reception sites and combining processing of the observations, but this greatly complicates the processes of signal reception and processing. Another drawback involves the constraints placed upon the transmitter and receiver sites by the necessity of observing a particular interesting region. Thus, to observe the territory of the Soviet Union, other countries have located transmitter sites for forward-scattering systems in Asia and the Pacific, in Japan, eastern Taiwan, and the Republic of the Philippines, and the receiver sites in Western Europe (Federal Republic of Germany, Italy, and United Kingdom) [9]. The need to place at least some of the equipment on foreign territory severely limits the ability to use forward-scattering systems.

A third problem is the difficulty in maintaining synchronous signal reception. Because the distance between the transmitting and receiving sites may exceed 10,000 km, it is difficult to achieve signal synchronization with communication lines. The receiver, therefore, includes a reference oscillator synchronized with the help of unique time signals [4, 15], which reduces the radar's reliability.

Without going into a more detailed discussion of the features of forward-scattering systems, the capabilities of which have not nearly been exhausted, we will simply note that in this book we will consider issues relating to HF radars using back-scattered ionospheric waves.

When waves propagate through a medium in which the refraction coefficient varies with height, the waves are refracted and the propagation path becomes curved. When propagating through the ionosphere, HF waves are refracted so strongly that they are reflected back to the surface, with the first hop occurring at approximately 3000 km. In backscatter OTH radars, the signals reflected from the earth's surface return to the transmitter site via the same path taken by the transmitted signal. In addition to single-hop propagation, multiple-hop paths may be used.

With typical local relief features (shorelines, mountains, islands), it is possible to form a radar map of a region far beyond the horizon from the received signals. This phenomenon in the HF band was first discovered and confirmed experimentally by N.I. Kabanov [2].

A reflected signal will be received from an object (target) which lies along the propagation path far beyond the horizon, so long as it has sufficient reflectivity. Together with this useful target signal, there will also arise powerful reflections from the surface and from ionospheric nonuniformities, which appear as passive interference when the signal is detected. It should be noted that in addition to the hopped propagation path, there is also a waveguide propagation mechanism ("ducting"), which will be examined in more detail in Ch. 3.

In a backscatter OTH radar system the transmitter and receiver may either be at the same site (forming a monostatic system) or separated by some small distance (forming a bistatic system). With a backscatter system it is possible not only to detect a target, but also to measure its coordinates. The range is determined by measuring the time delay of the two-way signal propagation; to determine the angular coordinates it is necessary to use a large antenna with sufficient directivity, as in the Wide Aperture Research Facility (WARF) OTH radar, for example (see Ch. 9, Ref. [18]).

A backscatter OTH radar operates on the following principle. Electromagnetic energy radiated by the transmitting antenna at a low angle to the horizon propagates until it reaches the ionosphere in the first hop region. After reaching the ionosphere, the radiated energy propagates either between the ionosphere and the surface or between ionospheric layers, with the operating frequency being chosen so as to minimize the propagation losses over a long path. If a target lies along the path, the reflected signal travels back to the radar and is captured by the receiving antenna.

There is much material on OTH principles and systems in the existing foreign and domestic literature. Over-the-horizon radar system design is emphasized in a number of works [4, 5, 7, 8, 14]. The principles of operation of OTH radars have been covered briefly by several authors [5, 12], and, in the opinion of US specialists, have been sufficiently verified.

Individual OTH issues have been addressed in several published works: for example, in the sources listed at the end of Ch. 4 of this book there are data on the effective cross section of a target at HF; and in [13] the choice of an optimum operating frequency for OTH radars is discussed. Among the works containing some degree of general material and treatment of the physics upon which OTH radar is based, two books [4, 5] are noteworthy.

Besides these larger works, there is an extensive amount of material which is valuable in the study of OTH radar. Foremost are those on the ionosphere, HF propagation in the ionosphere, various forms of interference in the HF band, and extraction of signals from interference. Works treating these questions are cited in the corresponding sections of this book.

Considering the huge number of works on the theory and practice of radar, it is natural to concentrate only on those questions with specific significance for OTH radar. We will, therefore, present a brief treatment of these subjects.

1.2 FEATURES OF OVER-THE-HORIZON RADAR

Just as with conventional radar, information about the parameters of a target's motion are obtained in an OTH radar by receiving and processing radio signals. There are certain features of over-the-horizon radar, however, which need to be examined in more detail (see [5]).

In comparison with the higher frequencies traditionally used in radar systems, the propagation conditions encountered by HF (short wave) signals change significantly; these changes are strongly dependent upon the ionosphere.

The high-frequency band is filled with many radio broadcasting stations, which become narrowband interference sources for OTH radar systems. In several portions of the band, the power of these interference sources may exceed the power of the reflected signal. Systems using the HF band are also subject to atmospheric interference, cosmic noises, and other active interference. Therefore, special methods are needed to minimize the effects of active interference on OTH radars.

Over-the-horizon radars may also be affected by passive interference originating from ground reflections. These interfering signals may also be several orders of magnitude stronger than the useful signal, and the radar will be inoperable unless this interference is reduced.

One of the common methods for achieving a long operating range in the presence of strong interference is the use of high radar power, which places specific requirements on the antenna and feeds, transmitter, and receiver of the radar. Sometimes it is sufficient simply to use a different band than that being affected by the interference. In this case, it is necessary for the transmitter and receiver to be capable of rapid tuning over a wide operating band, and for the antenna and feeds to maintain their performance over this band.

One of the primary factors determining the received target signal level is the effective scattering area of the target. Typical targets for an OTH radar may be sharply defined relief features such as mountains, cities, and islands; the sea surface; planes and ships; ascending ballistic missiles; nuclear blast regions; the aurora borealis; meteors, and other targets located at altitudes below the maximum ionization level [4, 5, 6, 11]. Data on objects observed with OTH radar are presented in Ch. 4.

The features of OTH radar outlined above are driving factors in the design of OTH radars used to detect targets. There are a number of works describing such radars [4, 5, 14].

The antenna in an OTH radar should have high directivity and be able to scan the entire azimuth sector rapidly. In addition, the antenna should be able to be steered in elevation and capable of transmitting high power signals. One of the most difficult requirements is that for operating the antenna over a wide band, which is required in order to be able constantly to tune away from interference, and also in order to select a propagation path in accordance with the state of the ionosphere. It is also desirable that the pattern sidelobes be as low as possible. This requirement, which is common to all antennas, is especially significant in OTH radars, due to the high interference levels at HF. As previously noted [5], interference from reflections from the aurora borealis and meteors may enter the receiver through the antenna sidelobes from great ranges. To obtain such a narrow beam, the antenna must have large dimensions; as a result, the length of the antenna array may reach several hundred meters (as, for example, in the US WARF OTH radar).

The receiver in an OTH radar must operate in extremely complex conditions. This is a consequence of the high levels of both active interference (from radio transmitters) and passive interference from ground and ionospheric reflections. In addition, there is the problem of fading. The design problems presented by these requirements are severely complicated by the further need for operation over a wide band of frequencies.

The continuously changing propagation conditions caused by ionospheric variations, and also the rapidly varying interference situation, make it virtually impossible to maintain any satisfactory signal-to-noise ratio (S/N) for a long period on a fixed frequency. As has been emphasized [5], a continuous description of real-time propagation conditions is needed for effective operation of an OTH radar. Furthermore, the transmitted signal and receiver signal processor must be matched to the ionospheric and interference situation at each moment. An over-the-horizon radar must therefore be an adaptive system.

Information about the current conditions may be obtained by using the same methods which are employed to maintain the best operating conditions for short-wave radio communications. These include vertical sounding of the ionosphere, oblique sounding, and acquiring current information on the presence of interference in the band which might be used to match the other conditions. A backscatter over-the-horizon radar possesses capabilities which cannot be matched by radio communication systems. By observing signals which have been reflected by the earth, it is possible to obtain information on the state of the ionosphere and the propagation conditions along the path. To accomplish this, however, the receiver must contain the components for selecting the band in which acceptable propagation losses are obtained, and also for choosing the operating channel with the minimal interference level.

1.3 CHARACTERISTICS OF OVER-THE-HORIZON RADAR SYSTEMS

As an example, we will examine the characteristics which may be achieved by HF OTH radars in detecting aircraft.

The operating range is taken to be 1000–4000 km; longer ranges may be used with multihop paths, but at the expense of poorer performance than is described below. The range resolution is 2 km or worse (values of 20–40 km are typical). The relative error in determining the range of one target relative to another with a single system is 2–4 km. The absolute range error is on the order of 10–20 km when the transmitted and received signal paths are determined correctly. The angular resolution is determined by the beamwidth; it may be smaller than 1°, which corresponds to a linear dimension of 50 km at a range of 3000 km.

Angle accuracy is achieved by forming several beams (up to 10) at a sufficiently high signal-to-noise ratio. Including ionospheric effects, the angular error may be several tenths of a degree. Target velocity resolution may be accomplished by extracting Doppler frequencies of 0.1 Hz and even lower. At a frequency of 20 MHz (15 m wavelength), a Doppler shift of 0.1 Hz corresponds to a relative velocity of 27 km/h.

The latest work on the construction of an OTH radar is described in [8]. The radar was constructed in Maine as part of an early warning system for approaching bombers and other aircraft. The transmitter and receiver sites are separated by 162 km, to lower the interaction between the high-power transmitters and the receiver.

The basic parameters of this radar are as follows [8]: it employs continuous wave (CW) frequency-modulated transmission; it has a dipole transmitter array 41 m high and 694 m long; the azimuthal scanning is electronically controlled by a computer, using additional frequency modulation in each of the radar's six operating bands; the computer also controls scanning in range. There are 21 transmitter modules, each with a power of 100 kW, of which seven may be transmitting simultaneously. The width of the receiving array is 1773 m; there are 96 superheterodyne receiver modules, and analog data is converted to digital form in the receiver. The receiver beam is formed digitally. The minimum operating range is 800 km, and the maximum is close to 3000 km.

Over-the-horizon radars may be applied in various ways, depending on the nature of the objects which it is designed to observe. The OTH system described by Headrick and Skolnik [5] is designed to monitor air movements over the sea and to observe the sea surface state. With such observations, it is possible to determine the direction of motion of the waves, estimate their height, and to estimate the strength of the winds which caused them. Recently, a great deal of attention has been given in the US to the question of using HF radars for observing sea state.

Radars operating in the HF band also find use in ionospheric studies. In 1976, for example, a radar operating at 7–29 MHz to study the electric fields and nonuniformities in the equatorial ionospheric current led to the discovery that the phase velocity and amplitude of reflected signals shifted sharply is dependent on range [10]. The results of these observations led to the conclusion that a low phase velocity and large reflected signal amplitude at large ranges is connected with the existence of horizontal plasma waves in the propagating layer of the ionosphere.

REFERENCES

1. Vasin, V.V. and Stepanov, B.M. *Spravochnik-zadachnik po radiolokatsii* (Radar Handbook). Moscow: Sovietskoe Radio, 1977. 317 pp.
2. *Discovery #1* (USSR) Byulletin Izobretenie (Patent Journal), 1959, no. 19, p. 8.
3. Lobanov, M.M. *Nachalo sovietskoy radiolokatsii* (*Early Soviet Radar*). Moscow: Sovietskoe Radio, 1975. 288 pp.
4. Mishchenko, Yu. A. *Zagorizontnaya radiolokatsiya.* Moscow: Voenizdat, 1972. 95 pp.
5. Headrick, D.M. and Skolnik, M.I. *Over-the-horizon radar in the HF band.* Proc. IEEE, vol. 62 (1974), no. 6, pp. 6–17.
6. Baghdady, E.J. and Ely, O.P. *Effects of exhaust upon signal transmission to end from rocket-powered vehicles.* Proc. IEEE, vol. 54 (1966), no. 9, pp. 1134–1146.

7. Desmond, S. *Over-the-horizon radar in defence of Australia.* Electronics Today, 1978, vol. 8, no. 2, pp. 35–40.
8. *Early-warning over-the-horizon radar being put together in Maine by G.E.* Electronics, 1977, vol. 50, no. 4, pp. 30–31.
9. Greenwood, T., *Reconnaissance and arms control.* Scientific American, vol. 228 (1973), no. 2, pp. 14–25.
10. *High frequency radar observations of horizontal plasma waves in the equatorial ionosphere.* Nature, vol. 277 (1979), no. 5693, pp. 203–204.
11. Jackson J.E., Whale, H.A., and Bauer, S.J. *Local ionospheric disturbance created by a burning rocket.* Geophys. Res., vol. 67 (1962), no. 5, pp. 2059–2061.
12. Mason J. and Sclater, N., *Over-the-horizon radars scan skies for FOBS.* Electronic Design., vol. 15 (1967), no. 26, pp. 25–28.
13. Ross, G.F. and Schwartzman, L. *Prediction of coverage for trans-horizon radar systems.* IRE Trans., vol. MIL-S (1961), no. 2, pp. 164–172.
14. Shearman, E.D.R. *Radar looks over the edge.* Spectrum, British Science News, vol. 67 (1969), no. 67, pp. 13–15.
15. Thomas, P.G. *Advanced ground radar.* Space-Aeronautics, vol. 44 (1966), no. 4, pp. 102–112.

Chapter 2
The Radar Equation

2.1 INTRODUCTION

The radar equation used in traditional radar applications describes the case when a target with effective cross section σ is located at a range R within direct sight of the radar, which has a transmitter power P_t, transmitter antenna gain G_t, and effective receiving antenna area A_r. This equation, which may be expressed in several forms, is used for above-the-horizon radar applications. [*Note:* The Russian term "beyond-the-horizon" is far less ambiguous than our "over-the-horizon," which actually refers to "under-the-horizon;" in any case, "above-the-horizon" should be interpreted "not over-the-horizon"–Tr.] In these equations, the propagation losses from the transmitter antenna to the target and from the target to the receiver are calculated. The propagation is assumed to take place along a direct line of sight, however, so that these equations take into account only those losses associated with the spherical spreading of the wave.

In the case of over-the-horizon radar, when the target is hidden beyond the limit of the horizon and the wave propagates at least partially in the ionized gases of the ionosphere, the propagation losses assume a much more complex character than spherical spreading, and the standard radar equation is inappropriate.

2.2 THE STANDARD RADAR EQUATION

The radar equation may be placed in the form of a ratio between the signal power received from a target P_r to the noise power P_n referenced to the input of the receiver:

$$P_r/P_n = P_t G_t A_r \, \sigma/(4\pi R^2)^2 P_n \tag{2.1}$$

This equation relates to the simple case, when reception and signal processing take place against a background of fluctuating noise, and the interfering action of passive interference is not considered. Nevertheless, we will limit ourselves to an examination of this simple case, in order more clearly to compare the standard and over-the-horizon equations.

With given requirements for probability of detection and false alarm rate, (2.1) is sometimes used to express the maximum detection range as a function of the radar energy parameters. The right side of (2.1) is set equal to some threshold value which provides the desired detection characteristics, and then transformed to obtain

$$R_{\max} = [PG_tA_r\sigma/(4\pi)^2P_n(P_r/P_n)_{thr}]^{1/4} \tag{2.2}$$

This form of the equation is usually referred to as the radar range equation.

With standard radars, the radio wave propagates along a line joining the radar and the target; the loss in electromagnetic energy due to spherical spreading in both directions is $W = (4\pi)^2R^4$. This is sometimes referred to as the radar loss in free space. In some cases it is necessary to introduce additional loss terms in the right side of (2.1), to take account of losses due to absorption or scattering in the atmosphere.

With over-the-horizon radar, when the target is hidden beyond the horizon and the radar signal propagates between the earth and the ionosphere, the signal path loss has a non-monotonic character and a complex form which depends on the target range and altitude, ionospheric characteristics, and the radar operating frequency.

2.3 THE GENERAL RADAR EQUATION

We will now examine the basic physical processes which govern the target signal power received by an over-the-horizon radar. The emphasis will be on developing a simple interpretation of these processes and using fairly simple mathematical descriptions of them.

Let a radar with directive antennas be located at point 1 on the earth's surface (see Fig. 2.1). We will assume that the beam is pointed directly at the target in azimuth. The elevation pattern is inclined at a low angle, and its width is such that the beam includes the critical elevation angle γ_{cr}, above which the radiated energy penetrates the ionosphere and is lost in space; for $\gamma < \gamma_{cr}$ the radiated energy is confined by the ionosphere and propagates further along the surface. The critical angle may be calculated as a function of the ionospheric parameters and the operating frequency f_{op} with the formula [7]:

$$\gamma_{cr} \approx \arccos[\sqrt{1 - (1 + z_n/a)^2(f_{cr}/f_{op})^2}]$$

where f_{cr} is the critical frequency associated with vertical transmission through the ionosphere; z_n is the altitude of maximum concentration in the *F*-layer of the ionosphere; and a is the radius of the earth.

With the limitations mentioned above, the power density in the region of a target located beyond the horizon at a range R and altitude z may be taken to be proportional to P_tG_t. The proportional relation to G_t may be lost when the critical elevation angle γ_{cr} lies outside the elevation pattern. The power density at the target range R may even decrease with an increase in G_t if the beam is too narrow in elevation. This situation does not ordinarily arise, however, and results only when the chosen operating frequency is not optimal, i.e., is substantially lower or higher than the best applicable frequency.

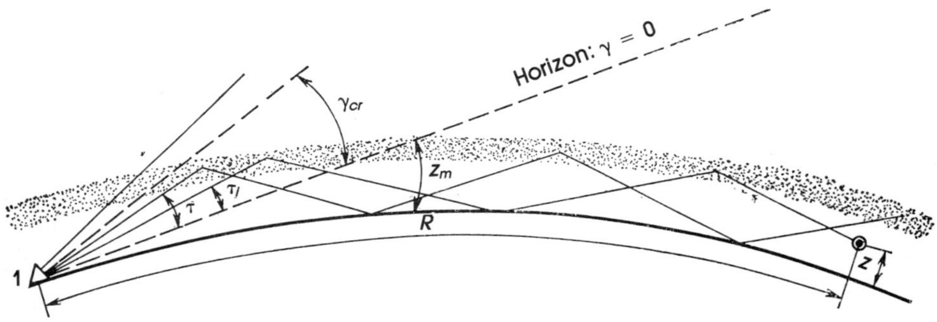

Fig. 2.1 Propagation of radio waves between the earth and the ionosphere.

In order to simplify the ensuing discussion, we will assume that the energy is radiated between the earth and the ionosphere along the elevation axis of the transmitted pattern. We will now study the propagation of electromagnetic energy radiated at some angle $\gamma < \gamma_{cr}$ through a differential beam, the dimensions of which are $\delta\gamma$ in elevation and $\delta\beta$ in azimuth. The solid angle of the differential beam is $\delta\Omega = \delta\gamma\delta\beta \cos \gamma \approx \delta\gamma\delta\beta$ and contains the power:

$$\delta P = P_tG_t\delta\Omega/4\pi \qquad (2.3)$$

We will find the power density in a cross section of a differential beam at a range R. Since all of the energy radiated in elevation from 0 to γ_{cr} is contained by the ionosphere, each differential beam radiated at an angle γ will be found at a range R at the altitude $z(\gamma, R)$. The quantity $z(\gamma, R)$ is thus the altitude at a range R from the radar at which a beam will arrive when transmitted at an angle γ. The character of this dependence for a fixed value of R changes significantly with ionospheric parameters and the operating frequency. The function $z(\gamma, R)$ may be calculated on a computer with numerical integration, and we will not discuss its calculation here, but just point out some of its properties.

First, this function is single-valued, that is, each differential beam radiated at an angle γ arrives at a single height z at a range R. [Note: In all that follows, "beam" should be understood to mean a very small element of the actual beam; multiple beams are not being transmitted –*Tr.*] The inverse function may be multivalued, meaning that at a given range R, there are many different angles from which a beam arriving at the height z could have been transmitted. Second, the interval within which the function z exists is limited by $(0, z_m)$, but it may have narrower limits. This results when the beam is trapped in an ionized duct between layers (E; F_1; F_2), in the earth/E-layer duct, and so on.

To calculate the absorption of electromagnetic energy in the ionosphere and the time delay of the radar signal when propagating from the radar to the target, we assign to the ith discrete beam an absorption coefficient Γ_i and time delay τ_i.

We will now calculate the spatial divergence of the ith beam, which will, in general, be different from that associated with spherical spreading. To do this we find the area of a normal cross section, having determined separately the spread in height δz_i and in azimuth δl_i. The cross section area will then be

$$\delta S_i = \delta z_i \delta l_i \tag{2.4}$$

The spread in azimuth will be the same for all beams and will be equal to

$$\delta l_i = \delta\beta_i a \sin(R/a) \tag{2.5}$$

where a is the radius of the earth.

For spherical spreading:

$$\delta z_i = \delta\gamma_1 R \quad \text{and} \quad \delta l_i = \delta\beta_i R \tag{2.6}$$

The power density of the radar field due to the ith beam in the region of the target may be calculated using the absorption coefficient $\Gamma_i = \Gamma_i(\gamma_i, R)$, and also (2.3)–(2.5):

$$B_i = \frac{\delta P_i}{\delta S_i}\frac{1}{\Gamma_i} = \frac{P_t G_t(\gamma_i)}{4\pi R a \sin(R/a)}\left(\frac{1}{R}\frac{\delta z_i}{\delta\gamma_i}\Gamma_i\right)^{-1} \tag{2.7}$$

We will use the notation:

$$g_i = (1/R)(\delta z_i/\delta\gamma_i) \tag{2.8}$$

The dimensionless quantity g_i may be called the *altitude factor,* characterizing the relative change in field intensity of the ith beam for a change in altitude δz_i in comparison with that due to spherical spreading, $R\delta\gamma$.

On the basis of (2.7) and (2.8), we have

$$B_i = P_t G_t(\gamma_i)/W_{12}(i) \tag{2.9}$$

where

$$W_{12}(i) = 4\pi Ra \sin(R/a)\, g_1 \Gamma_i \tag{2.10}$$

characterizes the weakening of electromagnetic energy, that is, the path loss of the radio wave in propagating from the radar to the target through the ith beam. If the transmitter power P_t is expressed in watts, and R and a in (2.10) are in meters, then in accordance with (2.9), the power density B_i will be in w/m^2 (the transmitter antenna gain G_t is a dimensionless quantity). The method just used for determining the path loss is based on the approximations of geometric optics and, as is well known, may not be used to estimate the power density in caustic regions, that is, at heights where $\delta z_i/\delta\gamma_i = 0$. In order to avoid this difficulty, the boundary increments δz_i are chosen to be larger than the caustic zones in height, and for the chosen values of δz_i the corresponding increments $\delta\gamma_i$ are determined. We assume that such an approach will give a correct estimate of the average power density in the caustic zones, but will not describe the fine structure of the fields in this height interval.

It should be understood that the summing of RF fields from different beams will occur with random phase. This means that the average power density at the target will be equal to the sum of the power densities of the component beams: $B = P_t G_t/W_{12}$. The quantity:

$$W_{12} = \left[\sum_i^n W_{12}^{-1}(i)\right]^{-1} \tag{2.11}$$

is the total loss in electromagnetic field power due to propagation from the radar to the target. We note that the field voltage in the region of the target has a random character, since the phases of the RF signals from the individual beams change slowly with changes in the time and location of the observation point. The distribution of the field voltage depends on the distribution of power in the beams and the number of beams. The spatial and temporal frequency of the voltage depends on the difference in the time delays of the beams, and the larger this difference, the more rapidly the fields fluctuate.

We will now examine the propagation of energy in the opposite direction—from the target to the receiving antenna of the radar. We will assume the target's backscatter coefficient is constant within the angular sector from which the beam arrives and that the effective cross section is equal to σ. Neglecting the effects of the earth's magnetic field when calculating the power density of the reflected signal at the radar site, we may assume that the signal propagates back to the radar along the same path taken by the corresponding discrete beam. Traveling back to the radar, this beam will experience the same amount of spatial spreading and time delay as on the forward leg of the path.

The reflected target signal at the radar is formed from n^2 beams. This is because the target may scatter the energy from the ith beam into the jth beam. We will call this signal path in the forward and backward directions the i, jth beam. At the point of reception, for each beam i, j there will be, in accordance with (2.9) and (2.10), a corresponding power density $B_r(i, j) = P_t G_t(\gamma_1)\sigma/[W_{12}(i)W_{21}(j)]$ and time delay $\tau(i, j) = \tau(\gamma_i, R) + \tau(\gamma_j, R)$. The attenuation factors are related as $W_{12}(i) = W_{21}(i)$ and $W_{21}(j) = W_{12}(j)$, and are determined by (2.10). Since the various beams are incoherent, the average received target signal power is given by the sum of the powers of all the beams:

$$P_r = \sum_i^n \sum_j^n B_r(i, j)\, A(\gamma_i)_r$$

Thus, the expression analogous to (2.1) for an over-the-horizon radar when $G(\gamma_i) = G$ and $A(\gamma_j) = A$ for all i, j is the signal-to-noise ratio formula:

$$P_r/P_n = P_t G_t A_r \sigma / P_n W \tag{2.12}$$

Here,

$$W = W_{12} W_{21} \tag{2.13}$$

where

$$W_{12} = \left[\sum_j^n W_i^{-1}(i)\right]^{-1} \quad \text{and} \quad W_{21} = \left[\sum_j^n W_j^{-1}(j)\right]^{-1}$$

Equation (2.12) is a generalization of the well-known radar equation (2.1) for the case when the received signal power is determined by the sum of the powers of individual components arriving at the receiver from different beams. This situation is characteristic of over-the-horizon radars.

The coefficient $W = W_{12}W_{21}$ appearing in (2.12) is the total path loss experienced by the radar signal along the combined path, that is, the attenuation occurring along the forward and backward paths.

As can be seen from (2.10) and (2.12), the loss of electromagnetic energy W_{12} which occurs as the wave propagates to the target depends on the range R. In addition to the direct dependence on R, it is also necessary to consider the contributions of g_i and Γ_i, which are related to R in a complex manner. An analogous dependence on R also appears in the backward loss coefficient W_{21}, i.e., that describing the loss from the target to the radar. This determines the character of the dependence upon the range R of the total loss W along the path. In addition, the coefficients g_i and Γ_i, and consequently W, depend on the angle of radiation γ, which is chosen according to the operating frequency and the state of the ionosphere.

Due to the complex dependence of the loss W on all of these factors, determining its value is a difficult problem, the more so because various propagation effects will come into play depending on the path length and the state of the ionosphere. Questions relating to determining W will be examined in more detail in Ch. 3. In any case, if the quantity W in (2.12) is understood to be the total loss, then this equation will be valid regardless of the particular propagation effects.

Equation (2.12) may be cast in other forms. We will assume that the radar illuminates a target with an effective scattering cross section σ. Then the reflected signal power P_r at the input of the radar receiver may be expressed as the product of three factors:

$$P_r = B_t\,(\sigma/W_{21})\,A_r \tag{2.14}$$

The first factor B_t determines the power density illuminating the target; the product of this first term and the second $B_t(\sigma/W_{21})$ is the power density of the reflected wave at the receiver site. Multiplying the resulting quantity by the effective receiving antenna surface area A_r, we find the power of the reflected signal entering the matched receiver input. Expanding (2.14) we obtain

$$P_r = (P_tG_t/W_{12})(\sigma/W_{21})\,A_r \tag{2.15}$$

Going from the receiver input power P_r to the power signal-to-noise ratio P_r/P_n, we obtain (2.12). Equation (2.12) may be used for monostatic radars and also for bistatic situations, when the transmitter and receiver are separated by some distance.

For a monostatic situation, we may use $W_{12} = W_{21} = \sqrt{W}$, where W is the resulting two-way loss over the propagation path. We will also assume that for a bistatic radar, the separation R_0 between the transmitter and receiver sites is small compared with the target range, so that $R_1 \approx R_2$. We may then still use $W_{21} = W_{12} = \sqrt{W}$, where W is the two-way signal loss experienced over the path.

Comparing (2.1) and (2.12), it may be seen that if the spherical spreading factor $(4\pi R^2)^2$ is substituted for $W = W_{12}W_{21}$ in (2.12), then (2.1) results.

2.4 ANOTHER FORM OF THE RADAR EQUATION

Equation (2.12) is a description of an ideal radar, and all of the parameters are optimal. In an actual radar there will always be losses associated with mismatched HF feed lines, nonoptimal signal processing, and other causes. These energy losses may be taken into account by including a loss term in (2.12). The equation will then take the form:

$$\frac{P_r}{P_n} = \left(\frac{P_t}{P_n}\frac{G_t A_r}{L}\right)\frac{\sigma}{W} \tag{2.16}$$

where L is the loss coefficient.

We will now transform this equation, introducing in place of the ratio P_r/P_n the ratio E_s/N_0, where E_s is the energy of the received signal, and $N_0 = P_n B_n$ (P_n is the root mean square noise power, and N_0 is its spectral density, that is, the noise power in a unit of the noise bandwidth B_n).

Considering all quantities appearing in (2.15) except P_t and P_r to be constant during the target observation time, we integrate the right and left sides of the equation over the duration of the illuminating signal t_d, and of the reflected signal t_r received from the target, for which we assume $t_d = t_r = t$. Then, keeping in mind that

$$E_s = \int_0^t P_r(t)\,dt, \quad E_t = \int_0^t P_t(t)\,dt \tag{2.17}$$

we obtain in place of (2.16):

$$E_s/N_0 = [(E_t/N_0)G_t A_r/L]\sigma/W \tag{2.18}$$

where E_s and E_t are the energies of the received and transmitted signals respectively.

Equation (2.18), giving the energy signal-to-noise ratio, is distinctive in that it does not depend on the waveform, i.e., on the form of the envelope and the method of intrapulse modulation.

2.5 THE SIGNAL-TO-INTERFERENCE RATIO

In Sec. 2.4, the noise power was denoted by P_n, and its spectral density at the receiver input by N_0. These values, however, are determined not just by noises, but also by external sources of interference (atmospheric, industrial, and so on). It is, therefore, possible to talk of "noises" and "interference", with the latter being an additive mixture of noise in the radar receiver and external interference.

We will now consider the signal-to-interference ratio $q = E_s/N_0$ which appears in (2.18). Due to signal fluctuations and a varying interference level, this ratio is a random quantity. We will examine briefly the manner in which the probability of detection D and probability of false alarm F depend on the quantity Q.

The probability of detection D is the probability that the receiver response will exceed a threshold in the presence of a useful signal. It may be expressed as

$$D = \int_{u_{thr}}^{\infty} p_{si}(u)\, du$$

where p_{si} is the probability density of the voltage u resulting from the mixture of signal and interference after coherent and incoherent processing at the input of the threshold device.

A false alarm may arise when there is no target signal, and an interference voltage spike exceeds the threshold level; in this case, the presence of a target is registered, when in fact none exists. The probability of false alarm F is the probability that the threshold u_{thr} will be exceeded due to interference:

$$F = \int_{u_{thr}}^{\infty} p_i(u)\, du$$

where $p_i(u)$ is the probability density of the interference voltage $u = u_n(t)$.

A relationship between the ratio q and the detection characteristics D and F may be established. We will use an idealized case, when the passive interference is weak and the active interference is fluctuating. Furthermore, we will be using a Neyman-Pearson receiver, which provides the maximum probability of detection D for a given q and fixed false alarm rate F. Analytical expressions for the relations between D, q, and F are rather cumbersome, and it is, therefore, more convenient to present the relations in the form of graphs called *detection characteristics* (see Fig. 2.2). The curves of Fig. 2.2 are for a Rayleigh-distributed signal. It is plain from the curves that with an increase in the signal-to-interference ratio and the incoherent integration factor S, for fixed false alarm rate F, the probability of detection D increases substantially. The function $D = \phi(q)$ depends on the character of the signal fluctuations, the characteristics of the predominant interference, and the chosen detection criteria (see Ch. 7 and Ch. 8).

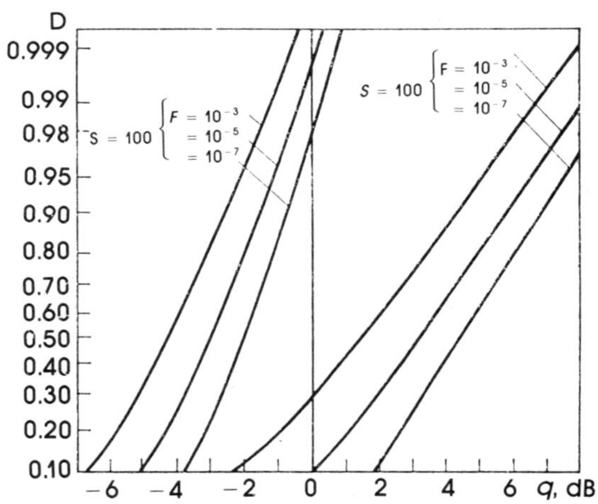

Fig. 2.2 Probability of detection D as a function of the signal-to-noise ratio q.

2.6 THE PERFORMANCE OF A RADAR SYSTEM

The performance of a radar may be assessed using the quantity:

$$Q = (E_s/N_0)\, G_t A_r / L \tag{2.19}$$

where $N_0 = P_n B_n$ is the spectral density of the interference resulting at the radar receiver input under actual operating conditions.

Because (2.19) uses the ratio of radiated energy E_s to N_0, the expression for Q is valid for any form of modulation. In addition to the overall performance index, it is convenient in some cases to consider the receiver performance index:

$$Q_r = (E_s/N_0')\, G_t A_r / L \tag{2.20}$$

In this case the interference spectral density $N_0 = N_0'$ is determined only by the receiver noises themselves: $N_0' = P_{n0} B_n$, where P_{n0} is the power of the receiver noises referenced to the receiver input. We will now examine the right side of (2.19).

2.6.1 The Radiated Energy

The radiated energy $E_s = E_p n$, where

$$E_p = \int_0^{\tau} p(t)\, dt = P_p \tau$$

where P_p is the pulse power. The pulse energy E_P is related to the average power P_{av} as $E_p = P_{av} T$, where T is the pulse repetition interval.

2.6.2 The Spectral Density of the Interference

The spectral density of the interference N_0 is the power of the resulting interference P_n per unit noise bandwidth B_n: $N_0 = P_n B_n = (P_{n \cdot ex} + P_{n0})\, B_n$, where $P_{n \cdot ex}$ is the interference power caused by external factors (galactic, atmospheric and other radio noises, passive interference, and so on), referenced to the input from the output of the linear portion of the receiver, and P_{n0} is the internal receiver noise power referenced to the receiver input.

2.6.3 The Noise Bandwidth

The noise bandwidth B_n is modeled as the rectangular spectrum with area equal to that of the amplitude noise spectrum (see Fig. 2.3):

$$B_n = \int_0^{\infty} \frac{1}{\sigma^2(f)}\, df$$

where $\sigma(f) = k_0/k_{\Delta f}$ is the ratio of the receiver gain when tuned to frequency f_0, to the gain at $f = f_0 + \Delta f$.

The problems of determining the noise bandwidth for an n-stage amplifier with identical stages and for a two-stage filter are solved in Gutkin

[2] and Kolosov [3]. For a narrowband amplifier, in which $f_0 \gg B_n$, we may use the relation $B_n = B_s\chi$, where χ is the correction coefficient, the value of which is as follows in Table 2.1 for an amplifier with identical stages at resonance:

Table 2.1

Number of Stages	*Coefficient* χ	*Number of Stages*	*Coefficient* χ
1	1.570	7	1.098
2	1.220	8	1.093
3	1.155	9	1.090
4	1.128	10	1.087
5	1.113	20	1.076
6	1.104		

For an amplifier with a two-stage filter, the value of the coefficient χ is smaller than for a single-stage amplifier. The closer the filter response is to rectangular, the closer χ is to 1.

If a multistage tuned amplifier is used in the receiver, then the noise bandwidth $B_n \approx B_s$, where B_s is the signal bandwidth.

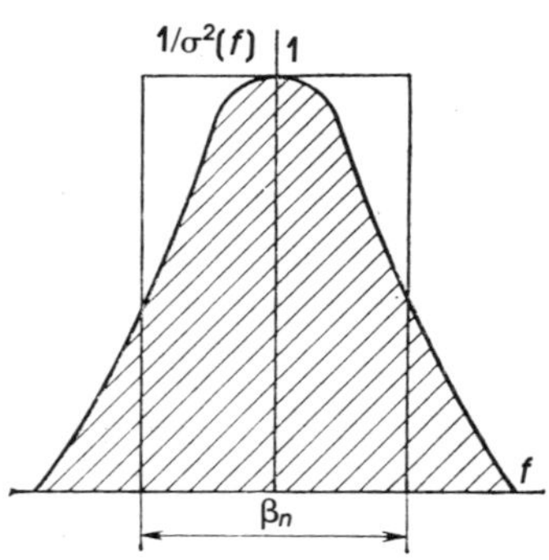

Fig. 2.3 Equivalent noise bandwidth.

2.6.4 The Antenna Gain

G_t represents the factor by which the input power increases for the directive antenna in relation to an ideal isotropic radiator, without losses and with the condition that the amplitude of the Poynting vector is constant at the observation point. The gain may be expressed in decibels or as a dimensionless quantity:

$$G_t = G_0 \eta \tag{2.21}$$

where G_0 is the directivity and η is the antenna efficiency.

The directivity G_0 characterizes the degree to which the radiated energy is concentrated in a given direction. This quantity may be determined experimentally by measuring the amplitude response *versus* direction. Usually, if it is not stated otherwise, the values G_t and G_0 correspond to the peak values of the gain and directivity. The antenna efficiency η is the ratio of the radiated power (including power radiated through the side lobes) to the power introduced at the antenna input. Thus, $\eta = P_\Sigma / P_{in} = P_\Sigma / (P_\Sigma + P_\Omega)$, where P_Σ is the radiated power; P_{in} is the total power at the antenna input; and P_Ω is the ohmic power loss in the antenna and matching devices. If the antenna efficiency is close to unity, then $G_t \approx G_0$. Feed losses are examined in Sec. 2.7.

2.6.5 The Effective Receiving Antenna Area

The quantity A_r is understood to be the surface area of the antenna in the plane of the wavefront, which intercepts and passes the received power to a matched load. The effective area of the receiving antenna A_r, in square meters, is related to the antenna gain:

$$A_r = \lambda^2 Gr/4\pi \tag{2.22}$$

In this case, the receiver antenna will extract the maximum power from the incident electromagnetic wave if the following conditions are met (see [4]): the antenna axis is aligned with the angle of arrival of the wave; the polarization of the antenna is matched to that of the signal; and the ohmic losses in the matching devices are kept to a minimum, that is, the antenna efficiency is maximized. If these conditions are not met, then various losses must be calculated (see Sec. 2.7).

The effective antenna area is directly related to the dimensions of its radiating surface, that is, $A_r = kS_a$, where S_a is the antenna area, and k is a coefficient representing the extent to which the antenna area is utilized. The value of k is always less than 1 for real antennas.

2.7 LOSSES DUE TO SYSTEM IMPERFECTIONS

2.7.1 The Loss Coefficient

The resulting loss coefficient L (see (2.19)), is the product of several components, each representing a different contribution, so that

$$L = L_1, L_2, L_3, \ldots, L_n \tag{2.23}$$

If the radar equation and corresponding radar performance index are expressed in logarithmic form, then the expression for the loss coefficient L in dB will have the form:

$$L = L_1 + L_2 + L_3 + \cdots + L_n \tag{2.24}$$

We will now examine each of the loss components contributing to L.

2.7.2 Feed Losses

The product $P_P\tau = E_p$ appears in the radar equation, where P_p is the radiated pulse power of the transmitter. The radiated power is usually measured at the last stage of the transmitter, so that a loss term L_f is necessary to take into account the loss in power occurring in the feed lines between the transmitter and the antenna. If the transmitter and antenna are well-matched to the feed lines, then this term is small, usually on the order of tenths of a decibel. If the radar operates over a wide band, however, it is difficult to obtain good matching at all the frequencies, as a result of which this term becomes substantially larger. The formula for the antenna gain, (2.21), holds for an antenna matched to its feed. If the antenna and feed are mismatched, then the expression for the gain takes the form [4]:

$$G_{equiv} = G(1 - |\Gamma|^2) = G_0\eta(1 - |\Gamma|^2)$$

where $|\Gamma|$ is the magnitude of the reflection coefficient connected with the mismatch. Thus, the loss due to mismatched feed lines may be represented by the coefficient $L_f = 1/(1 - |\Gamma|^2)$. Losses in the receiver feeds are not taken into account.

This treatment of the losses in the antenna (the coefficient η) and in the feed line (L_f) is based on a very simple idealization. The important object of this discussion has been to emphasize the necessity to take into account the indicated losses, especially in the case of wideband operation when the mismatch loss may be significant. For large multi-element antennas with complex feed networks, it is extremely difficult to estimate the antenna-feed losses, which depend on the specific construction of the antenna. The loss term will be $L_{af} = P_{rad}/P_{out}$, where P_{rad} is the power radiated from the antenna, and P_{out} is the power introduced at the input to the feed system by the transmitter.

2.7.2 Polarization Losses

Polarization losses arise when the polarization of the signal incident upon the receiving antenna differs from the polarization of the antenna by an angle ϕ. There may be different causes for this polarization mismatch, but in the HF band, the primary cause is the Faraday effect. As a plane-polarized wave propagates through the ionosphere, the magnetic field of the earth splits the wave into ordinary and extraordinary components, which in general will have elliptical polarizations with opposite directions of rotation. These two components propagate through the ionosphere at different speeds, as a result of which the phase relation between them changes along the path. Upon leaving the ionosphere, the waves recombine to form another plane-polarized wave which is shifted by some angle relative to the original polarization plane (the Faraday effect). This introduces a shift ϕ between the polarization of the receiving antenna and the polarization of the arriving wave. The loss resulting from this polarization mismatch may be estimated with the coefficient $L_\phi = \sec^2 \phi$ (see [5]).

2.7.3 Signal Processing Losses

The powers of the received signal, internal system noise and external noises referenced to the receiver input depend on the signal processing, and therefore so does the signal-to-interference ratio q, calculated at the output of the receiver circuit just preceeding the detector. The signal-to-interference ratio q will be largest when optimum coherent signal processing is used; in all other cases, q will be smaller.

If optimum signal processing is not used, then a loss term L_s characterizing the resulting loss needs to be introduced into (2.23) and (2.24). The term L_s will depend on: the form of the frequency response of the RF (predetector) portion of the receiver; the amount by which the bandwidth differs from the optimum; the imperfections in the video portion of the receiver; the use of incoherent integration in place of coherent integration; and other factors.

2.7.4 The Optimal Form of the Radar Frequency Response and Losses Due to a Mismatched Passband

The selection of the radar receiver's frequency response is a very important step in the system design, because it is used to obtain the best filtering of signals against a background of interference.

The issue of optimal filtering in the presence of fluctuating interference has been studied in a number of works (see, for example, [1]). There are several possible approaches to the selection of the structure and parameters of the filter, depending on the demands which it must meet. If target detection is the primary concern, then the filter should provide the highest signal-to-interference ratio possible; this is the case we will consider.

The power signal-to-interference ratio is $q = r^2$, where r is the voltage signal-to-noise ratio. The largest value $r_{\max}$, and accordingly $q_{\max}$, will be obtained when an optimum filter is used in the receiver. Analysis of this issue (see [1, 5]) indicates that the value of q reaches $q_{\max}$ for an optimum filter whose response is matched to the signal as a function of time.

At the output of the linear filter, $q_{\max} = \alpha E_s/N_0$, where E_s is the signal energy, and N_0 is the interference spectral density. The coefficient α depends on what is meant by the quantity E_s, which is given by $E_s = P_r t$, where P_r is the received signal power. If the signal power is taken to be the maximum power reached for the maximum amplitude of the RF oscillations, then $\alpha = 2$. If the signal power is taken to be the effective RF power, that is, the power averaged over an RF cycle, then $\alpha = 1$.

The quantities P_t and P_r appearing in the radar equation (2.15) are average powers. The ratio E_s/N_0 appearing on the left side of (2.18) is thus the signal-to-noise ratio q.

The transfer function of an optimum filter must satisfy the condition:

$$k(i\omega) = aS^*(i\omega)\exp(-i\omega t_0)$$

where a is a constant coefficient and $S^*(i\omega) = S(-i\omega)$ is the complex conjugate of the signal spectrum $S(i\omega)$. From the preceding relation it follows that $|k(i\omega)| = a|S(i\omega)|$, i.e., to within a constant, the frequency response of an optimum filter should be equal to the signal's spectrum amplitude.

If the chosen bandwidth is not the optimum, then the signal-to-noise ratio q will be reduced. The probability of detection is therefore also reduced when nonoptimum filtering is used.

In order for the probability of detection to have a definite relation to the signal-to-interference ratio q, we will assume that the receiver bandwidth B_s is optimal. If the bandwidth is not optimal, then in the radar equation and the formula for the performance index Q there needs to be added an additional loss term taking this factor into account [5]:

$$L_\tau \approx (B/4B_{\text{opt}})(1 + B_{\text{opt}}/B)^2$$

If, for example, $B = 0.5B_{\text{opt}}$, then $L_\tau = 1.125$, or approximately 0.5 dB. For $B = 0.1B_{\text{opt}}$, $L_\tau = 3.03$, or about 4.8 dB. Thus, with a comparatively small difference between the actual and the optimal bandwidth, the losses are small. The losses due to any mismatch of the video bandwidth may be neglected.

2.7.5 Losses Due to the Use of Incoherent Instead of Coherent Integration

The curve in Fig. 2.4 [6] may be used for a rough estimate of these losses. The graph was plotted for $q = 13.5$ dB and gives the loss in decibels resulting from the use of incoherent integration of n pulses in comparison with coherent integration. Optimum intrapulse signal processing is assumed with both incoherent and coherent integration. It is clear from the graphs that with 10 pulses in a group, for example, the loss due to incoherent integration will be about 1.5 dB.

In Secs. 2.3–2.7, the basic parameters appearing in the radar equation were examined. We will not delve into a detailed analysis of the issues relating to determining signal thresholds, however, because these are covered quite thoroughly in the literature (see, for example, [6]).

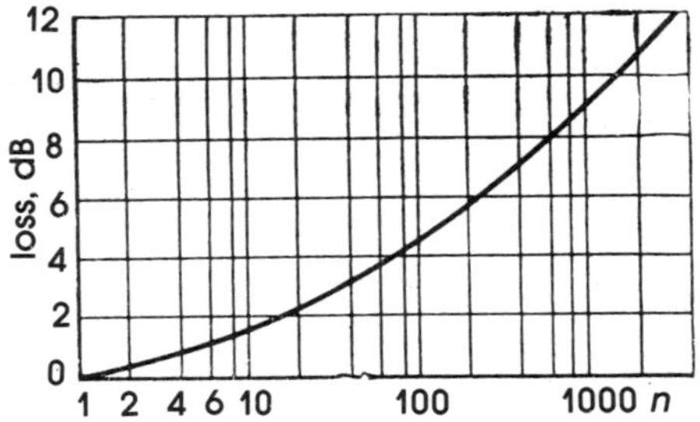

Fig. 2.4 Loss resulting from use of noncoherent integration, as compared with use of coherent integration.

REFERENCES

1. Gutkin, L.S. *Teoriya optimal'nykh metodov radiopriema pri fluktuatsionnykh pomekhakh (The theory of optimum radio reception in fluctuating interference)*. Moscow: Sovietskoe Radio 1972. 448 pp.
2. Kolosov, A.A. *Polosa shumov mnogokascadnykh rezonancnykh usiliteley (The noise band in multi-stage tuned amplifiers)*. DAN SSSR (Rep. of the USSR Acad. of Sci.) vol. 12 (1948), no. 4, pp. 473–475.
3. Kolosov, A.A. *Rezonancnye sistemy i rezonancnye usiliteli (Tuned systems and tuned amplifiers)*. Moscow: Svyaz', 1949. 559 pp.
4. Markov, G.T. and Sazonov, D.M. *Antenny (Antennas)*. Moscow: Energiya, 1975. 528 pp.
5. Skolnik, M.I., ed., *Radar Handbook,* New York: McGraw-Hill, 1970. 1520 pp.
6. Shirman, Ya.D., Golikov, V.N., Busygin, I.N. *et. al.* (Ya.D. Shirman, ed.). *Teoreticheskie osnovy radiolokatsii* (*The theoretical fundamentals of radar*). Moscow: Sovietskoe Radio, 1970. 559 pp.
7. Cherniy, F.B. *Rasprostranenie radiovoln* (*Radio wave propagation*). Moscow: Sovietskoe Radio, 1972. 463 pp.

Chapter 3
Methods for Calculating the Path Loss

3.1 INTRODUCTION

The operation of an over-the-horizon radar is governed by the physical processes by which the radar waves interact with the ionosphere to reach areas beyond the horizon. This interaction of the wave with the ionosphere, which in principle permits probing of regions beyond the horizon, has a significant influence on the energy characteristics of the transmitted signals, and is largely determined by the state of the ionosphere along the propagation path. The greatest influence is the change in the optimum radio frequencies for propagation, that is, the frequencies resulting in the least loss of radar power over the path. This leads to a requirement for adaptive tuning of the transmitted signal to match the changing ionospheric conditions.

The design of such an adaptive frequency capability, which is concerned primarily with the realization of the radar's full capabilities over the frequency band, presupposes the development of means to obtain real-time estimates of the optimum frequencies and resulting losses. The fundamental connection between the path loss and the radar frequency, the dependence of the minimum loss and the optimum frequency on the state of the ionosphere along the whole path, and the resulting strong dependence linking the radar's detection capabilities to these factors, all contribute to greater demands on the ability to determine the optimum frequency than are encountered in radio communications systems. This requirement necessitates a detailed analysis of the issues relating to the propagation of high-frequency radio waves in the ionosphere.

Numerous theoretical works in the field of HF ionospheric propagation have, to a significant extent, spread the understanding of the processes governing the energy characteristics of HF radiation in the far zone. A detailed bibliography of such works appears in Gurevich [17] and Kravtsov [23].

There are other questions, however, such as the role of upper ionospheric ducts, the function of ionospheric absorbing layers in the formation of fields along paths of varying extent, the formation of vertical field distributions, and so on, which are not individually treated in the literature. There is also a lack of adequate models of the global distribution of ionospheric characteristics affecting the propagation of radio waves. These deficiencies make the task of realizing theoretical methods in engineering practice more difficult. This has led to the need to develop semiphenomenological and statistical approaches to propagation problems (see [17, 28]). In these approaches, the basic functional connections beteween the field characteristics and the state of the ionosphere are obtained initially by one or another approximate solution (as opposed to explicitly solving or simply giving the characteristics of the medium). These estimates are rounded out with the addition of certain phenomenological constants having a statistical or deterministic character, which are determined experimentally. For the solution to be reasonable, the set of constants should be limited, and each of the constants should be associated with some easily grasped physical entity. The resulting model will be acceptable if it explains all the existing data with the proper constants. The constants which are determined by this approach may then be used to estimate the significance of one or another effect in forming the fields along actual paths, in those cases where more analytical estimates of the effects are not to be found in the theoretical works.

This approach clearly presupposes the existence of well-developed theoretical methods and a sufficiently complete set of independent experimental data describing the ionosphere for widely varying conditions. The experimental data may be obtained from studies of HF propagation characteristics using monostatic or bistatic oblique sounding along paths of varying length and orientation, on the basis of studies of round-the-world signals, and also using special measurements providing information on vertical field distributions. These subjects will be discussed in this chapter.

3.2 THEORETICAL METHODS FOR STUDYING THE LONG-RANGE PROPAGATION OF HIGH-FREQUENCY RADIO WAVES

3.2.1 The Wave Equation

The propagation of electromagnetic waves may be examined using the equations describing the electrodynamics of continuous media (Maxwell's equations [17, 36]):

$$\nabla \times \mathbf{H} = (i\omega/c)\tilde{\epsilon}\mathbf{E} + (4\pi/c)\mathbf{j}$$
$$\nabla \cdot \mathbf{H} = 0$$
$$\nabla \times \mathbf{H} = -(i\omega/c)\mathbf{H}$$
$$\nabla \cdot (\tilde{\epsilon}\mathbf{E}) = 4\pi\rho \tag{3.1}$$

Here $\tilde{\epsilon} = \epsilon + i(4\pi/\omega)\sigma$ is the complex dielectric permittivity of the medium, $\mathbf{j}$ is the current density, and $\rho = (i/\omega)\nabla \cdot \mathbf{j}$ is the charge density of an arbitrary source distribution. The magnetic permeability of the medium is taken to be unity. If there are no external currents or sources, then $\mathbf{j} = 0$ and $\rho = 0$, and we arrive at the vector wave equation [13]:

$$\Delta\mathbf{E} + k_0^2\tilde{\epsilon}\mathbf{E} = \nabla(\nabla \cdot \mathbf{E}) \tag{3.2}$$

where $k_0 = \omega/c$.

If the variation of $\tilde{\epsilon}$ is sufficiently smooth, which is true for a wide range of actual conditions, the right side of the last equation may be set to zero. The propagation of electromagnetic waves is then described by the vector form of the Helmholtz equation:

$$\Delta\mathbf{E} + k_0^2\tilde{\epsilon}(\mathbf{r})\mathbf{E} = 0 \tag{3.3}$$

In the general case, when $\tilde{\epsilon} = \tilde{\epsilon}(\mathbf{r})$, obtaining a solution to Maxwell's equation (3.1) or the vector equations in (3.2) and (3.3) becomes rather difficult.

If certain symmetries are assumed, however, the vector equations become somewhat simpler, and may be decomposed into independent scalar equations for the field components, similar to the Debye decomposition for homogeneous media (see [26, 34, 42]). This simplifies the problem to that of solving an equation like (3.3) for the scalar potential. However, the problem is still difficult to solve for the general case. This circumstance leads to the use of approximate solutions (see [2–8, 13, 17, 30]).

The character of the approximation depends on the conditions of the problem and is determined by the relative degree of variability in the medium, which in our case relates to fluctuations in the permittivity $\tilde{\epsilon}(\mathbf{r})$, the relation between the wavelength and the spatial scale of the inhomogeneities, and also the global geometry of the problem, including the path length, the width of the wave bundle, and so on. If the fluctuations in the medium are insignificant, then the scattering caused by these fluctuations will also be weak in comparison with the main wave. It is sensible in this case to use an approximation with just the primary scattering. As the fluctuations grow, it becomes necessary to take multiple scattering into account. If the scale l_ϵ of the inhomogeneities in the permittivity are much larger than the wavelength. $l_\epsilon \gg \lambda$, the scattering occurs primarily in a direction close to that of the incident wave, and three main approximation methods are used: the method of geometric optics and the more general asymptotic theory of diffraction, the second-order method, and the method of continuous perturbations (see [30]).

For a medium with small-scale variations in $\tilde{\epsilon}$, comparable to the wavelength ($l_\epsilon \approx \lambda$), a number of perturbation theories, primarily developed in quantum field theory, are applied for both weak and strong fluctuations.

3.2.2 The Qualitative Characteristics of Ducts

As is known from numerous experimental data, the ionosphere is a very inhomogeneous medium. This inhomogeneity manifests itself functionally as a dependence of the electron concentration N_e on altitude (see Fig. 3.1(a)). The electron concentration N_e in turn enters the equation for the permittivity:

$$\tilde{\epsilon}(\mathbf{r}) = 1 - \frac{4\pi e^2}{m\omega}\left(\frac{N_e(\mathbf{r})}{\omega - i\nu_e(\mathbf{r})}\right) \tag{3.4}$$

where e and m are the charge and mass of the electron; $\nu_e(\mathbf{r})$ is the frequency with which electrons collide in the ionosphere. Issues related to the inhomogeneous structure of the ionosphere are treated in Ginzburg [13], Gurevid [16], and Kolosov [22]. Without dwelling on these questions, we will note that studies of the structure of the ionosphere, carried out with radars and geophysical rockets, have shown that the ionosphere has a pronounced layered structure in altitude, the basic parameters of which are functions of latitude and longitude, and also exhibit seasonal and daily variations determined largely by the zenith angle of the sun.

In undisturbed conditions, the ionospheric structure in height may be represented by two regions, F and E; in daytime (especially in the summer or in the middle latitudes) the F layer is separated into the F_1 and F_2 layers, and in the lower layers an absorbing region called the D layer arises. Although, in general, $\tilde{\epsilon}(\mathbf{r})$ depends on all three coordinates, the most important factor is the height dependence—the variation in a horizontal direction is much slower than that in altitude:

$$\tilde{\epsilon}(\mathbf{r}) = \tilde{\epsilon}(x; y; z); \quad \left|\frac{\partial\tilde{\epsilon}/\partial x}{\partial\tilde{\epsilon}/\partial z}\right| \approx 10^{-2}$$

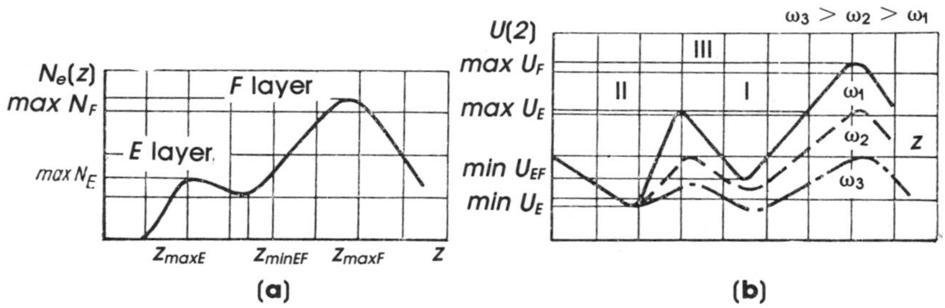

Fig. 3.1 Distribution of $N_e(z)$ and $U(z)$ in height.

The inhomogeneous structure of the ionosphere determined by the zenith angle of the sun has traditionally been considered to be regular. In addition, within the limits of a relatively small region near the earth's surface in which the sun's zenith angle may be considered constant, there are random inhomogeneities in the ionosphere which add to the regular structure. These variations may be considered to be small when compared with the large-scale inhomogeneities characterizing the regular structure of the ionosphere.

When solving an equation such as (3.3), the inhomogeneous structure of the ionosphere may be treated as follows. The permittivity $\epsilon(\mathbf{r})$ is written in the form:

$$\tilde{\epsilon}(\mathbf{r}) = \tilde{\tilde{\epsilon}}(\mathbf{r}) + \delta\tilde{\epsilon}(\mathbf{r}) \tag{3.5}$$

where $\tilde{\tilde{\epsilon}}(\mathbf{r})$ is the ensemble average over the entire medium; $\delta\tilde{\epsilon}$ is the fluctuation in the permittivity caused by the irregular structure. When describing the fields caused by the regular structure, for $\lambda \ll l_\epsilon$, the methods of geometric optics or second-order approximations are often used. The effect of the small-scale irregular structure of the ionosphere ($\lambda \sim l_\epsilon$) is determined for simple scattering or with the help of combinatorial methods and various perturbation theories, depending on the strength of the fluctuations. Before turning to a discussion of the results of studies of HF propagation using the methods just presented, we will formulate the basic questions addressed by those studies.

3.2.3 Ionospheric Ducts

The long-range propagation of radio waves is conditioned by the existence of ionospheric ducts, which arise due to the spherical shape of the earth and the layered structure of the ionosphere. The existence of duct propagation is easily established using the method of normal waves (see [24, 41, 42]) for a spherically symmetric ionosphere. In this case, (3.1) is reduced to a scalar equation for the Debye potential $U(\mathbf{r})$.

If the source is a vertical dipole, then considering the azimuthal symmetry of the system, (3.1) becomes

$$\tilde{\epsilon} \frac{\partial}{\partial r}\left(\frac{1}{\tilde{\epsilon}}\frac{\partial U}{\partial r}\right) + \frac{1}{r^2}\frac{1}{\sin\theta}\frac{\partial}{\partial \theta}\left(\sin\theta \frac{\partial U}{\partial \theta}\right) + k_0^2\,\tilde{\epsilon} U = -\frac{4\pi}{c}\mathbf{j} \tag{3.6}$$

where r and θ are the polar coordinates in a coordinate system with the origin at the center of the earth and the polar axis passing through the source.

Equation (3.6) lends itself to solution by separation of variables; the solution is found in the form $U(r, \theta) = \sqrt{\tilde{\epsilon} r}\,\Psi(r)T(\theta)$; (see [17]). For the radial function $\Psi(r)$ and the angular function $T(\theta)$, we find that in any region with no sources:

$$\frac{d^2\Psi}{d\eta^2} + [W(\eta) - \chi]\Psi = 0$$

$$\frac{1}{\sin\theta}\frac{d}{d\theta}\left(\sin\theta\frac{dT}{d\theta}\right) + \chi T = 0 \tag{3.7}$$

Here

$$W(\eta) = k_0^2\tilde{\epsilon}r^2 - \sqrt{\tilde{\epsilon}r}\frac{d^2}{d\eta^2}\left(\frac{1}{\sqrt{\tilde{\epsilon}r}}\right), \quad \chi = p(p + 1)$$

is the separation constant; $\eta = \ln r$.

The radial equation in (3.7) determines the profile of the normal wave, and the eigenvalues χ determine the square of the wave number. The angular equation is a canonical Legendre equation. The general solution of (3.7) has the form

$$U(r, \theta) = \sum_n C_n\Psi_n(r)P_n(\cos\theta) \tag{3.8}$$

where $P_n(\cos\theta)$ is the Legendre polynomial.

Considering that the ionosphere is a thin layer (in comparison with the radius of the earth, a), and making the substitution $r \to z = (r - a)$, $z \ll a$, the first equation in (3.7) may be placed in the form:

$$\frac{d^2\Psi}{dz^2} + k_0^2 (E - U(z))\Psi = 0 \tag{3.9}$$

Here $U(z) = -\tilde{\epsilon}(z)(1 + 2z/a)$.

Equation (3.9) is analogous to the Schrödinger equation of quantum physics, which describes a particle moving in a potential field $U(z)$. The quantity E plays the role of the energy level in the quantum mechanical equation. If there are minima in $U(z)$, then *potential wells* are formed, in which the particle may be captured. In the problem being considered, this means that if there are minima in $U(z)$, then there will be ducts within which electromagnetic waves may propagate.

The qualitative form of the function $U(z)$ for a two-layer ionosphere is shown in Fig. 3.1(b). For the case illustrated in this figure, it is possible to distinguish three regions over different height intervals within which electromagnetic waves may propagate. The region between the earth and the E layer is the E duct, for which the energy E is limited by $\max(U_E)$ and $\min(U_E)$; the region between the F and E layers is the FE duct, for which $\max(U_E) > E \geq \min(U_{FE})$; and the region between the earth and the F layer is the F duct, for which $\max(U_F) > E > \max(U_E)$.

For a source located on the surface of the earth, the value of E lies in the limits $0 \geq E \geq -1$ and is connected with the angle α_0 between the vector normal to the wavefront and the horizon through the relation $\cos^2 \alpha_0 = -E$.

As analysis shows (see [17]), the ducts as calculated above possess different absorbing characteristics. The total absorption of HF waves along the path from θ_0 to θ is determined by the equation: (see [17]):

$$A(\theta, \theta_0) = \int_{\theta_0}^{\theta} K(\theta)\, d\theta = \frac{4\pi e^2}{mc\omega^2} \int_{\theta_0}^{\theta} \frac{d\theta}{\Theta} \int_{z_{\min}}^{z_{\max}} \frac{N_e(z)\nu_e(z)}{\sqrt{E - U(z)}}\, dz \tag{3.10}$$

Here

$$K(\theta) = \left(\frac{\overline{d\Gamma}}{d\theta}\right) = \frac{4\pi e^2}{mc\omega^2\Theta} \int_{z_{\min}}^{z_{\max}} \frac{N_e(z)\, \nu_e(z)}{\sqrt{E - U(z)}}\, dz \tag{3.11}$$

is the absorption, averaged over one oscillation period:

$$\Gamma = \frac{2\pi e^2}{mc\omega^2} \int_{z_{\min}}^{z_{\max}} \frac{N_e(z)\ \nu_e(z)}{\sqrt{E - U(z)}}\, dz \tag{3.12}$$

is the absorption coefficient:

$$\Theta = \frac{2}{a} \int_{z_{\min}}^{z_{\max}} \frac{dz}{\sqrt{E - U(z)}} \tag{3.13}$$

is the period of oscillations; $z_{\min}$ and $z_{\max}$ are the turning points, corresponding to the level E and determined by the equation:

$$E - U(z) = 0 \tag{3.14}$$

As follows from (3.10), the absorption of the radio waves is determined by the imaginary component of the permittivity, and exhibits a marked altitude dependence, with the maximum values occurring in the regions of the lower ionospheric layers *E* and *D*. The radio waves contained within the various levels (referred to as propagation modes) therefore undergo varying absorption depending on the intensity of their interaction with the absorbing layers of the ionosphere.

An example illustrating the absorption of various modes, belonging to the ducts labeled *E, F,* and *FE*, is shown in Fig. 3.2 (see [17] for more detail).

As may be seen in Fig. 3.2, the various modes experience nearly identical absorption at night, but significantly different absorption levels during the day. This is due to the fact that the absorbing layers *E* and *D* are, for the most part, present during the day and absent at night. With long-range and extremely long-range propagation, the radio wave may pass through differing portions of the ionosphere, including both day and night regions. As has been noted, the absorbing properties of the ionospheric ducts in these regions may be quite different. Of importance in extremely long-range propagation is the between-layer *FE* duct, which exhibits relatively low absorption. The role of this duct, in forming the energy characteristics of a field at long ranges from the sources, is determined not only by its absorption properties, but by its ability to capture waves radiated from the earth's surface. In the final analysis, the role of the various ducts, including the *FE* duct, in forming fields at a given range, is thus determined by the quantitative absorption characteristics of the ducts and the quantitative characteristics of their ability to capture waves or be excited by a source located on the ground. The study of these

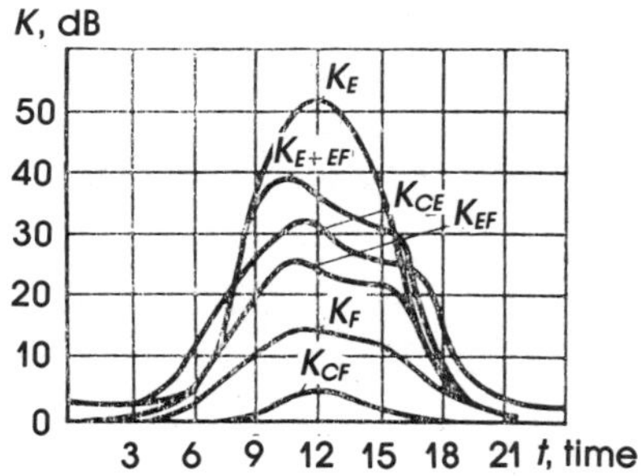

Fig. 3.2 Diurnal variation in the absorption coefficient K in various ducts.

questions and especially the issue of exciting the upper ionospheric ducts makes up a large fraction of the theoretical investigations of the propagation of *HF* waves over long paths.

The quantitative characteristics are highly dependent upon the mechanisms by which the waves are captured: refraction at the horizontal gradients in the ionosphere, diffraction effects, and scattering at irregular ionospheric inhomogeneities. The study of these mechanisms requires the application of various approximation methods. Comparative qualitative estimates of the effectiveness of the various capture mechanisms may be used to discern the important effects which govern the energy characteristics of fields in long-range propagation, and also to decide which effects may be neglected. It is thus possible to choose only those methods which are convenient to use in solving practical propagation problems.

3.2.4 Solving the Wave Equation for a Regularly Varying Medium

Referring to a number of works [2, 8, 13, 17, 23, 25, 30, 36, 37, 41] for details, we will present a brief treatment of the approximation methods most often used in solving the wave equation for a medium with regular characteristics. The *method of geometric optics* offers simplicity and ease of visualization, and is therefore often used in propagation studies. Its major drawback is the fact that it does not cover diffraction effects, owing to which its application is limited. In those areas where it is used, however, it has definite advantages over methods which do consider diffraction effects, enabling comparatively simple studies of such effects as the influence of regular refraction on the capture of waves in an ionosphere with horizontal inhomogeneities, and the amplification of field fluctuations in the neighborhood of a caustic surface.

Using the geometric optics approximation, the Helmholtz equation for the case of a smoothly varying medium in which $\tilde{\epsilon}$ changes only slightly over a wavelength ($\lambda \cdot |\nabla \tilde{\epsilon}| \ll \tilde{\epsilon}$), is solved by the *Debye method*. The solution is found in the form of the series:

$$U = \left(\sum_{n=0}^{\infty} \frac{A_n}{(ik)^n} \right) \exp(ik\Phi) \tag{3.15}$$

Using (3.15) in the Helmholtz equation and setting expressions of identical degree k to zero, we obtain the system of equations:

$$\begin{aligned} (\nabla \Phi)^2 &= \tilde{\epsilon} \\ 2(\nabla \Phi \nabla A_0) + A_0 \nabla \Phi &= 0 \\ 2(\nabla \Phi \nabla A_1) + A_1 \nabla \Phi &= -\Delta A_0 \\ \cdots\cdots\cdots \\ 2(\nabla \Phi \nabla A_n) + A_n \nabla \Phi &= -\Delta A_{n-1} \end{aligned} \tag{3.16}$$

Here ∇ denotes the gradient; Δ is the Laplacian operator. The first equation in (3.16) is called the characteristic equation, and the function Φ is the characteristic function. The equations for the amplitudes A_k are called the transfer equations. Analytical studies, as a rule, are limited to the zeroth term, so that in (3.15) only the zero term $A_0 \exp(i\Phi)$ remains.

This approximation is valid when the radius of the first Fresnel zone $\sqrt{\lambda L}$ (L is the distance traveled by the wave) is small in comparison with the scale of the inhomogeneities: $l_\epsilon \gg \sqrt{\lambda L}$. The characteristic equation corresponds to the rays for which the functional $\int \sqrt{\tilde{\epsilon}}\, dS$ is an extremum (the Fermi principle). The beam equations may be placed in the form:

$$\frac{d\mathbf{r}}{dS} = \mathbf{t}, \quad \frac{d\mathbf{t}}{dS} = \frac{1}{2\tilde{\epsilon}} [\nabla \tilde{\epsilon} - t(t, \nabla \tilde{\epsilon})] \tag{3.17}$$

where dS is an element of the beam length and $\mathbf{t}$ is the unit vector tangent to the beam.

If the beam equations are solved, then the characteristic and transfer equations may be integrated along the path of the beams:

$$\Phi = \int_0^S \sqrt{\tilde{\epsilon}}\, dS$$

and the amplitude A_0 is found from the condition of conserving the intensity in an infinitely thin beam of cross section $d\Sigma$;

$$\sqrt{\tilde{\epsilon}}\, A_0^2 d \sum = \text{const}$$

For the case of an inhomogeneous medium in which $\tilde{\epsilon}(\mathbf{r})$ varies randomly, an analytical solution for (3.16) is impossible. In this case perturbation methods must be used (see the works [2, 3, 30] for details).

Where the methods of geometric optics may be used, it is comparatively simple to make use of analytical solutions or numerous experiments to study the propagation of short waves. This approximation is sufficiently close to reality in describing such characteristics of the propagation of HF waves as absorption, time delay, spatial distributions, and the amplitude-frequency characteristics of oblique sounding signals along 1- and 2-bounce paths, so long as the ionospheric models used are sufficiently accurate. When using this method to model longer propagation paths, the results are not as close to experimental observations. This has the greatest impact on the description of the frequency dependence of the path loss, and especially in the upper portion of the observed signal band, and results, for example, in the calculated values of the maximum observed frequencies (MOF) of round-the-world signals being lower than those actually observed (see [17]).

Furthermore, geometric optics cannot explain the reception of round-the-world signals from a ground source at frequencies exceeding the maximum usable frequency (MUF), in the region of the initial entry of the wave into the ionsophere, which is regularly observed at midlatitudes during summer nights (see [28] and Sec. 3.4). This phenomenon is a manifestation of the part played by the upper ionospheric layers in forming the signals; analysis carried out using geometric optics indicates that such capture of waves in the upper ionospheric ducts is impossible in these conditions.

Using the *adiabatic approximation,* developed in Gurevich [17], is the next step after the geometric optics approximation. Being, in fact, a variation of the latter method, this method has several advantages connected with the simplicity of using it in the analysis of long-range propagation. We use the adiabatic approximation (or the method of adiabatic invariance) when the characteristics of the ducts change slowly, or "adiabatically," over the length of one wave oscillation, i.e.:

$$\mu = \frac{\Theta}{2\pi P_i}\frac{dP_i}{d\theta} \ll 1 \tag{3.18}$$

Here Θ is the period of oscillation of the beam (3.13); P_i is some characteristic of the waveguide duct, for example, its width or height; and μ is the adiabatic parameter.

If the condition of (3.18) is satisfied, it is possible to derive approximate integrals which give the path of the beams with sufficient accuracy. These are called adiabatic invariants.

As was noted above, if the beam varies symmetrically as $U = U(r)$, the beam will experience oscillations within the channel, characterized by the constant E. In the case $U = U(r, \theta)$ the quantity E is a function: $E = E(\Phi)$:

$$\frac{dE}{d\theta} = \left.\frac{dU}{d\theta}\right|_{z=z(0)} \tag{3.19}$$

The quantity $E(\theta)$ contains a component which varies smoothly with θ, and a component which oscillates rapidly (with period Θ) as well. In order to be considered adiabatic, the amplitude of the oscillations in E must be small. In this case it is convenient to extract the quantity averaged over the oscillations, $\overline{E}$, which varies smoothly with changes in θ. Then the trajectory of the beam will be characterized by the adiabatic invariant:

$$I = \frac{4}{a}\int_{z_{min}}^{z_{\max}} \sqrt{\overline{E} - U}\, dz \tag{3.20}$$

It may be shown that in adiabatic conditions, $dI/d\theta \approx 0$. For this, to exponential accuracy, $\Delta I \approx \exp(-C/\mu)$, $C \approx 1$. Thus, the adiabatic invariant I is the approximate integral of the trajectory equation.

Using the invariance in a weakly inhomogeneous ionosphere, it is possible to determine practically all of the characteristics of the trajectory of the beam. The path of the beam $\overline{E} = \overline{E}(\theta)$ is found in the form $\overline{E}(\theta) = \overline{E}(\theta, I_0)$ from the equation:

$$I(\overline{E}, \theta) = I_0 \tag{3.21}$$

where I_0 is the invariant at the initial point of the path. In (3.21) and in what follows, the average notation "–" will be dropped.

The dependence of the height of the beam z on θ is determined by the relation:

$$\theta = \theta_0 + \frac{1}{a}\int_{z_0}^{z} \frac{dz}{\sqrt{\overline{E}(\theta) - U(z, \theta)}} \tag{3.22}$$

The period of oscillations of the beam Θ and the number of oscillations of the beam P in the duct along the path from the point θ_0 to θ are found from the expressions:

$$\Theta = \frac{2}{a}\int_{z_{\min}}^{z_{\max}} [E - U(z, \theta)]^{-1/2}\, dz = \frac{dI}{dE} \tag{3.23}$$

$$P = \int_{\theta_0}^{\theta} \frac{d\theta}{\Theta} \tag{3.24}$$

If P is known, it is possible to determine the phase of the oscillations at the point θ:

$$\Phi = 2\pi(P - [P])$$

where $[P]$ is the integral part of the number P.

The group delay time of the signal pulses is given by

$$\tau = \left(\frac{a}{c}\right)\int_{\theta_0}^{\theta} (-E)^{-1/2}\left[1 + \frac{2z(\theta)}{a}\right] d\theta \tag{3.25}$$

where

$$\bar{z}(\theta) = \frac{2}{a\Theta}\int_{z_{\min}}^{z_{max}} [E - U(z)]^{-1/2}\, z\, dz \tag{3.26}$$

Finally, the absorption of the radio waves may be determined using (3.12). Thus, the adiabatic approximation may be used to determine all of the basic parameters of the path of the beams.

As has been noted, this approximation is also very useful in predicting the propagation of radio waves over long and extremely long distances, both below and above the maximum transmitted frequency. It also makes it comparatively simple to formulate the conditions under which waves are captured, transmitted, and released by ionospheric ducts. Compared with the monthly method which is usually used, and which takes into account only propagation by hops, the adiabatic invariant method provides much more information about the characteristics of radio wave propagation, making it possible to analyze the possible paths of the beams in the various ionospheric channels.

The basis of such analysis is the comparison of the invariant I_0, characterizing the path chosen for the analysis in the duct of interest (E, F, or FE) with the so-called maximum invariant of this channel:

$$I_m = \frac{4}{a}\int_{z_{\min}}^{z_{\max}} [U_m - U(z)]^{1/2}\, dz \tag{3.27}$$

where $U_m = E_m$ is the value of the potential $U(z)$ at its maximum, characterizing particular points, called separators (with corresponding modes), which separate the various ducts (see Fig. 3.1). In Fig. 3.1 the separator I divides channels F and FE, separator II divides channels E and F, and separator III determines the trajectory supported in the F layer.

The adiabatic invariant of any path contained within the given duct is less than I_m. The maximum invariant changes along the path: $I_m = I_m(\theta)$. The beam characterized by the invariant I in a given duct is supported in the duct so long as the condition:

$$I < I_m(\theta)$$

is satisfied.

If this condition is disrupted, the beam leaves the layer, the path of the beam nears the separator, where the adiabatic approximation breaks down: the adiabatic invariant changes sharply at this point. This necessitates the derivation of boundary conditions governing the propagation past a special point (see [17]). It should be noted that the quantity $E(\Theta)$ does not exhibit a jump, and the path of the beam remains continuous.

The construction of a map of the change in the maximum invariants along the path under consideration, at the various frequencies, makes it possible to establish a general idea of the propagation along the path, to examine the propagation of waves at various frequencies, and the conditions which are favorable for capturing waves in an ionospheric duct.

The condition for capturing a wave is expressed in terms of the invariant as follows:

$$dI_m(\theta)/d\theta > 0 \tag{3.28}$$

This condition is realized, as a rule, at the transition from day to night or from night to day, and corresponds to the effective increase in "volume" of the duct characterized by the potential function $U(z, \theta)$ in the direction of propagation.

The second-order method. The variations of the method of geometric optics, examined above, make it possible to treat refraction in the propagation of radio waves. Diffraction effects, however, remain outside the scope of these methods, which are limited by the condition $(\lambda L)^{1/2} \ll l_\epsilon$. If this condition is not satisfied, then the diffraction effects become substantial. One of the methods used to analyze the effects of diffraction, in both regular and random inhomogeneous media, is the second-order approximation. In deriving the second-order equation, we obtain from the Helmholtz equation:

$$\Delta U + k_0^2\bar{\bar{\epsilon}}\,(1 + \delta\tilde{\epsilon}/\bar{\bar{\epsilon}})U = 0 \tag{3.29}$$

where $\delta\tilde{\epsilon}/\bar{\bar{\epsilon}}$ is the relative variation of the dielectric permittivity; $\overline{\delta\tilde{\epsilon}/\bar{\bar{\epsilon}}} = 0$; $k_0\sqrt{\bar{\bar{\epsilon}}} = k$.

Let the inhomogeneous medium occupy the half-space $z > 0$ and let the plane wave described by $U = A_0 \exp(ikz)$ be incident upon it. We may express the field in the medium in the form:

$$U(\rho, z) = \vartheta(\rho, z)\exp(ikz) \tag{3.30}$$

Here $\vartheta(\rho, z)$ is the amplitude of the wave. In an inhomogeneous medium where the scale of the variation $l_\epsilon \gg \lambda$, the function $\vartheta(\rho, z)$, and its derivative with respect to z, change slowly over a wavelength:

$$\lambda\left|\frac{\partial\vartheta}{\partial z}\right| \ll \vartheta, \quad \lambda\left|-\frac{\partial^2\vartheta}{\partial z^2}\right| \ll \left|\frac{\partial\vartheta}{\partial z}\right| \tag{3.31}$$

The conditions in (3.31) are met when the back-scattered field is small. Placing (3.30) in (3.29) and noting (3.31), we arrive at the second-order equation for the amplitude:

$$2ik\frac{\partial\vartheta}{\partial z} + \frac{\partial^2\vartheta}{\partial x^2} + \frac{\partial^2\vartheta}{\partial y^2} + k^2\frac{\delta\epsilon}{\bar{\bar{\epsilon}}}\vartheta = 0 \tag{3.32}$$

The most rigorous discussion of the second-order equation is given in Rytor et al. [30] and Fok [37]. It is convenient to write (3.32) in spherical coordinates when analyzing the propagation of a wave in a medium with permittivity ϵ. In the case of azimuthal symmetry, (3.6), the equation takes the form (see [29])

$$\epsilon\frac{\partial}{\partial r}\left(\frac{1}{\epsilon}\frac{\partial\vartheta}{\partial r}\right) + i\frac{2k_0a}{r^2}\frac{\partial\vartheta}{\partial\theta} + \left(k^2 - \frac{k_0^2a}{r^2}\right)\vartheta = 0 \tag{3.33}$$

Equations (3.32) and (3.33) have physical significance, in that they express the slow transverse diffusion of energy of the wave with an increase in z or θ. The second-order equation was first used to solve the problem of radio wave diffraction by M. A. Leontovich [26]. The effectiveness of this method when applied to ionospheric propagation problems was shown in the works [29, 36–39]. Without dwelling on the details of the results of the enumerated works, we will note the following. The analytic and numerical results obtained in the works [29, 38, 39] in the study of the field intensity distribution along extended paths are in good agreement with the results obtained using the geometric optics approximation, outside the caustic zones and the regions of reflection from the upper layers of the ionosphere and the earth. The preliminary estimates performed in this work indicate that diffraction effects exert a slight influence (on the order of 10^{-4}) on the excitation of weakly damped propagation modes in the spherically symmetric *FE* layer of the ionosphere.

3.2.5 Methods of Solving Problems Involving a Medium With Irregular Structure

These include the small perturbation method, and the multiple scattering methods. The *method of small perturbations* (see [30]) is used when the fluctuations $\delta\tilde{\epsilon}$ are weak ($\delta\tilde{\epsilon}/\bar{\bar{\epsilon}} \ll 1$). The solution of (3.29) is found in the form of a power series for $\delta\tilde{\epsilon}$. To accomplish this, it is necessary to go from the differential equation (3.29) to the equivalent integral equation, with the help of the following procedure.

Let $U_0(\mathbf{r})$ be the field of the primary wave, satisfying the Helmholtz equation for $\delta\tilde{\epsilon} = 0$. Then

$$\Delta U_0(\mathbf{r}) + k_0^2\bar{\bar{\epsilon}}\, U_0(\mathbf{r}) = 0 \tag{3.34}$$

where $k_0 = \omega/c$.

We will denote the unperturbed Green's function through $G(\mathbf{r}, \mathbf{r}')$, for which

$$\Delta G(\mathbf{r}, \mathbf{r}') + k_0^2\bar{\bar{\epsilon}} G(\mathbf{r}, \mathbf{r}') = \delta(\mathbf{r}, \mathbf{r}') \tag{3.35}$$

where $\delta(\mathbf{r}, \mathbf{r}')$ is the delta function. The field $U_0(\mathbf{r})$ and the Green's function $G(\mathbf{r}, \mathbf{r}')$ both satisfy the required boundary conditions. Placing (3.24) in the form:

$$\Delta U(\mathbf{r}) + k_0^2\bar{\bar{\epsilon}}\, U(\mathbf{r}) = -\, k_0^2 \delta\tilde{\epsilon}\, U(\mathbf{r}) \tag{3.36}$$

and expressing the solution of the inhomogeneous equation (3.36) with the Green's function, the following integral equation is obtained:

$$U(\mathbf{r}) = U_0(\mathbf{r}) - k_0^2 \int G(\mathbf{r}, \mathbf{r}')\, \delta\tilde{\epsilon}(\mathbf{r}')\, U(\mathbf{r}')\, d^3\mathbf{r}' \tag{3.37}$$

The integration in (3.37) is over the volume V of the inhomogeneities $\delta\tilde{\epsilon}$.

The series solution is constructed by iterated application of (3.37):

$$\begin{aligned} U(\mathbf{r}) &= U_0(\mathbf{r}) - k_0^2 \int G(\mathbf{r}, \mathbf{r}')\delta\tilde{\epsilon}(\mathbf{r}')\, U_0(\mathbf{r}')\, d^3\mathbf{r}' \\ &\quad + k_0^4 \int G(\mathbf{r}, \mathbf{r}')\delta\tilde{\epsilon}(\mathbf{r}')\, d^2\mathbf{r}' \int G(\mathbf{r}', \mathbf{r}'') \\ &\quad \delta\epsilon(\mathbf{r}'')\, U_0(\mathbf{r}'')\, d^3\mathbf{r}'' - \cdots \\ &= U_0(\mathbf{r}) + U_1(\mathbf{r}) + \cdots + \mathbf{U}_k(\mathbf{r}) + \cdots \end{aligned} \tag{3.38}$$

This series is called a Neyman series, or the Bohr expansion. The second term in the series (3.38), linear in $\delta\tilde{\epsilon}$, describes the first scatter of the field and is called the Bohr approximation; the kth term describes the kth scatter. In the Bohr approximation, the fluctuation $\delta\tilde{\epsilon}$ is considered to be sufficiently small that (3.38) may be limited to just the first term U_1. If this is not the case, it is necessary to use additional terms.

Knowledge of the Green's function for the unperturbed equation and the correlation function of the permittivity fluctuations:

$$\Psi_\epsilon(\mathbf{r}, \mathbf{r}') \equiv \overline{\tilde{\epsilon}(\mathbf{r}')\tilde{\epsilon}(\mathbf{r})}$$

enable us to determine the basic quantities characterizing the scattering, such as the average intensity of the scattered field I_1, the average current density of the scattered energy J, and the effective scattering cross section $\sigma(\Theta, \phi)$.

We will consider the case when the Green's function $G(\mathbf{r}, \mathbf{r}')$ describes the field of a point source in an unbounded homogeneous medium:

$$G(\mathbf{r}, \mathbf{r}') = -\frac{1}{4\pi} \frac{\exp(ik|\mathbf{r} - \mathbf{r}'|)}{|\mathbf{r} - \mathbf{r}'|}$$

and the field of fluctuations $\delta\tilde{\epsilon}$ is statistically quasihomogeneous, i.e., the correlation function for $\tilde{\epsilon}$ has the form $\Psi(\mathbf{r}, \mathbf{r}') = \Psi(\rho, \mathbf{R})$, where $\rho = \mathbf{r} - \mathbf{r}'$, $\mathbf{R} = (\mathbf{r} + \mathbf{r}')/2$. $\Psi(\rho, \mathbf{R})$ thus changes slowly with $\mathbf{R}$ (the scale L_ϵ of the variation of I with $\mathbf{R}$ is substantially larger than the scale l_ϵ of the variation with ρ).

The average intensity of the scattered wave is determined with the expression (see [30]):

$$\overline{|U(\mathbf{r})|^2} = \bar{I}(\mathbf{r}) = \frac{\pi k_0^4}{2} \int_{-\infty}^{\infty} d^3\mathbf{R}\, I_0(\mathbf{R})\, \Phi_\epsilon(\mathbf{q}, \mathbf{R})(\mathbf{r} - \mathbf{R})^{-2}$$

where

$$\Phi_\epsilon(\mathbf{q}, \mathbf{R}) = \frac{1}{8\pi^3} \int \Psi_\epsilon(\rho, \mathbf{R})\exp(-\mathbf{i}\mathbf{q}\rho)\, d^3\rho$$

$I_0(\mathbf{R})$ is the intensity of the initial wave; $\mathbf{q}(\mathbf{R}) = k[\mathbf{n}_s(\mathbf{R}) - \mathbf{n}_1(\mathbf{R})]$ is the scattering vector; $\mathbf{n}_i$ and $\mathbf{n}_s$ are the unit vectors characterizing the direction of the incident and scattered waves, respectively.

The average energy current density is found from the relation:

$$\bar{J} = \frac{q}{2ik} \overline{(U_\nabla^* U - U_\nabla U^*)} = a\mathbf{n}_s \bar{I}(\mathbf{r})$$

where a is a normalization factor. Knowing the average energy current density, we find the differential scattering cross section $\sigma(\mathbf{n}_s)$ of a unit of volume in a unit of solid angle $d\Omega$ with direction $\mathbf{n}_s$:

$$\sigma(\mathbf{n}_s) = \frac{1}{V\,|J_0|} \frac{d\bar{P}}{d\Omega} = \frac{1}{2} \pi k_0^4 \Phi_\epsilon(\mathbf{q}) \tag{3.39}$$

Here V is the scattering volume; J_0 is the energy current density in the initial wave; $d\bar{P}$ is the average power scattered in the solid angle $d\Omega$.

It is possible to show that in the case of small-scale inhomogeneities, when the correlation function $\Psi_\epsilon(\rho)$ is nonzero in a small region $\rho < l_\epsilon \ll \lambda$, the function $\Phi_\epsilon(\mathbf{q})$ is independent of $\mathbf{q}$, and the scattering is isotropic, so that:

$$\sigma(\mathbf{n}_s) = \tfrac{1}{2}\, \pi k_0^4\, \Phi_\epsilon(0)$$

i.e., does not depend on the scattering angle.

In the case of large-scale inhomogeneities ($l_\epsilon \gg \lambda$), the spectral density $\Phi_\epsilon(\mathbf{q})$ decreases rapidly as $\mathbf{q}$ grows, i.e., with an increase in the scattering angle, which results in preferred forward scattering. The range of angles β, in which the scattering is concentrated, may be estimated from the condition $\beta \leqslant 1/kl_\epsilon$.

Knowing the differential scattering cross section, we find the total scattering cross section:

$$\sigma_0 = \oint \sigma(\mathbf{n}_s)\, d\Omega = \frac{1}{2} \pi k_0^4 \oint \Phi_\epsilon(\mathbf{q})\, d\Omega \tag{3.40}$$

and determine the average current density of the scattered energy and the average intensity:

$$\bar{J} = \frac{n_s}{r^2} |J_0| \, V_\sigma(\mathbf{n}_s), \quad \bar{I} = \frac{1}{r^2} I_0 V_\sigma(\mathbf{n}_s)$$

The total scattering cross section σ_0 is connected with the attenuation of the initial wave due to scattering, i.e., $|J_0(l)| \sim \exp(-\sigma_0 l)$.

This single-scatter approximation is valid with the assumption that the amplitude of the primary wave U_0, and, consequently, the current density J_0, are practically constant within the scattering volume $\sigma_0 V \approx \sigma_0 L^3 \ll L^2$, i.e., when the total cross section of the whole volume $\sigma_0 L^3$ is small in comparison with the area L^2 presented by the volume to the incident wave.

Multiple-scatter methods. When the single-scatter approximation is not applicable, multiple scattering must be considered. In addition to the methods described above—geometrical optics, the second-order equation, smooth perturbations, and the approximation of multiple scattering as a Markov random process (see [30]), methods for approximating the series (3.38) for the field $U(\mathbf{r})$ from perturbation theory are also used. In comparison with the methods enumerated above, these methods may be used for cases of both large- and small-scale inhomogeneities.

With these methods, all equations obtained in the various approximations are derived, and the limits of their application are indicated. In particular, in the so-called "scalariform" approximation, a diagrammatic technique and use of the Bethe-Solpiter equation gives a "statistical-wave" basis for the empirical transport equation, and establishes the diffraction character of this equation (see [30]):

$$\begin{aligned} \mathbf{n}\nabla j(\mathbf{R}, \mathbf{n}) = & -\Gamma j(\mathbf{R}, \mathbf{n}) \\ & - \oint \frac{\pi k^4}{2} \Phi_\epsilon \, [k, (\mathbf{n} - \mathbf{n}')] \, d\Omega(\mathbf{n}') j(\mathbf{R}, \mathbf{n}) \\ & + \oint \frac{\pi k^4}{2} \Phi_\epsilon \, [k, (\mathbf{n} - \mathbf{n}')] j(\mathbf{R}, \mathbf{n}') \, d\Omega(\mathbf{n}') \end{aligned} \tag{3.41}$$

Equation (3.41) characterizes the change in the current density:

$$|dJ| = j(\mathbf{R}, \mathbf{n}) \, d\Omega$$

in the solid angle $d\Omega$ with vertex at the point $\mathbf{R}$, as the current propagates through the medium. The quantity $j(\mathbf{R}, \mathbf{n})$ is called the radiant intensity. The first and second terms in the right side of (3.41) describe the attenuation due to absorption (Γ is the absorption coefficient) and scattering.

The third term describes the increases in the current in the direction **n** due to the scattering of currents initially propagating in different directions. The transport equation (3.41) is widely used to calculate the energy characteristics of radiation in scattering media.

To study the energy characteristics of ionospheric propagation, it is convenient to start with the radiation transport equation written in spherical coordinates. In the case of azimuthal symmetry, the equation for the distribution of intensity $j = j(\Theta; z; \alpha)$ takes the form (see [17]):

$$\frac{1}{a}\frac{\partial j}{\partial \theta} + \alpha\frac{\partial j}{\partial z} - \frac{1}{2}\frac{\partial U}{\partial z}\frac{\partial j}{\partial \alpha} = -\Gamma j - \sigma_0 j + \int \sigma(\alpha') j(\theta, z, \alpha) d\alpha \qquad (3.42)$$

Here σ_0 and $\sigma(\alpha)$ are the total (3.40) and differential (3.39) scattering cross sections, α is the angle with the horizontal, and z = constant:

$$U = -\bar{\bar{\epsilon}}\,(1 + 2z/a)$$

The left part of (3.42) describes the variation of the quantity $j(z, \theta, \alpha)$ along the ray trajectories of the beams, and the right part describes the change due to scattering and absorption. If the function j_0 is known at the initial point, its form may be determined at any point θ, using this equation to take into account refraction and scattering effects.

If the radiator is located at the point $\theta = 0$; $z = z_0$, then at the initial point $j = j_0(\theta, z_0, \alpha_0) = (P_0/c)G(\alpha_0)\delta(z_0)\cos\alpha_0$, where P_0 is the total radiated power; $G(\alpha)$ is the antenna pattern. Scattering leads to diffuse spreading of the function $j(\theta, z, \alpha)$ in the angle α.

Equation (3.42) may then be approximated in the form:

$$\frac{1}{a}\frac{\partial j}{\partial \theta} + \alpha\frac{\partial j}{\partial z} - \frac{1}{2}\frac{\partial U}{\partial z}\frac{\partial j}{\partial \alpha} = -\Gamma j + \frac{\partial}{\partial \alpha}\left(D_\alpha \frac{\partial j}{\partial_\alpha}\right)$$

where D_α is the diffusion coefficient in the vertical plane.

The transport equation takes an even simpler form when it is written in terms of the adiabatic invariant I (see ([17]):

$$\frac{\partial \bar{j}}{\partial \theta} = \frac{\partial}{\partial I}\left(\overline{D}\Theta\,\frac{\partial \bar{j}}{\partial I}\right) - 2\overline{K}(I)\bar{j} \qquad (3.43)$$

where

$$\overline{D} = 8 \int_{z_{\min}}^{z_{\max}} D_{\alpha}[E(I) - U(z)]^{1/2}\, dz$$

$$\overline{K}(I) = \frac{4\pi e^2}{mc\omega^2\Theta} \int_{z_{\min}}^{z_{\max}} \frac{N_e(z)\, \nu_e(z)}{[E(I) - U(z)]^{1/2}}\, dz$$

and the bar on K indicates the average over the period of oscillation Θ. The limiting cases of (3.43) occur for the total current energy at the edge of the layers ($I \rightarrow I_m$), and when the current density at the bottom of the layers is 0 ($I \rightarrow 0$).

3.2.6 The Mechanism by Which Radio Waves are Captured by Ionospheric Ducts

With the methods outlined above, it is possible to examine all of the currently known mechanisms by which HF waves propagate. The most important of these are skip propagation, sliding (anti-waveguide) propagation, and propagation through ionospheric ducts. The possible mechanisms for the capture and propagation of radio waves in ionospheric ducts is presented in the books [17, 23] with a detailed bibliography, and we will follow that presentation in this section.

The propagation of ground waves may be described in several ways. The field in the channel between the earth and the refraction layer may be taken to be the sum of normal waves, propagating along the channel, or a summing of plane waves alternately reflected from the earth and the ionosphere (see [8, 18, 21, 24, 38]). In the latter case, one speaks of a series of hops. In the short wave asymptotic limit, both of these models approach the geometrical optics approximation, which, encompassing the most significant aspects of radio wave propagation, is very convenient when considering horizontal inhomogeneities in the ionosphere. While explaining a large number of experimental observations, usually over short paths, this approximation is not always satisfactory for long-range propagation, in which the observed field strengths and transmission frequencies are greater than those predicted. This circumstance has led to the proposed existence of other propagation mechanisms, differing from skipped propagation. One of these mechanisms is "sliding" or "anti-waveguide" propagation (see [11, 19, 31, 32, 35]).

In an ionospheric layer having a minimum permittivity (max N_e), a sliding propagation mechanism may arise, by which one of the beams asymptotically approaches the maximum of the smooth N_e profile, while the neighboring beams, initially close to one another, gradually "fall off" from the vertex. The flatness of the layer at the maximum $N_e(z)$ facilitates confinement of the wave. These beams behave in a manner directly opposite to waves in a waveguide, hence the name "anti-waveguide" propagation. In the neighborhood of the outer beams the absorption of radio waves is less than the absorption experienced by skipped waves in the lower layers of the ionosphere. The strong divergence of the beams, however, leads to an exponential decrease in the field strength with distance, which significantly affects the range of sliding propagation. In addition, the sliding beams are very sensitive to random perturbations. This leads to the conclusion that, on the whole, it does not contribute fundamentally to the formation of fields at long ranges from the source, but it may play some role in ionospheric ducts.

Propagation through ionospheric ducts is very stable (see [4–7, 10, 11, 14, 17, 23, 38, 41]). From the point of view of minimizing absorption, HF waves propagate the best through the below-layer F duct at night, and in the between-layer FE duct during the day, inasmuch as they are located in height intervals where the product $N_e \nu_e$ is small. If the source is located on the ground, however, some sorts of capture mechanisms are required to capture the waves in these ducts. The most effective mechanism by which ionospheric ducts are excited is refraction at the horizontal inhomogeneities of the ionosphere.

Excitation of ducts through refraction is possible most of all in regions where there are regular horizontal longitudinal (at the day-night border) and latitudinal gradients in the electron concentration. Horizontal gradients below the maximum of the F layer are essential for exciting the below-layer F duct, and horizontal gradients in the E layer are required to excite the FE duct. In the last case, energy can propagate only for a well-defined relation between the electron concentration gradients in the E and F layers, which does not always exist in actual conditions. This is the apparent explanation for the fact that propagation through a twilight zone is observed to be significantly worse from the night side than from the day side. Numerical and analytical estimates (see [4, 7, 17, 18, 21, 33, 47]), performed using the methods of ray optics and the adiabatic invariant, have shown that the excitation of ionospheric ducts by refraction at horizontal gradients is extremely effective.

With narrow transmitted beams, a substantial fraction (10–20%) of the transmitted energy may fall in these ducts. However, large horizontal gradients in the electron concentration do not always exist. The experimental data on long-range propagation therefore suggest that there are less effective but more regular excitation mechanisms, including refraction at large-scale inhomogeneities, diffusion of beams in a randomly inhomogeneous ionosphere, excitation with the help of sliding propagation, and diffraction mechanisms, including sub-barrier seepage of fields through the wall of the duct, scattering at small-scale random inhomogeneities, and also polarization capture.

Refraction at large-scale ionospheric inhomogeneities is conditioned by the existence in the ionosphere of moving inhomogeneities on the order of 100–500 km in size, connected with sporadic type E_s formations and localization of the dimensions of inhomogeneities to several tens of kilometers. Although these inhomogeneities are characterized by smaller dimensions than those associated with twilight zones, they may in some cases be rather effective in exciting ionospheric ducts, and especially against a background of moderate regular gradients. Estimates performed using ray optics and perturbation methods show that the sectors of the beams which are captured are characterized by a fraction of a degree (with sufficiently narrow transmitter antenna patterns, this may consist of $10^{-2} - 10^{-3}$ of the radiated energy; see [22]).

Diffusion of beams in a randomly inhomogeneous ionosphere may be considered to result from the existence of a large number of randomly located large-scale inhomogeneities in the ionosphere. The diffusion approximation for beams (see [17] and Sec. 3.2) is statistical in nature, describes the refraction of waves at random inhomogeneities and takes into account multiple scattering processes. The diffusion of beams leads to the diffusion of energy across the boundaries, owing to which the layers are excited. Even with a slow diffusion process, such excitation may be rather strong, accumulating along extended paths.

The role of sliding propagation in exciting ionospheric ducts. The sliding mechanism may play an important role in the excitation of ducts in combination with one or another perturbations. Numerical modeling, performed for the case of a horizontally inhomogeneous ionosphere (see [32, 33]), has shown that sliding beams may excite both sub-layer and between-layer ducts; the ducts are fed by waves incident from within an angular sector estimated to be a fraction of a degree. The capture effectiveness is typically on the order of $10^{-2} - 10^{-3}$. Despite this, over long paths the combination of the sliding mechanism plus ionospheric ducting may compete with skipped propagation, because the loss experienced in excitation is compensated by a decrease in absorption for high-altitude waves captured in the ionosphere.

The role of sub-barrier seepage of a field through the wall of a duct may be estimated with the method of phase integrals (see [17]). Since the depth of the barrier exceeds the wavelength, the fraction of captured energy is small. Energy which has seeped into the duct gradually leaks back out of it due to the same effect. Calculations (see [7]) show, that in a between-layer ionospheric duct, only waves from a narrow sector of angles in the neighborhood of the separator effectively penetrate the duct, and the relative quantity of captured energy is approximately 10^{-3} for a radiation pattern of near 1°. Results on the same order are given by numerical field calculations using the parabolic equation method (see [29, 38, 39]). On the whole, the phenomenon of sub-barrier penetration is less effective than refraction mechanisms in ionospheric ducts, but, on the other hand, it is always present. There are additional capture mechanisms at natural and artificial inhomogeneities in the ionosphere. We will limit ourselves to those discussed above, referring the reader to Kravtsov et al. [23].

There is evidently a rather wide class of differing mechanisms for exciting ionospheric ducts. In the actual ionosphere all of these effects are apparently at work, but only those exhibiting minimal losses will predominate.

Dealing with the entire range of propagation mechanisms is a fundamental issue in deriving a completely satisfactory theory and in developing adequate methods for calculating the attenuation of radio waves over extended paths. Another difficulty encountered when developing such techniques is the absence of a model of the propagation medium which reflects with sufficient accuracy the true ionosphere, especially when describing irregular structures, which, as was noted above, exert significant influence on the excitation of ionospheric ducts. These difficulties may be avoided to some extent with the help of so-called semiphenomenological approaches (see [28]), which occupy a middle ground in relation to purely phenomenological descriptions.

The essence of such approaches is that the uncertainties associated with the choice of one or another approximate solution of the wave equation, and also with incomplete knowledge of the characteristics of the propagation medium, are handled with the help of constants, which are calibrated using experimental data. Such an approach will be used in Sec. 3.5 to construct a model for calculating the attenuation of transmitted signals over long paths. These phenomenological constants are calibrated, and the model verified, using experimental data on the frequency dependence of the energy of round-the-world signals examined in Sec. 3.4. Before turning to these questions, we will present several examples of the calculation of the field energy characteristics for oblique sounding of the ionosphere, using the ray optics approximation.

3.3 CALCULATING THE SPATIAL FIELD ENERGY CHARACTERISTICS WITH THE RAY OPTICS APPROXIMATION

The ray optics approximation is very often used in numerical analyses of the spatial energy characteristics of the propagation of HF waves, especially for relatively short (3–4 $\times$ 10^3 km) single-skip paths.

Numerical methods for calculating such characteristics based on ray optics have been widely developed in response to the recent interest shown by many investigators in oblique sounding of the ionosphere (see [40]). This method has unique capabilities, providing information on the state of the ionosphere out to several thousand kilometers in any direction, and also about the basic propagation characteristics, the information being contained in signals which are reflected from the ground at long ranges from the source. The oblique sounding method is apparently the most informative for optimizing the frequency and angle of radiation in OTH radars, allowing the propagation losses along the radar-target-radar path to be minimized. The intensity of the signals, characterizing the HF propagation conditions, may be measured as a function of the time delay, antenna polarization, carrier frequency, Doppler shift of the carrier frequency, azimuth and elevation angles for transmission and reception, and so on. Of all of these quantities, the most sensitive to the state of the ionosphere is the signal intensity. The complex character of these dependencies, however, often complicates the interpretation of the observed results. It is not currently possible to determine the structure of the ionosphere with oblique sounding signal data. The reverse procedure, at least, is possible—the expected characteristics of oblique sounding signals for rather complex ionospheric profiles may be calculated. Analysis of these calculated signal forms facilitates the understanding of the causes for their variations, and leads to a more accurate interpretation of observed data (see [44] and the references listed therein).

There is currently one realistic method for interpreting the observed characteristics of oblique sounding signals. This method is based on creating a data base of the calculated signal characteristics, corresponding to actual expected physical situations. The received signal data is compared with that stored in the computer, and on the basis of the similarities observed, it is possible to draw conclusions about the physical state of the ionosphere (see [44)].

A large number of methods have been developed for calculating the spatial and energy characteristics of the propagation of HF radio waves, such as the trajectory of the beams, the sounding signal characteristics, the vertical field distribution, and others, all based on the ray optics approximation (see [11, 13, 17, 36]). These methods have reached the level where they may be used to obtain synthetic data for any model of the ionosphere, for an actual or hypothetical transmission system. The adequacy of the synthesized data is actually determined by the quality of the model of the medium, and the limitations associated with using the data in computer calculations. The main difficulty encountered in the current development of numerical techniques is the absence of realistic models of the ionosphere, especially when attempting to describe the field characteristics at long ranges from the source. Numerical results, obtained with ray optics approximations, are therefore used most often for calculating the field characteristics over single-skip paths, because the problem of determining a description of the ionosphere to match experimental conditions is then greatly simplified, especially if data have been obtained from vertical sounding near the peak of the skip. Referring the reader to the works [11, 13, 17, 36], we will not present the detailed standard formulas for calculating the trajectory of the radio beams and the oblique sounding signal characteristics. We will only consider, as an illustration, several examples of calculating the field characteristics of HF fields for single- and multiple-skip paths.

The signal characteristics over a single-skip path for a single-layer model of the ionosphere are presented in Fig. 3.3. Similar results are shown in Fig. 3.4 for a two-layer spherically symmetric ionosphere. The indicated drawings show examples of the calculation of the distance-frequency characteristics, the dependence of the length of the radio path on the angle of radiation α in the vertical plane (Figs. 3.3(b), 3.4(b)) given a model of the distribution of the plasma frequencies f in height (Figs. 3.3(d)), 3.4(d)), and the radiation pattern $G(\alpha)$ in the vertical plane (Figs. 3.3(c), 3.4(c)). In contrast with the situation shown in Fig. 3.3(a), two oblique sounding signal paths are visible in Fig. 3.4(a) (1), (2), caused by reflection from the *E* and *F* layers with subsequent reflection from the earth's surface. A method laid out in Chernov [40] was used to construct the curves shown in Figs. 3.3 and 3.4. In order to exclude the influence of the scattering properties of the earth's surface, from which the oblique sounding signals are reflected, a normalized unit A/A_m is used along the ordinate of the frequency response graphs.

These predicted curves for the characteristics of oblique sounding signals may be used to analyze the distinguishing features of signal formation with a change in the character of the vertical distribution of the

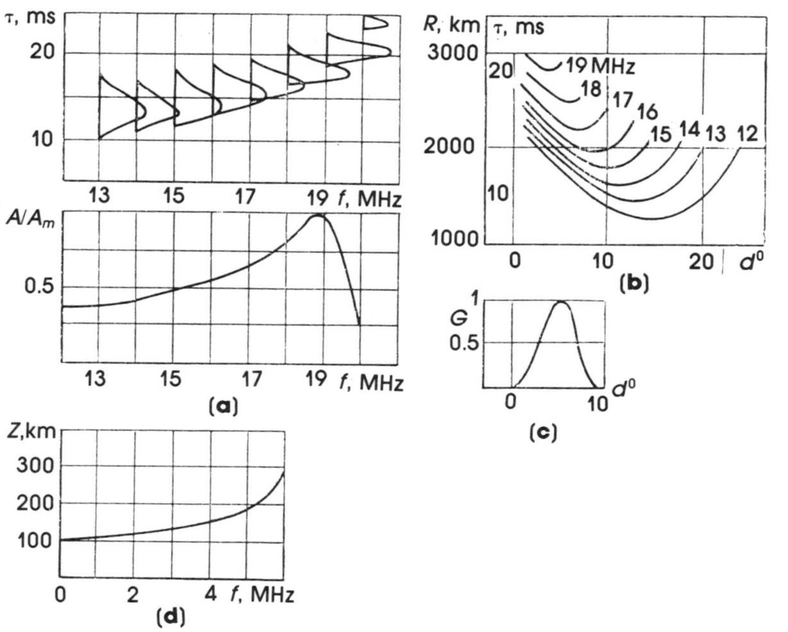

Fig. 3.3 The predicted characteristics of reflected oblique sounding signals, for a homogeneous single-layer ionosphere.

(a) spatial-frequency and spectral characteristics

(b) the function $R(\alpha)$ and the group delay τ (α) at the frequencies 12–19 MHz and various values of the angle α

(c) the elevation antenna pattern $G(\alpha)$

(d) the function $z(f)$, the change in the plasma frequency with height

electron concentration for both single-layer (Fig. 3.3(d)) and two-layer (Fig. 3.4(d)) models of the ionosphere. An analysis of the influence of the vertical distribution of the ionospheric electron density on the range- and amplitude-frequency responses of reflected oblique sounding signals is presented in Croft [44].

Figures 3.5 and 3.6 show the experimental and corresponding predicted distributions of reflected oblique sounding signals as a function of group delay τ. The experimental data were collected along a midlatitude path at an oblique sounding station (see [21]). The predicted curves were constructed on the basis of a model of the ionosphere (see [45]), corrected at the peak of the first skip with data obtained from vertical sounding. As is evident from the curves, the predicted responses satisfactorily describe the character of the amplitude distribution of the oblique sounding signals

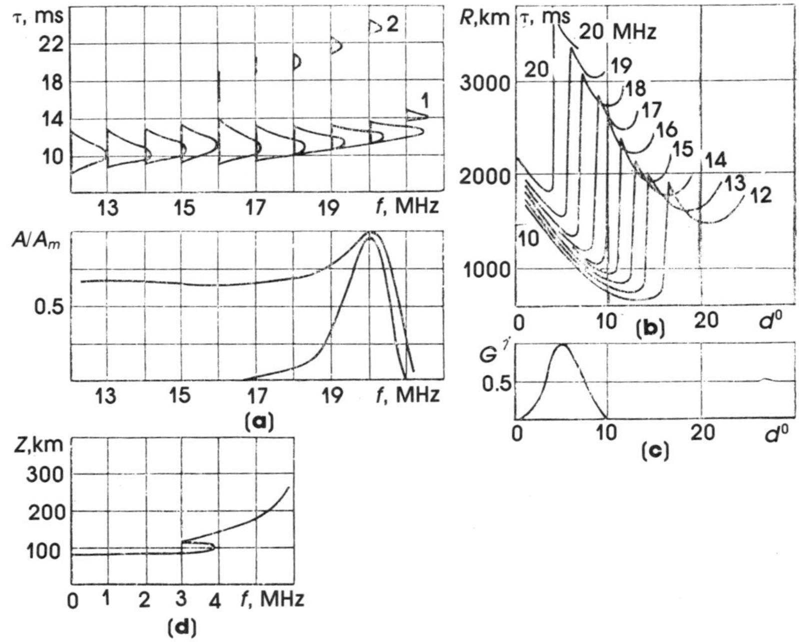

Fig. 3.4 The predicted characteristics of reflected oblique sounding signals for a two-layer ionosphere.
(a) spatial-frequency and spectral characteristics
(b) the group delay τ for various frequencies and angles α
(c) the elevation antenna pattern
(d) the change in the plasma frequency with height

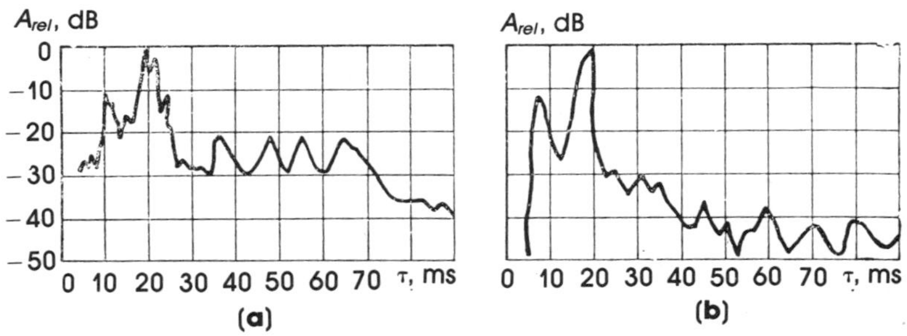

Fig. 3.5 Experimental (a) and predicted (b) distributions of oblique sounding signals *versus* the group delay τ for a midlatitude path (f = 22 MHz, winter, midday).

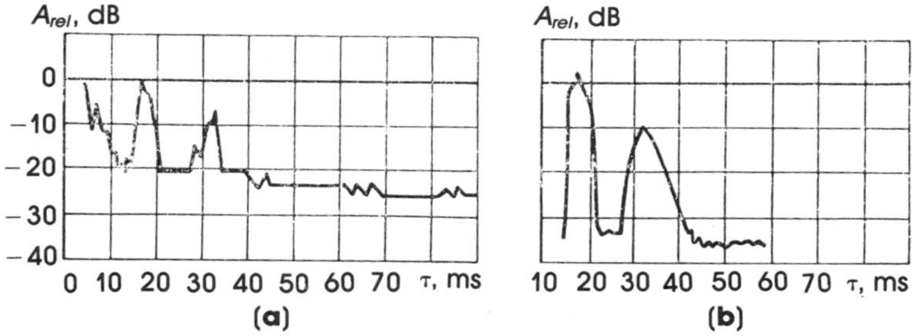

Fig. 3.6 Experimental (a) and predicted (b) distribution of oblique sounding signals *versus* the group delay τ for a midlatitude path (f = 10 MHz, winter, morning).

at ranges from one to two skips ($\tau < 40$ μs). The correspondence between the predicted and actual results grows substantially weaker at longer ranges. This is apparently connected with the errors in describing the ionosphere at long ranges from the source, along with errors in describing the scattering properties of the earth's surface, which were taken to be constant at all ranges in these calculations. In addition, the effects of scattering from irregular ionospheric structures were neglected, and, as was noted above, these effects may exert a significant influence on long-range propagation. A comparison of the experimentally obtained and the predicted diurnal distribution of signal delays for single-, double-, and triple-skip reflected oblique paths ROP-I, ROP-II, and ROP-III is presented in Fig. 3.7. The experimental data were obtained from observations along a midlatitude sounding path with the help of an ionosonde, and represent the median value of the observed delays, with the indicated intervals covering the range 0.1 to 0.9. The prediction of the diurnal variations in the sounding signal delays was carried out on the basis of the model in Ching and Chiu [45], using the antenna pattern used in the sounding system.

As may be seen in Fig. 3.7, the prediction reflects the actual character of the delays in ROP-I and ROP-II (single- and double-skip paths) rather well. As the range (delay) increases, the relation is upset.

An example of a modeled amplitude distribution $A(\tau)$ for oblique sounding signals is shown in Fig. 3.8, along with the vertical dependence of the field attenuation $W(z)$ at ranges between ROP-I and ROP-II, at the range ROP-II, and at a range exceeding ROP-III. It is apparent that backscatter energy is observed out to the range $R \approx 15{,}000$ km, and that out to 7500 km there are local increases in energy, corresponding to the first (ROP-I), second (ROP-II), and third (ROP-III) hops. The vertical

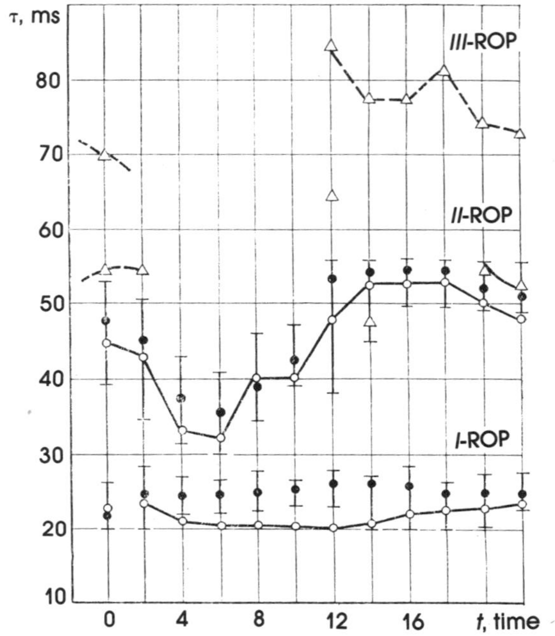

Fig. 3.7 Experimentally observed and predicted diurnal variations in the group delay for the signals I-ROP, II-ROP, and III-ROP, for a midlatitude path (winter).

distributions of the field strength decay at the ranges 4000 and 5500 km exhibit marked nonuniformities in the field strength, illustrating the irregular character of the fields in the near zone. The distribution $W(z)$ at the range
$R \approx 7500$ km already has a quasi-uniform character, which is a consequence of the fact that the energy radiated by a directive source in the elevation plane is almost uniformly distributed in the earth-ionosphere duct at ranges exceeding 7500 km, and is no longer irregular.

A comparison of the experimental and predicted spatial-energy characteristics of the fields from an HF source, based on these illustrations, indicates that they are very similar over ranges from one to two hops ($R \approx 4–5 \times 10^3$ km). In the opinion of many investigators, this extension of the methods of ray optics to the calculation of the field characteristics is the most simple and promising for the solution of engineering problems, so long as the effects which are not covered by the approximation may be neglected.

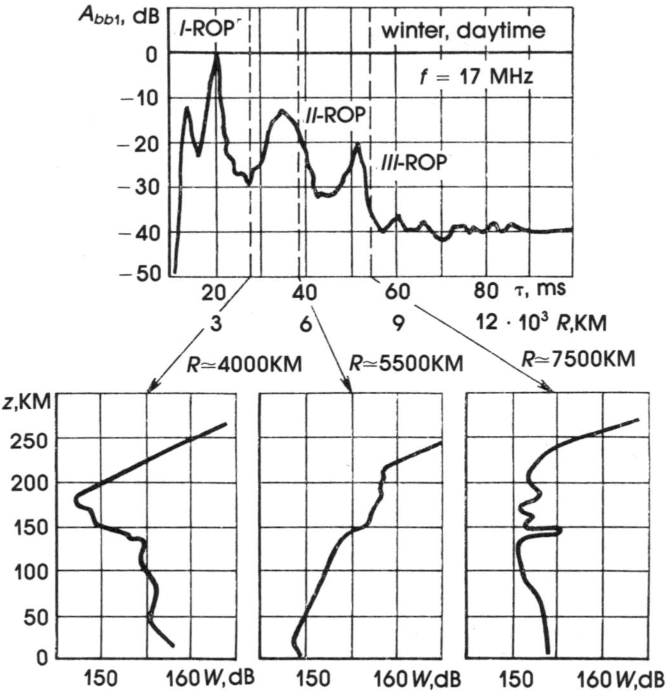

Fig. 3.8 Predicted field characteristics (distribution of oblique sounding signals and the vertical distribution of field attenuation), obtained using the ray optics approximation for a midlatitude path (winter, daytime, $f = 17$ MHz).

3.4 THE RESULTS OF EXPERIMENTAL INVESTIGATIONS INTO THE SPECTRAL CHARACTERISTICS OF ROUND-THE-WORLD SIGNALS

Round-the-world (RTW) signals carry information about the state of the ionosphere along propagation paths. The character of the frequency dependence of the overall attenuation of RTW signals is indicative of the state of both the lower and upper layers of the ionosphere, and of the extent to which they affect the field energy characteristics. The propagation paths of RTW signals exhibit characteristics which vary over a wide range, both in the lower absorbing layers, and in the electron density in the *F* layer. Experimental data on RTW signals thus contain fairly complete information about the processes governing long-range propagation of HF waves, and may be used both for verifying and calibrating models.

The traditional methods of registering RTW signals with the help of ground stations do not give direct information about the vertical field distribution, which, in light of the issues discussed in the previous section, is extremely valuable. Several results relating to the vertical field distribution have been obtained on the basis of measurements carried out with the help of geophysical rockets (see [9]). In addition, the necessary data may be extracted by studying the propagation of radio waves along significantly inhomogeneous paths, RTW paths often falling into this category, and also by analyzing the absorption of RTW signals and variations in the lower layers of the ionosphere.

Among the basic characteristics of round-the-world signals which are studied in experiments are the following: the preferred band of observed frequencies, the optimal azimuth angles for propagation, the time-dependent band of optimum RTW frequencies for various seasons, the variation in the absorption of RTW signals, and so on (see [14, 15, 17]).

To calibrate the model it is possible to use the results of RTW signal experiments carried out from 1974 to 1977 at a midlatitude observation point (see [28]), which basically confirmed the results obtained earlier by other authors (see [14, 15]). Below we present some data from Alebastrov *et al.* [28], used later to calibrate a model for the formation of the energy characteristics of long-range HF propagation, and to verify the extent to which it explains experimentally observed results.

The experiments were carried out using receiving-transmitting systems with antenna patterns which permitted the angle of arrival of the RTW signals to be selected in azimuth. The basic measurement methodology was based on a cycle which typically lasted 20 minutes, during which the spectral and azimuth responses of round-the-world signals were registered.

The seasonal-diurnal variations in the observability of RTW signals of the first and second frequency are shown in Fig. 3.9. Figure 3.10 presents the experimental data relating to the time dependence of the optimal azimuth angle of arrival, the minimum attenuation W and the time-dependent frequency band of RTW signals—with the median values of the maximum observed and lowest observed frequencies (MOF, LOF)— for three seasons. In Fig. 3.11 are presented the median values of the lower (LOB) and upper (UOB) limits of the optimum frequency band for RTW signals, measured at a level 6 dB down from the maximum of the spectral response, the solid and dashed lines corresponding to the $\frac{1}{2}F$ MUF (maximum usable frequency) and $\frac{1}{2}E$ MUF at the top of the first half-hop and in the region in which the waves first enter the F and E layers, respectively.

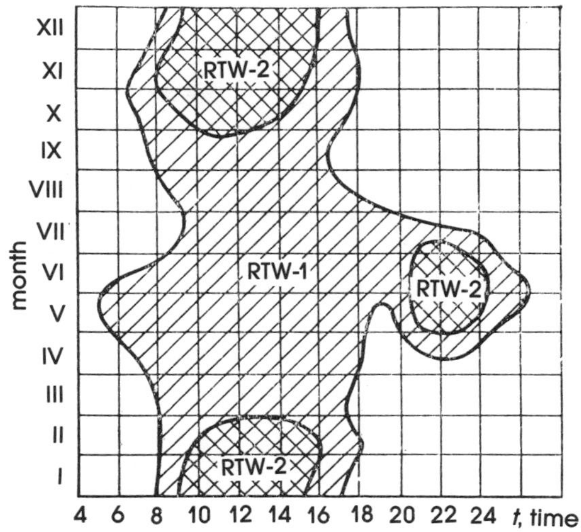

Fig. 3.9 Diagram of times during which round-the-world signals are observed.

[*Note: The notation for the various frequencies is that adopted by the National Bureau of Standards, wherein the first (optional) number indicates the number of hops, the italicized letter (possibly with subscript) indicates the layer in which the signal is reflected, and the ensuing letters indicate the nature of the frequency*–Tr.] The monthly variation in the value of the MUF in the *F* layer over the RTW propagation path is shown with the punctuated line. The following properties attract attention in Figs. 3.10 and 3.11. As a rule, the MOF-RTW (maximum observed frequency of round-the-world signals) $\leqslant \frac{1}{2}F$ MUF, and is 1.5 to 2 times greater than the minimum of *F* MUF (see [28]), except during the summer months, when the MOF exceeds $\frac{1}{2}F$ MUF, which may be attributed to the combined effects of a horizontal gradient in the *F* layer electron density at the peak of the first half-hop, along with scattering at irregular inhomogeneous structures. The value of LOF-RTW, as a rule, lies below $\frac{1}{2}E$ MUF in the region where the radio waves first enter the ionosphere. The band of optimum RTW frequencies, corresponding to minimal attenuation, satisfies the condition $\frac{1}{2}E$ MUF $\leqslant f_{\text{opt}} \leqslant \frac{1}{2}F$ MUF, except during summer days, when $f_{\text{opt}} > \frac{1}{2}E$ MUF. This bears witness to the fact that the characteristics of the *E* layer in the region of the initial entry of radio waves into the ionosphere do not always play a role in determining the optimum RTW

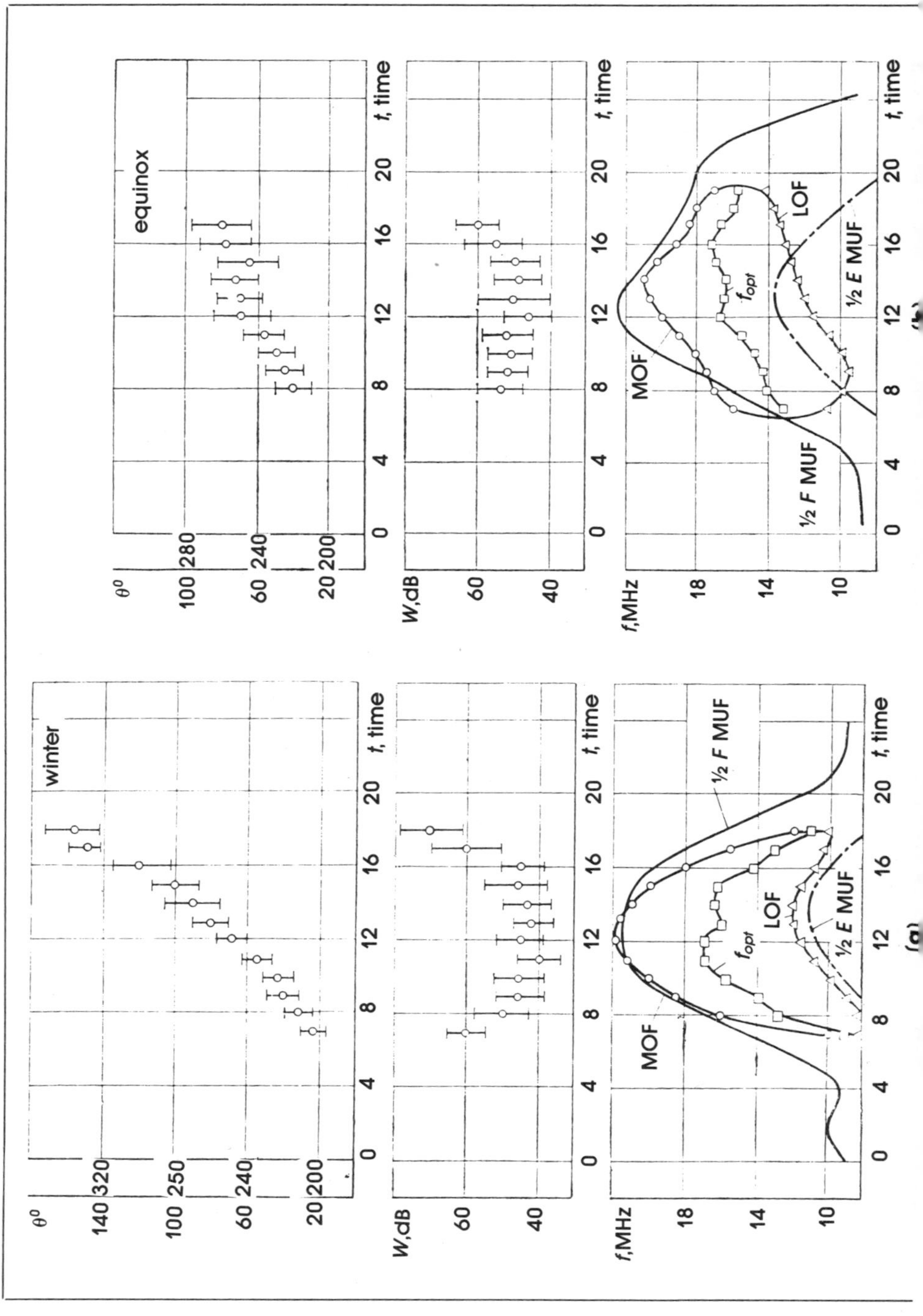
winter
equinox
θ°
W,dB
f,MHz
t, time
MOF
LOF
f_{opt}
½ F MUF
½ E MUF

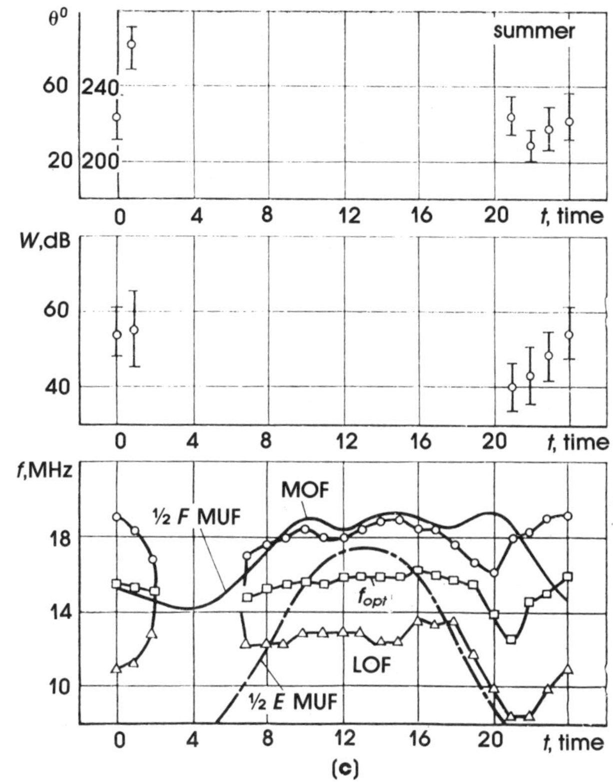

Fig. 3.10 Azimuth, energy, and frequency-time characteristics of round-the-world signals.
(a) winter (b) equinox (c) summer

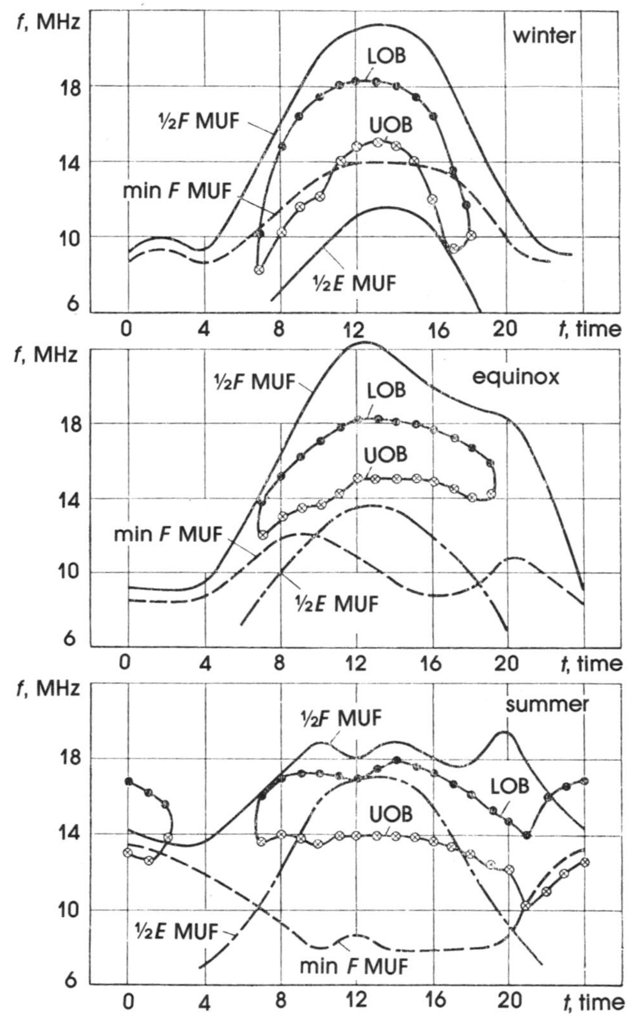

Fig. 3.11 Time dependence of the band of optimum frequencies for three seasons.

frequency, that is, the optimum frequency is in general a rather complex function of the distribution of critical frequencies of the *E* and *F* layers along the entire propagation path (see [28]).

The frequency dependence of the ionospheric losses W_i experienced by RTW radio waves is determined on the basis of the measured spectral characteristics of the RTW signals with the formula W_i =

$10 \log(E_0^2/E^2)$, where E_0 is the voltage calculated for free space at a range equal to a half-hop; E is the voltage at the reception point.

Typical curves for $W_i(f)$ for RTW signals are shown in Fig. 3.12, along with plots of the MUF for the F and E layers, plotted using monthly predictions. As may be seen in the drawings, the frequency responses $W_i(f)$ for RTW signals were obtained for paths which vary significantly in ionospheric characteristics. Later, precisely these characteristics will be used to calibrate the constants in the model.

The frequency dependences $W_i(f)$ for RTW signals, shown in Fig. 3.12, may be divided into two basic types: parabolic and flat. The flat curves are characterized by a weak dependence of attenuation on the lower frequencies, while the parabolic curves exhibit clearly defined minimum attenuation levels. The parabolic type results when it is close to midday

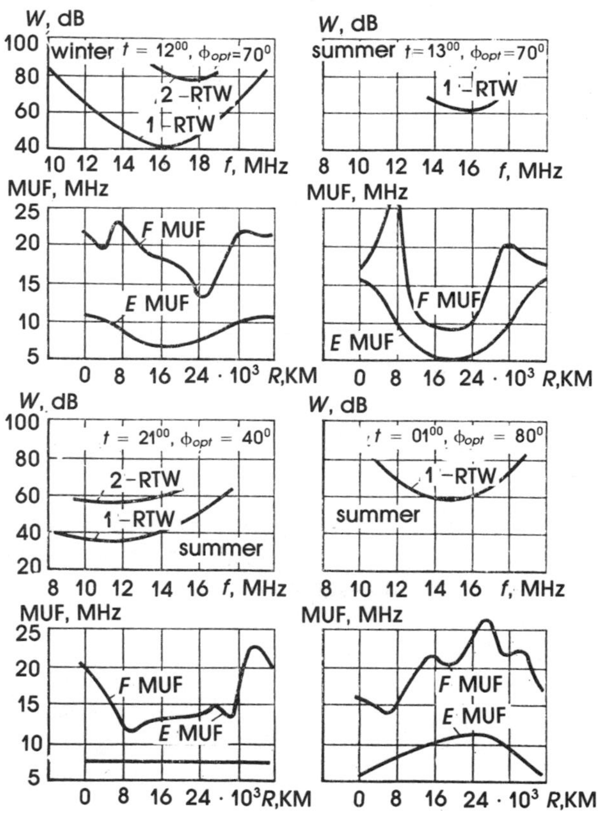

Fig. 3.12 Predicted frequency dependence of round-the-world signal attenuation for various paths.

or midnight at the observation point. The flat type is observed when it is near twilight, at azimuths close to the terminator. The slope of the low-frequency portion of the attenuation curves varies in the range 0–10 dB/MHz, at frequencies below optimal. The slope of the high-frequency portion varies in the limits 10–15 dB/MHz (see [28]). This is difficult to explain in terms of a model of a regular ionosphere with fairly weak random inhomogeneities.

The similarity of the frequency responses of the RTW attenuation for winter day and summer night conditions should be noted. In addition to 1-RTW signals, during winter daytime hours and summer nighttime hours, multiple RTW signals are observed (2-RTW). [*Note: The number preceding RTW is the number of "trips" taken by that signal*–Tr.]

The frequency-time dependences of the MOF and LOF for multiple RTW signals are shown in Fig. 3.13. If 1-RTW and 2-RTW signals are received simultaneously, it is possible to measure the attenuation due to the round-the-world propagation in the wave duct (removing the contribution of absorption on the way into and out of the duct):

$$\Delta W_{1\text{-}2} = 20 \log(A_{1\text{-RTW}}/A_{2\text{-RTW}}),$$

where $A_{1\text{-RTW}}$ and $A_{2\text{-RTW}}$ are the measured amplitudes of the multiple-hop signals; this is one of the most important propagation characteristics. In principle, by analyzing the frequency dependence of the attenuation, in conjunction with the characteristics of the lower absorbing layers along the path, one may estimate the extent of the interaction of the mode with the lower *E* and *D* layers, thus gaining information on the role of these or other ionospheric layers in determining the energy characteristics of fields along extended paths. Thus, low values of the absorption $\Delta W_{1\text{-}2}$ with a weak frequency dependence indicates the apparent advantage of propagating signals in the upper ionospheric layers, where the absorption is low. Increased values of $\Delta W_{1\text{-}2}$ with pronounced frequency dependence may point to the role of layers in which the radio waves interact fairly strongly with the lower absorbing layers.

Analysis of the frequency dependence shows that in the fall, winter and spring days, and also during summer nights, the quantity $\Delta W_{1\text{-}2}$ decreases sharply with increasing frequency, and is well approximated by the dependence C/f^2 (see Fig. 3.14). A completely different frequency dependence is observed during the twilight hours for RTW paths, passing close to the terminator. Under these conditions, $\Delta W_{1\text{-}2}$ is practically independent of frequency (Fig. 3.15).

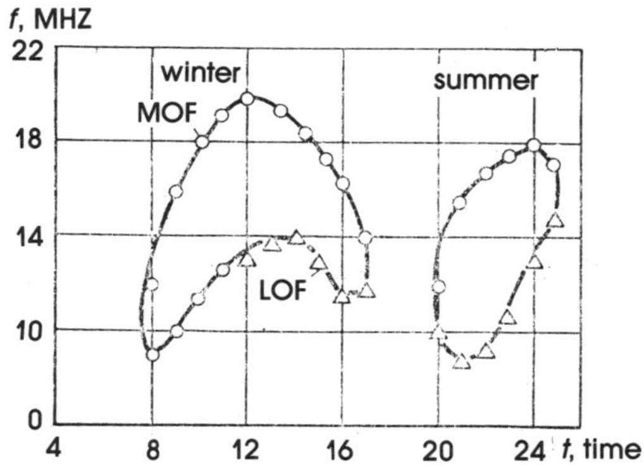

Fig. 3.13 Observed frequency-time ranges for 2-RTW in winter and summer.

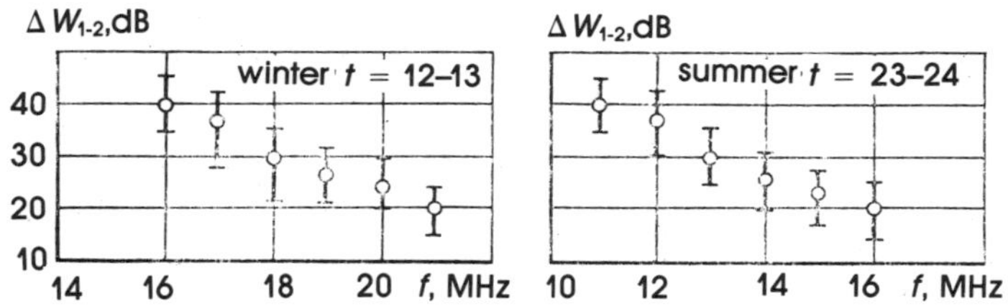

Fig. 3.14 Frequency dependence of the attenuation $\Delta W_{1\text{-}2}$ for multiple round-the-world paths for winter midday and summer night.

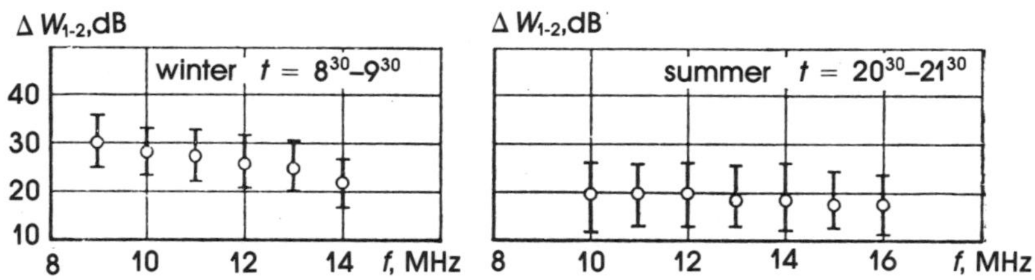

Fig. 3.15 Frequency dependence of the attenuation $\Delta W_{1\text{-}2}$ for multiple round-the-world paths, at winter and summer twilight.

The time dependence of absorption for a number of frequencies is shown in Fig. 3.16. As is evident, the average value of the absorption $\Delta W_{1\text{-}2}$, measured by simultaneous observation of multiple-trip RTW signals, varies in the range 15–45 dB, reaching its maximum value at noon in winter and at midnight in summer (see [28]).

Information about the primary effect of one or another propagation duct in forming the energy characteristics of the fields may be obtained indirectly from an analysis of the optimum RTW frequencies as a function of the inhomogeneity of the path. In particular, it is possible to assess the group of modes forming RTW signals by comparing the optimum frequencies of signal propagation with the minimum value of *F* MUF on the path. Thus, if the optimum RTW frequency is near the minimum value of *F* MUF for the path, then the RTW field will be formed by modes lying in the region $E \sim 1$; that is, the *F* duct plays the determining role. If the optimum frequency significantly exceeds the minimum value of *F* MUF at the path, then the field will be formed by *FE* modes. The experimentally observed relations between the optimum RTW frequency as a function of the inhomogeneity of the path and the critical *F* layer frequency are shown in Fig. 3.17.

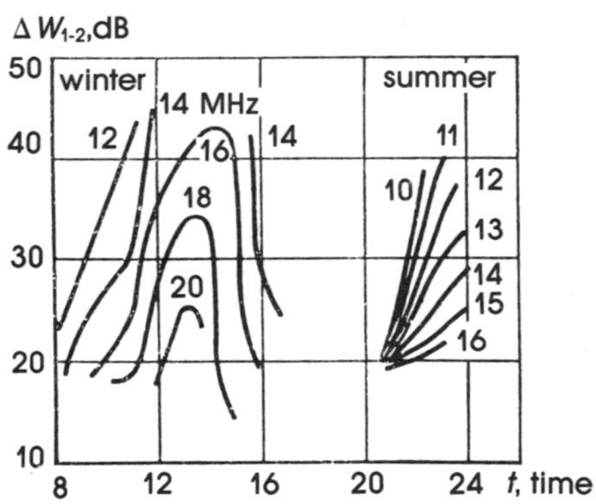

Fig. 3.16 Time dependence of the attenuation $\Delta W_{1\text{-}2}$ for multiple RTW paths, at various frequencies.

The degree of inhomogeneity of the path in Fig. 3.17 is characterized by the parameter $\chi = \min F\ \text{MUF}/\frac{1}{2}F\ \text{MUF}$. Here min *F* MUF is the minimum value of *F* MUF along the RTW path, determined by long-term prediction: $\frac{1}{2}F$ MUF is the maximum usable frequency in the *F* layer at the peak of the first half-hop.

As follows from Fig. 3.17, for the general inhomogeneous path, the optimum RTW signal frequency satisfies the condition min *F* MUF $< f_{opt}$

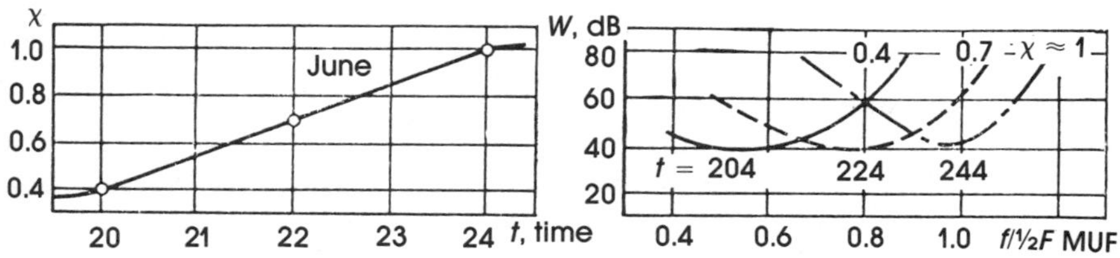

Fig. 3.17 Relation between the optimum RTW frequency and $\frac{1}{2}$ F MUF depending on the degree of inhomogeneity along the path.

$< \frac{1}{2}F$ MUF, somewhat exceeding the value min F MUF. For a homogeneous path ($\chi = 1$), $f_{\text{opt}} = \min F$ MUF $= \frac{1}{2}F$ MUF. This fact is an indication that the energy of RTW signals is formed by a group of modes, lying to the right of the mode for which the mode number $E = 1$.

It should be noted that RTW signals are recorded up to frequencies approximately two times greater than min F MUF for an inhomogeneous path ($\chi < 1$), and exceeding F MUF by a factor of 1.2–1.5 for a homogeneous path ($\chi = 1$), with the signal level lower than it would be at the optimum frequency (see [28]). This bears witness to the fact that at the indicated frequencies, RTW signals are propagated through modes which belong to the upper layer of the ionosphere. The distribution of field intensity by modes is characterized by a reduction in energy with an increase in the mode number $E \rightarrow \min U(z)$, compared to the energy in the $E = 1$ mode, and by a reduction in the RTW signal with increasing frequency, compared to the optimum frequency level.

The propagation of RTW signals at frequencies exceeding F MUF at the peak of the first half-hop is regularly observed at night in the summer, which testifies to the substantial contribution to the feeding of an ionospheric duct made by the effects of scattering at irregular ionospheric structures.

In order to shed light on the particular influence of the focusing region on the frequency characteristics of RTW signal attenuation, for reception at the point of radiation, these responses were measured simultaneously at a removed point. A comparative analysis showed that there were no noticeable differences in the frequency responses of RTW signals with identical receiving conditions at the transmission point and at the removed point. In addition, these experiments produced data on the frequency dependence of RTW signal attenuation at the transmission point and a removed point for the case where the critical frequency of the F layer was substantially different at these points. These data enabled an assessment to be made as to the change in the field level at the earth's surface due to a break in the duct in a region of reduced electron density in the F layer (at the twilight transition from day to night).

An example of such measurements is shown in Fig. 3.18. The frequency dependence of the RTW signal attenuation at the radiation point is plotted with a solid line, and at the other reception point with a dashed line. F MUF at the radiation point was 21 MHz, while at the removed point it was approximately 8–10 MHz. As may be seen, in these cases the transmission and reception of RTW signals are also observed at frequencies higher than F MUF at the removed point of reception.

Experimental data obtained from simultaneous measurement of the frequency dependence $W_i(f)$ for RTW signals at the point of radiation and a removed point, similar to that shown in Fig. 3.18, may serve as the experimental basis for estimating the effects of scattering and sub-barrier penetration in the excitation of, and release of energy by, ionospheric ducts. The spectral characteristics of RTW signals which have been derived are used to calibrate and verify models for estimating the attenuation of HF waves, as discussed in the following section.

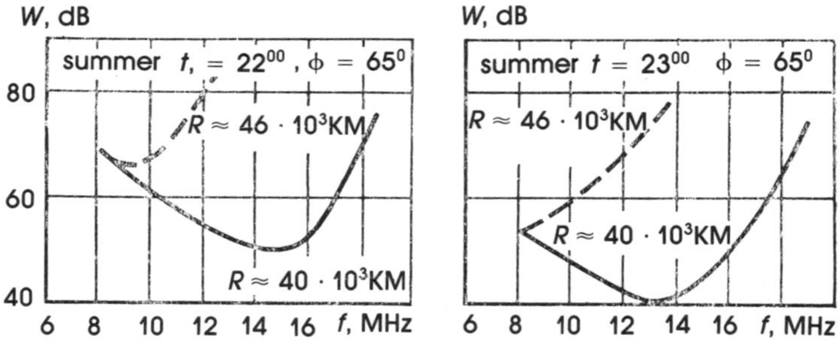

Fig. 3.18 The relative frequency dependence of RTW attenuation for signals received simultaneously at the transmitter site and a distant site.

3.5 ESTIMATING ATTENUATION USING A SEMIPHENOMENOLOGICAL MODEL

3.5.1 Design of the Model

We will consider one of the possible models explaining the energy characteristics of long-range propagation, developed on the basis of the semiphenomenological approach discussed above. The following scheme was adopted in constructing the model. The ray optics approximation was used as the foundation, describing the effects of regular ionospheric structures in forming the electromagnetic fields. The influence of irregular structures was modeled as a process of diffusion of the beam intensity obtained from the geometric optics analysis. This diffusion leads to a constant inter-layer exchange, which, on the one hand, excites the upper ionospheric ducts, and on the other, weakens the energy density in one or another layer due to "flourescence" of the energy in the neighboring ducts and free space.

The coefficient of diffusion is determined by a phenomonologically simplified functional dependence on the frequency of radiation, the geometry of the beam, and the parameters of the ionosphere along the propagation path. The difficulties associated with a detailed description of ionospheric irregularities may thus be avoided, and the problem is limited to that of the regular structure, which is taken care of with the help of existing models (for example, see [16, 45]).

We note that the inter-layer exchange, caused by diffusion, has a significant effect on the absorption of electromagnetic energy propagating in one or another duct. This is connected with the fact that the absorption of the modes, contributing to the field in a given duct, is determined not only by the absorbing properties of that duct, but also by the absorbing properties of the neighboring ducts, depending on the previous history of each of the modes and the strength of the diffusion, which is determined by the inter-layer exchange. Thus, the absorption of modes, contributing to the field in each channel at a given range from the source, should be described by functions which depend on the intensity of the diffusion and the absorbing properties of all the neighboring ducts.

With a given coefficient of diffusion of the beam intensity, the absorption may be found along the trajectory of each mode. This, however, requires a detailed description of the electron concentration $N_e(z)$ and the collision frequency $\nu_e(z)$ along the entire path.

3.5.2 Equation for the Distribution of Intensity by Mode Spectrum

The approach outlined above may be realized on the basis of a radiation transfer equation, the mathematical tools for which are rather well developed (see [30]). In the two-dimensional case, the transfer equation for the beam intensity may be written in the form (see [17]):

$$\frac{1}{a}\frac{\partial j}{\partial \theta} + \alpha\frac{\partial j}{\partial z} - \frac{1}{2}\frac{\partial U}{\partial z}\frac{\partial j}{\partial \alpha} = -\Gamma j + \frac{\partial}{\partial \alpha}\left(D_\alpha \frac{\partial j}{\partial \alpha}\right) \qquad (3.44)$$

We will call the function j of θ and α the intensity distribution in α.

Here we have: α is the angle made by the beam to horizontal; a is the radius of the earth; θ is the polar angle in a geocentric coordinate system; z is the height relative to the earth's surface:

$$U(z) = -\epsilon(z)\left(1 + \frac{2z}{a}\right)$$
$$= -\left(1 - \frac{4\pi e^2}{m\omega^2} N_e(z)\right)\left(1 + \frac{2z}{a}\right)$$

is the potential function; $\Gamma = 2\pi e^2 N_e(z)\nu_e(z)/mc\omega^2$ is the absorption coefficient; and D_α is the coefficient of diffusion for the angle α.

The left part of (3.44) describes the transfer of field intensity along the beam trajectories, and the right part describes its change due to diffusive spreading and ionospheric absorption. It is convenient to transform the transfer equation to a simpler form, averaging it in z over a period of beam oscillation, and choosing new independent variables $E = U(z)\cos^2 \alpha$ and $x = \theta a$. In this case the transfer equation may be written in the form (see [28]):

$$\frac{\partial \bar{j}}{\partial x} = \frac{\partial}{\partial E}\left(\overline{D}_E \frac{\partial \bar{j}}{\partial E}\right) - \overline{C}_E \frac{\partial \bar{j}}{\partial E} - \overline{\Gamma}_F \bar{j} \qquad (3.45)$$

In contrast with (3.43), (3.45) contains a "shift" term, corresponding to a redistribution in the field intensity spectrum of E, which occurs with the existence of regular gradients in the distribution of electron concentrations along the propagation path.

The solution of (3.45), given concrete functions for the coefficients $\overline{D}_E$, $\overline{C}_E$, $\overline{\Gamma}_E$, is, in principle, readily solved with numerical techniques. In particular, when the coefficients do not depend on x or E, the solution to this equation in the absence of boundaries is known to be

$$F(E, x) = \frac{1}{2\sqrt{\pi \overline{B} x}} \exp[-(E - cx)^2/4\overline{B}x - \overline{\Gamma}x] \qquad (3.46)$$

It is easy to see that (3.45) describes the evolution of the spectral density $j(E, x)$ with a change in x. We note that if $\overline{C}_E$ and $\overline{\Gamma}_E$ are equal to 0, then (3.45) describes the diffuse spreading of the field intensity by the mode spectrum. When $\overline{D}_E$ and $\overline{\Gamma}_E$ are equal to 0, the solution to the equation describes the mode transformation with a change in x, caused by horizontal electron concentration gradients. Finally, the case with $\overline{D}_E = 0$

and $\overline{C}_E = 0$, corresponds to the change in intensity due to ionospheric absorption. If the functional dependence of the initial field intensity distribution over E is known,

$$\bar{j} = j(0, E) = P_0 G(0, E) \tag{3.47}$$

where P_0 is the radiated power, and $G(0, E)$ is the radiation pattern, the general solution is found in the form:

$$\bar{j}(x, E) = \int \bar{j}_0(0, E')F(E - E', x)\, dE'$$

Taking account of the pattern $G_r(E)$ associated with the receiving equipment, the power P_r received at the reception point is described in the form:

$$P_r(x) = \int \bar{j}(x, E) G_r(E) dE$$

The power loss is $W = -10 \log(P_r/P_0)$.

If the coefficients $\overline{D}_E$, $\overline{C}_E$, and $\overline{\Gamma}_E$ are variable, an explicit solution to (3.45) cannot be found in the form of (3.46). In order to solve the equation, the coefficients must be expressed as specific functions of E, and the boundaries of the ionospheric ducts must be taken into account. To simplify the problem, we will consider three ionospheric ducts—*F, FE,* and *E*—each of which is characterized by its own set of coefficients $\overline{D}_i$, $\overline{C}_i$, and $\overline{\Gamma}_i$ $(i = F, FE, E)$, which are the averages over the spectrum of modes forming the ducts. We will use the indices "0" and "c" to denote the lower and upper boundaries of the *F, FE,* and *E* ducts (see Fig. 3.19). At the common boundary of the ducts:

$$E = E^c_{FE} = E^c_E = E^0_F$$

Equation (3.45) may be written for each of the ducts *F, FE,* and *E*:

$$\frac{\partial \bar{j}_i}{\partial x} = \frac{\partial}{\partial E}\left(D_i \frac{\partial \bar{j}_i}{\partial E}\right) - \overline{C}_i \frac{\partial \bar{j}_i}{\partial E} - \overline{\Gamma}_i \bar{j}_i \tag{3.48}$$

Joining the solutions to (3.48) at the duct boundaries $E = E^0_F = E^0_{FE} = E^c_{FE} = E^c_E$ and $E = E^c_F$, we introduce the function:

$$J_i(x) = \int \bar{j}_i(x, E)\Big|_{\overline{\Gamma}=0} dE, \quad B_i = \frac{\partial}{\partial x} J(x) \tag{3.49}$$

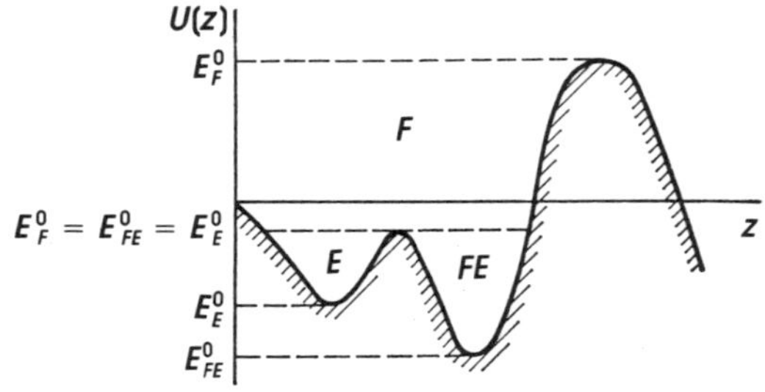

Fig. 3.19 Boundaries of the ionospheric ducts *F, FE,* and *E.*

describing the total field power in the given duct, with no absorption. Here $i = 0, F, FE, E,$ and J_0 is the total power radiated from the duct into free space for $E \geqslant E_F^c$. The integration is performed at the boundaries of the corresponding ducts. From (3.49) it is not difficult to obtain

$$\frac{\partial J_0}{\partial x} = - \sum_{k=F,FE,E} \left[\overline{D}_k \frac{\partial \bar{j}_k(x, E)}{\partial E} - \overline{C}_k \bar{j}_k(x, E) \right] \Bigg|_{E=E_F^c}$$

$$\frac{\partial J_i}{\partial x} = \sum_{k=F,FE,E} \left[\overline{D}_k \frac{\partial \bar{j}_k(x, E)}{\partial E} - \overline{C}_k \overline{J}_k(x, E) \right] \Bigg|_{E-E_i^c}$$

$$- \left[\overline{D}_k \frac{\partial \bar{j}_k(x, E)}{\partial E} - \overline{C}_k \bar{j}_k(x, E) \Bigg|_{E-E_i^0} \right] \quad (i = F, FE, E)$$

The solutions to the system of equations in (3.48) are joined at the boundaries of the ducts by requiring the total radiated power to be conserved, as follows:

$$\sum \frac{\partial J_i(x)}{\partial x} = 0 \quad (i = 0, F, FE, E) \tag{3.50}$$

An approximate solution to (3.48), convenient for numerical analysis, is found using (3.50):

$$\bar{j}_F(x + \Delta x, E)$$

$$= \int_{E_F^0}^{E_F^c} dE' \{ \bar{j}_F(x; E') \left[F_F(E - E'; \Delta x) \right.$$

$$\left. + \frac{F_F(2E_{FE}^0 - E + E'; \Delta x) \Delta E_{FE} + F_F(2E_E^0 - E + E'; \Delta x) \Delta E_E}{\Delta E_{FE} + \Delta E_E} \right]$$

$$+ \bar{j}_{EF}(x; E')[F_{FE}(E - E'; \Delta x) + F_{FE}(2E_{FE}^0 - E + E'; \Delta x)]$$

$$+ \bar{j}_E(x; E')[F_E(E - E'; \Delta x) + F_E(2E_E^0 - E + E'; \Delta x)]\}$$

$$\text{for } E_F^0 \leqslant E \leqslant E_F^c$$

$$\bar{j}_i(x + \Delta x; E)$$

$$= \int_{E_i^0}^{E_i^c} dE' \{ \bar{j}(x; E')$$

$$\times \left[\frac{F_F(E - E'; \Delta x) + F_F(2E_{FE}^0 + E' - E; \Delta x)}{\Delta E_{FE} + \Delta E_E} \Delta E_i \right]$$

$$+ \bar{j}(x; E')[F_i(E - E'; \Delta x) + F_i(2E_i^0 - E + E'; \Delta x]\}$$

$$(i = FE, E) \quad E_i^0 \leqslant E \leqslant E_i^c \qquad (3.51)$$

In this equation:

$$\Delta E_{FE} = E_{FE}^c - E_{FE}^0, \quad \Delta E_E = E_E^c - E_E^0$$

In the notation of (3.51), the redistribution of power, in the form of radiation from the F duct into the FE and E ducts, is taken to be proportional to ΔE_{FE} and ΔE_E, respectively. The value of $j(x, E)$ in any of the ducts at any distance $x \neq 0$ may be found by iterating (3.51), using an initial distribution $\bar{j}_0(0, E)$ in accordance with (3.47).

3.5.3 Choosing the Functional Dependencies of the Phenomenological Model Parameters

In order to solve (3.51) numerically, it is necessary to assign concrete functions for the coefficients $\overline{D}_i$, $\overline{C}_i$, $\overline{\Gamma}_i$ and the duct boundaries E_E^0, E_{FE}^0, E_{FE}^c, E_F^c along the path. The duct boundaries may be determined with the help of any model of the electron concentration distribution. The model in Ching and Chiu [45] was used in this capacity, supplemented in the inter-layer region with the value for the concentration N_d at a height of approximately 130 km:

$$N_d = \begin{cases} N_e \dfrac{\cos \chi}{0.1 + \cos \chi} 0.9 & \text{for } \cos \chi > 0 \\ 0 & \text{for } \cos \chi \leqslant 0 \end{cases} \qquad (3.52)$$

Equation (3.52) corresponds to the fact that at a transition from day to night conditions, the depth of the inter-layer trough grows sharply.

In constructing the functional dependence for the coefficients of diffusion $\overline{D}_i$, it is considered that $\overline{D} \sim (\Delta E)^2$, $\Delta E \sim \Delta U$, and $\Delta U \sim \Delta N_e/f^2$, where ΔN_e corresponds to the electron density functions. The coefficient of diffusion is accordingly given in the form $D_i = (\eta_i/f^2)^2$ ($i = E, FE, F$). With the assumption that the inhomogeneity is uniformly distributed in height from the E layer to the F_2 layer, the values η_i in the different channels are given by the relations $\eta_E = 0$, $\eta_F = \eta_{FE} = \eta$. This corresponds to the absence of diffusion in the E duct and identical diffusion in the F and FE ducts, the value of the coefficient η being chosen experimentally.

The absorption in the F layer is used as a reference for choosing the functions $\overline{\Gamma}_i$. The concrete function for $\overline{\Gamma}_F$ is formed using the Kazantsev method for calculating the absorption. Using this method, for typical hop lengths (~3000 km) and radiation angles (0°–10°), we obtain

$$\overline{\Gamma}_F = \gamma \frac{f_{0E}^2(x)}{f^2}, \quad \gamma = 0.1 \text{ dB/km}$$

The absorption in the E duct is taken to be two times greater than in the F duct: $\Gamma_E = 2\Gamma_F$.

To determine the function Γ_{FE} in the FE duct, an analysis was carried out on the results of calculations on the absorbing properties of various ionospheric ducts [17] at various latitudes in night and twilight conditions. With these data, the following function is obtained:

$$\overline{\Gamma}_{FE} = \left(\beta_0 + \beta_f \frac{N_d}{N_E} \frac{f_{0E}}{f} \right) \overline{\Gamma}_F$$

where N_d and N_E are the electron concentrations in the inter-layer region and at the maximum of the E layer; β_0 for a given choice of the model N_d (3.52) is the residual absorption in the FE duct in twilight (at the terminator) and night conditions; and β_f is the frequency dependence of the absorption due to the upward shift of the "bottom" of the FE duct at higher frequencies.

As a result of the analysis of typical electron concentration distributions in height, and the relations of the beam paths in various regions of the ionosphere, the following approximate expressions for the shift coefficients (describing the effects of horizontal gradients) were obtained:

$$\overline{C}_E = 0$$

$$\overline{C}_{FE} = \frac{\partial E^0_{FE}(x)}{\partial x}$$

$$\overline{C}_F = 0.2 \frac{\partial E^c_F}{\partial x} + 0.7 \frac{\partial E^c_E}{\partial x}$$

The coefficients η, β_0, and β_f were introduced above as constants of the model; we will now determine their values. We note that diffusion influences the effective absorption in each duct. In this sense the diffusion and absorption constants are interdependent, which significantly complicates the problem. In selecting the values for η, β_0, and β_f, we made use of the frequency dependencies of the attenuation $W(f)$ for round-the world signal propagation, which were presented in the preceding section.

It is expedient to start by determining the coefficient η with the high-frequency branch of the function $W_i(f)$. With this in mind, the most suitable function $W(f)$ is that for summer night conditions, when it is possible to receive RTW signals with $F > \frac{1}{2}F$ MUF at the peak of the first hop. The value which is closest to the experimental results is $\eta = 1.2\ \text{MHz}^2 \times 1000^{-1/2}$ km (Fig. 3.20(a)). This value is used in the following calculations.

The value of β_0 is determined from the frequency dependencies $W_i(f)$ and $\Delta W_{1\text{-}2}(f)$ over the round-the-world path, oriented along the terminator in summer twilight. The influence of variation in B_0 on the modeled characteristics of RTW signals, for these conditions, is shown in Fig. 3.20(b) (the attenuations $W_{1\text{-RTW}}$ and $W_{2\text{-RTW}}$ are shown for single and double RTW signals). It was determined that the value which best matches experimental results is $\beta_0 = 0$–0.1, which corresponds to absorption in the FE duct of less than 10 dB for $f = 10$ MHz (the missing portion of the RTW attenuation $\Delta W_{1\text{-}2}$ is caused by diffusion from the FE duct).

The coefficient β_f is determined by comparing the results of modeling the frequency dependence of the RTW attenuation $W(f)$, with the results of winter daytime measurements. It was observed that for small values of

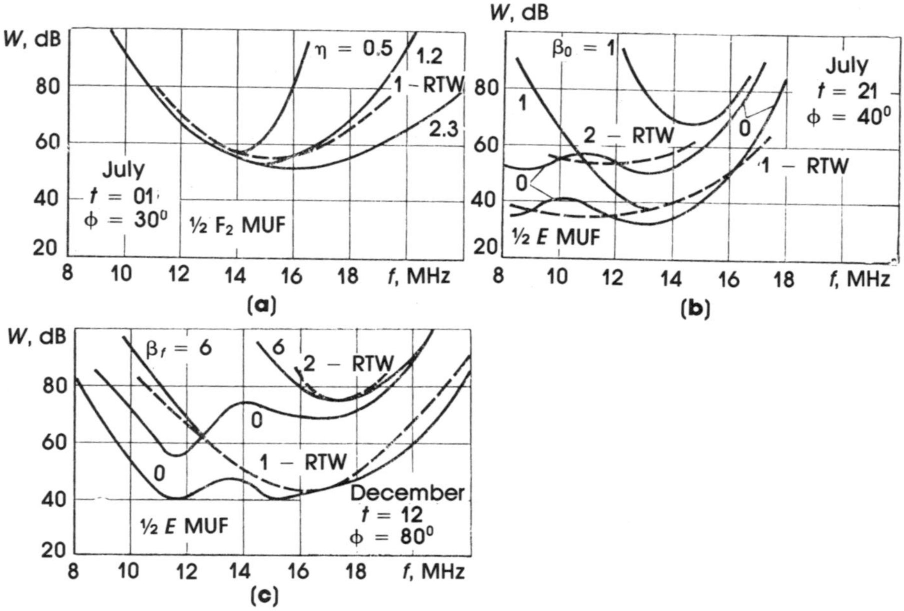

Fig. 3.20 The frequency dependence of RTW signal attenuation for various values of the model constants η (α), $\beta_0(\alpha)$, and $\beta_f(\alpha)$.

β_f, there appears a steep minimum in $W_i(f)$ at f = E MUF. As a result, $W_i(f)$ has two minima (Fig. 3.20(c)), which is never noticed experimentally under the specified conditions. This puts the value of β_f in the range 6–8, and corresponds to the fact that in daytime conditions at the frequency E MUF, $\Gamma_F = \Gamma_{FE}$, i.e., the absorption in the F duct is near that in the FE duct. The values of the phenomonological constants thus determined are not changed in what follows.

3.5.4 Results of Modeling the Frequency-Energy Characteristics of RTW Signals.

Using the model described above, the RTW attenuation was calculated at 1 MHz steps in frequency, and 10° steps in azimuth, for fixed moments in time. The optimum azimuth Φ_{opt} and the optimum frequency f_{opt}, corresponding to the minimum attenuation $W_{\min}$, were determined. The limits of the frequency and azimuth bands were determined by the attenuation level, according to the actual sensitivity of the RTW signal measuring equipment. A diurnal dependence was constructed from the following data: the optimum azimuth and width of the acceptable range of azimuth values, the optimum frequency and width of the band of acceptable frequencies within which RTW signals were collected, and the value of minimum attenuation. The following months were taken to be representative of the various conditions: for winter—December, for equinox—September, and for summer—July. An example of modeled frequency dependences for the ionospheric loss $W_i(f)$ experienced by RTW signals, for conditions corresponding to those under which experimental data was obtained (see Fig. 3.12), is presented in Fig. 3.21. Results of the calculations of the diurnal dependence of the optimum azimuth Φ_{opt}, the minimum attenuation $W_i(t)$ and the RTW frequency band $f(t)$ is shown in Fig. 3.22. Experimental measurements are also shown, for comparison. As reflected in the curves, there is good agreement between the modeled and experimental data, that is, at points where there is a large amount of experimental data, the predicted results are close to the measured results.

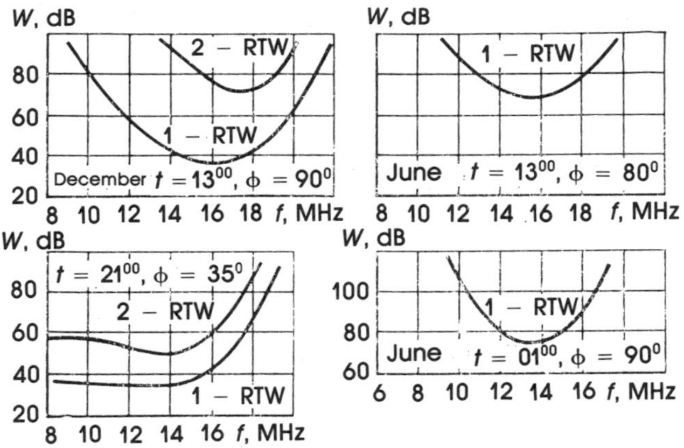

Fig. 3.21 The predicted attenuation of RTW signals for various observation conditions.

The model reflects the following important features of round-the-world signals:

1. The azimuth sector within which RTW signals are received, all day during winter and at night during summer, is narrow (about 20° at the 6 dB level), and close to the direction of minimal path illumination. During summer, fall, and spring days, there is no clear azimuth dependence in the absorption, and RTW signals are best propagated at azimuths along which F_2 MUF is greatest.

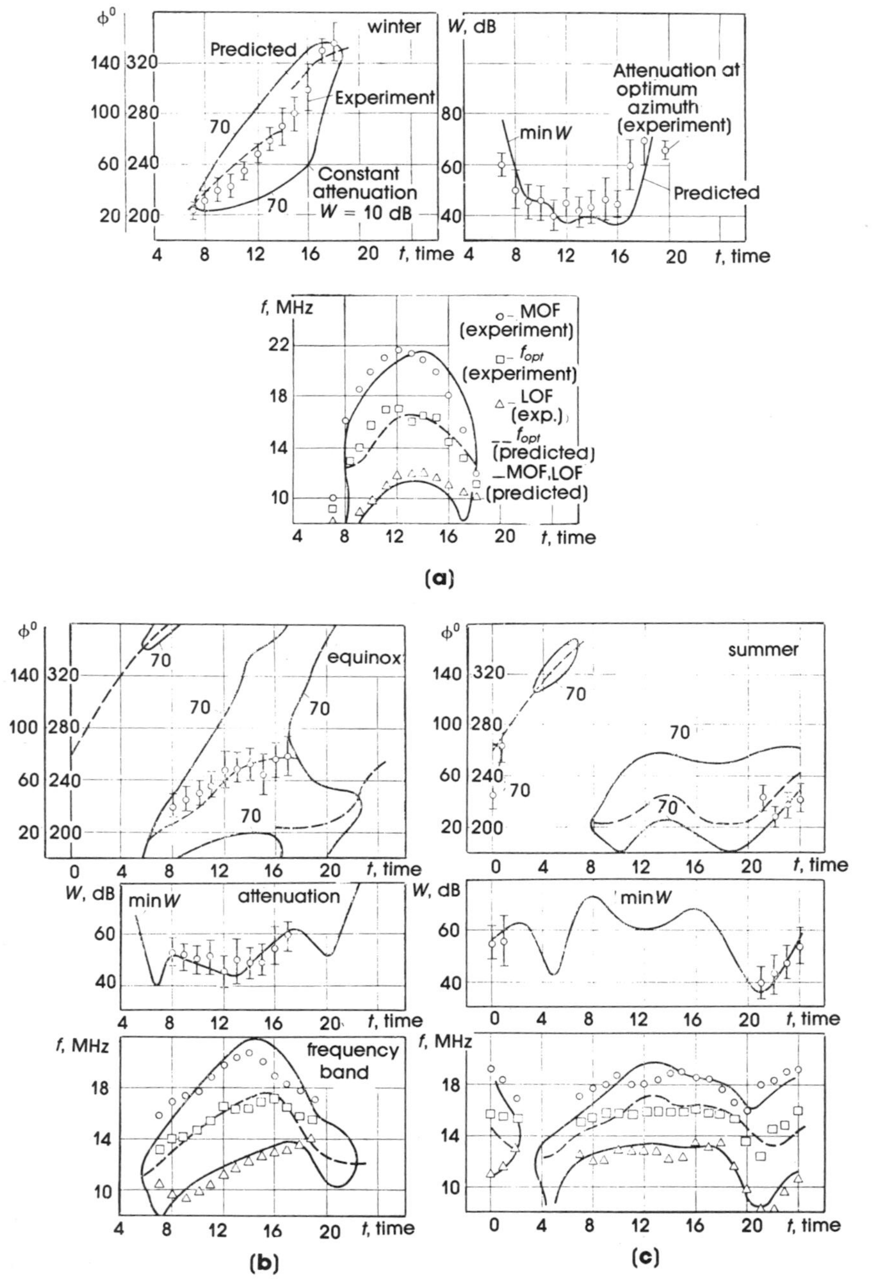

Fig. 3.22 The predicted and experimentally observed characteristics of RTW signals.
(a) winter (b) equinox (c) summer

2. The minimum attenuation W_i occurs for RTW signals during winter days and summer nights, which results in the model due to high values of F_2 MUF and relatively low absorption levels on the RTW path.
3. The lowest RTW signal frequency is observed along the terminator, and is due to the significant influence of the *FE* duct.
4. The ability to observe RTW signals at frequencies higher than F_2 MUF for the path is described in the model, in a natural way, through the effects of diffusion.

The predicted frequency dependencies of the ionospheric loss $W_i(f)$ in the ground-ionosphere duct are shown in Fig. 3.23, for a midlatitude path, 6000 km in length.

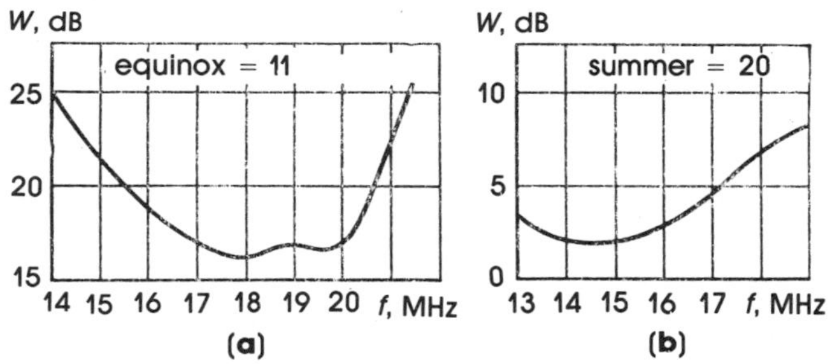

Fig. 3.23 The frequency dependence of the average attenuation of HF waves in the earth-ionosphere duct.
(a) equinox (b) summer

3.5.5 Further Development of the Model

A comparison of the model's predictions for the frequency-energy characteristics of long-range HF propagation of radiowaves, radiated from a source on the ground, with experimentally observed data, shows that the results coincide satisfactorily, which justifies the use of the model in estimating the radar attenuation experienced by sounding signals.

The most interesting, and possibly problematic, result obtained within the scope of the model, is when the contribution of a duct in forming the spectral characteristics of the RTW signal change dramatically with the orientation of the propagation path relative to the terminator line. The contribution of the *FE* duct is greatest for paths oriented at a minimum angle to the terminator line, and decreases monotonically as this angle increases. In particular, for paths passing through the point under the sun, the contribution of the *FE* duct in forming RTW signals may be neglected at frequencies $f \leqslant f_{opt}$. This result may be explained by the fact that the model is calibrated with experimental data on RTW signal measurements taken on the earth's surface. In addition, it may depend on the use in the model of the simplistic assumption concerning the independence of the diffusion coefficient D on the mode numbers in the *FE* duct. Thus, the possiblity is not excluded that the experimentally observed characteristics of RTW signals may be produced with another choice for the functional dependence of the diffusion coefficient D on the various modes forming the *FE* duct.

REFERENCES

1. Alebastrov, V.A., Belkina, L.M., Bocharov, V.I. *et al. Prostranstbenno-bremennye, energeticheskie i chastotnye kharaketeristiki signalov VNZ* (*The space-time, energy and frequency characteristics of obliquely reflected signals*). 13-ya Vsyecoyuznaya conferentsiya po rasprostraneniyu (The 13th All-Union Conf. on Propag.): Tez. Dokl. Ch. 1. Moscow: Nauka, 1980, pp. 203–204.
2. S.V. Butanovoy, ed. *Analiticheskie metodi v teorii difraktsii i rasprostraneniya voln* (*Analytic methods in the theory of wave diffraction and propagation*). Moscow: AN SSSR, 1970. 450 pp.
3. Baranov, V.A. and Kravtsov, Yu. A. *Metod vozmyshcheniy dlya luchey v neodnorodnoy srede* (*A method for exciting beams in a nonhomogeneous medium*). Izv. Vuzov SSSR. Radiofizika, vol. 18 (1975), no. 1, pp. 52–62.
4. Baranov, V.A., Popov, A.V., and Cherkashin, Yu.N. *Vliyanie rasseyaniya na krupnomasshtabnykh neodnorodnostyakh na zakhvat radiovoln v ionosferniy volnovodniy kanal (The effects of scattering at large irregularities on the capture of radio waves in ionospheric ducts.* Difraktsionnye Effekty Dekametrovykh Radiovoln v Ionosfere (Diffraction Effects for HF Waves in the Ionosphere). IZMIRAN SSSR. Moscow: Nauka, 1977, pp. 43–53.
5. Borisov, N.D. and Gurevich, A.V. *Raspredelenie polya korotkikh radiovoln v ionosfernykh volnovodnykh kanalakh* (*The field distribution of short radio waves in ionospheric ducts*). Issledovanie Sverkhdal′nevo Rasprostraneniya Korotkikh Radiovoln (Investigating Long-Range Propagation of Short Radio Waves in the Ionosphere IZMIRAN. Moscow: 1975, pp. 3–14.

6. ——, *Volnovaya teoriya rasprostraneniya v gorizontal'no neodnorodnoy ionosfere i zakhvata v kanaly korotkikh radiovoln* (The wave theory of propagation in a horizontally inhomogeneous ionosphere and the capture of short waves in ducts). Sverkhdal 'nee Rasprostranenie Radiovoln i Modeli Ionosfery (Very Long Range Radio Wave Propagation and Models of the Ionosphere). IZMIRAN. Moscow: 1977, pp. 13–29.
7. ——, *Vozbuzhdenie ionosfernykh volnovodov istochnikami, raspolozhennymi na zemle* (*The excitation of ionospheric ducts by sources located on the earth*). Izv. Vuzov SSSR. Radiofizika, vol. 20 (1977), no. 12, pp. 1876–1886.
8. Brekhovskikh, L.M. *Volny v sloistykh sredakh* (*Waves in layered media*). Moscow: Nauka, 1973. 343 pp.
9. Brylev, I.S., Kalinin, Yu.K., Kolosov, A.A., *et al. Difraktsionniy zakhvat korotkikh radivoln ionosfernym volnovodom i evo kharakteristiki, poluchennye s pomoshch'yu geofizicheskikh raket* (*The diffractive capture of short radio waves in an ionospheric duct and its characteristics, obtained using geophysical rockets*). DAN SSSR (Rep. of the USSR Acad. of Sci.), vol. 235 (1977), no. 4, pp. 802–804.
10. Budnov, V.A., Golyan, S.F., and Shlionskiy, A.G. *Osobennosti zatukhaniya sverkhdal'nikh signalov* (*Long-range signal attenuation*). Ionosfernye Rasprostraneniya Korotkikh Radiovoln (Ionospheric Propagation of Short Radio Waves). IZMIRAN. Moscow: 1975, pp. 90–96.
11. Voyk, E.E. *Formula radiotraektorii* (*The radio trajectory formula*). Geomagnetizm i Aeronomiya, vol. 14 (1974), no. 1, pp. 57–66.
12. Vinogradov, A.G., Kravtsov, Yu.A., and Feyzulin, Z.I. *Vliyanie zemnoy atmosfery na tochnost' radiotekhnicheskikh izmereniy* (*The effects of the earth's atmosphere on the accuracy of radio measurements*). Radiotekhnika, vol. 36 (1981), no. 12, pp. 21–31.
13. Ginzburg, V.L. *Teoriya rasprostraneniya radiovoln v ionosfere* (*The theory of radio wave propagation in the ionosphere*). Moscow-Leningrad: Gostekhizdat, 1949. 350 pp.
14. Golyan, S.F. and Shlionskiy, Sh.G. *Rasprostranenie krugosvetnykh ekho-signalov* (*The propagation of round-the-world echo-signals*). Geomagnetizm i Aeronomiya, vol. 11 (1971) no. 1, pp. 98–101.
15. Golyan, S.F. *Ob optimal'nykh ucloviyakh dal 'nevo rasprostraneniya korotkikh radiovoln* (*The optimum conditions for the long-range propagation of short radio waves*). Ionosfernoe Rasprostranenie Korotkikh Radiovoln (Ionospheric Propagation of Short Radio Waves). IZMIRAN SSSR. Ch. 2. Moscow: 1974, pp. 30–51.

16. Gurevich, A.V., Fishchuk, D.I., and Tsedilina, E.E. *Trekhmernaya analiticheskaiya model' raspredeleniya elektronnoy kontsentratsii spokoynoy ionosfery* (*Three-dimensional analytic model of the electron density distribution in a calm ionosphere*). Geomagnetizm i Aeronomiy, vol. 13 (1973), no. 1 p. 31.
17. Gurevich, A.V., and Tsedilinea, E.E. *Sverkhdal'nee rasprostranenie korotkikh radiovoln* (*Long-range propagation of short radio waves*). Moscow: Nauka, 1979. 245 pp.
18. Davis, K. *Radiovolny v ionosfere* (*Radio waves in the ionosphere*). Moscow: Mir, 1973. 502 pp.
19. Kazantsev, A.N. and Lukin, D.S. *Napryazhennost' polya korotkikh radiovoln, izluchaemykh iskusstvennym sputnikom Zemli* (*The field strength of short radio waves, radiated by an artificial earth satellite*). Kosmicheskie Issledovaniya, vol. 4 (1966), pp. 221–238.
20. Kalinin, Yu.K. and Ruchkin, A.D. *K metodike vydeleniya paketa normal'nykh voln, obladayushchikh minimal'nym dekrementom zatukhaniya, v sluchae tochechnovo istochnika, nakhodyashchevosya v sfericheskoy kvasisloistoy srede* (*On a method for extracting a packet of normal waves, which are minimally attenuated, for a point source located in a spherical quasi-layered medium*). Izv. Vuzov SSSR. Radiotekhnika, vol. 19 (1976), no. 2, pp. 240–243.
21. Kerblay, T.S. and Semenova, E.M. *O traektoriyakh korotkikh radiovoln v ionosfere* (*On trajectories of short radio waves in the ionosphere*) Moscow: Nauka, 1974, 160 pp.
22. Kolosov, A.A. *Struktura ionosfery* (*The structure of the ionosphere*). Moscow: Dep./NIIEIR, No. 3-6025. 1979. 69 pp.
23. Kravtsov, Yu.A., Tinin, M.V., and Cherkashin, Yu.N. *O vozmozhnykh mekhanizmakh vozbuzhdeniya ionosfernykh volnovodnykh kanalov* (*On the possible mechanisms by which ionospheric ducts are excited*). Geomagnetizm i Aeronomiya, vol. 19 (1979) no. 5, pp. 769–787.
24. Mrasnyshkin, P.E. *Metod normal'nykh voln v primenenii k probleme dal'nikh radiosvyazey* (*The method of normal waves applied to the problem of long-range radio communications*). Moscow: MGU, 1974. 52 pp.
25. Kravtsov, Yu.A. and Orlov, Yu.I. *Granitsy primenimosti metoda geometricheskoy optiki i smezhnye voprosy* (*The limits of applicability of the ray optics approximation and related questions*). UFN, vol. 132 (1980) no. 3, pp. 475–496.

26. Leontovich, M.A. *Ob odnom metode resheniya zadach rasprostraneniya elektromagnitnykh voln vdol' poverkhnosti Zemli* (*A method for solving problems concerning the propagation of radio waves along the earth's surface*). Izv. AN SSSR, Ser. Fizicheskaya (J. of the USSR Acad. of Sci., Phys. Ser.), vol. 8 (1944), no. 1, pp. 16–22.
27. Malykov, A.A., Orlov, V.N., and Popov, V.N. *Uravneniya traektorii rasprostraneniya korotkovolnovovo signala* (*Equations for the propagation paths of shortwave signals*). Geomagnetizm i Aeronomiya, vol. 15 (1975), no. 2, pp. 370–373.
28. Alebastrov, V.A., Kolosov, A.A., Kubov, V.N., *et al. Polufenomenologicheskaya model' formirovaniya chastotno-energeticheskikh kharaketeristik krugosvetnykh signalov* (*A semiphenomenological model of the formation of the frequency-energy characteristics of round-the-world signals*). Moscow: IZMIRAN, 1982. Preprint 7(732).
29. Kuz'minskiy, F.A., Kovaleb, L.K., Sakharov, I.V., *et al. Primenenie metoda parabolicheskovo uravneniya k raschetu volnovykh poley v ionosfere.* Moscow: IZMIRAN, 1979. Preprint 12(241). 54 pp.
30. Rytov, S.M., Kravtsov, Yu.A., and Tatarskiy, V.I. *Vvedenie v statisticheskuyu radiofiziky* (*Introduction to statistical radiophysics*). Part 2. Moscow: Nauka, 1978. 463 pp.
31. Sazhin, V.I. and Tinin, M.V. *O dal'nem rasprostranenii posredstvom lucha Pedersona* (*On long-range propagation through a Peterson beam*). Geomagnetizm i Aeronomiya, vol. 15 (1975), no. 4, pp. 748–749.
32. Sazhin, V.I. and Tinin, M.V. *Rol' skol'zyashchevo mekhanizma rasprostraneniya v vozbuzhdenii ionosfernykh volnovodov* (*The role of the sliding propagation mechanism in the excitation of ionospheric ducts*). Izv. Vuzov SSSR. Radiofizika, vol. 18 (1975), no. 9, pp. 1389–1393.
33. Sazhin, V.I., Semeney, Yu.A., and Tinin, M.N. *Nekotorye effekty vliyaniya gorizontal'nykh gradientov elektronnoy kontsentratsii na rasprostranenie korotkikh radiovoln* (*Some effects of horizontal gradients in the electron density on the propagation of short radio waves*). Issledovaniya po Geomagnetizmy, Aeronomii i Fizike Solntsa. Moscow: Nauka, 1974, no. 32, pp. 53–56.
34. Stretton, D.N. *Teoriya elektromagnetizma* (*The theory of electromagnetism*). Moscow: Gostekhizdat, 1948. 539 pp.

35. Tinin, M.V. *Rol' lucha Pedersena i svyazannykh s nim naprablyaemykh voln pri rasprostranenii voln vdol' parabolicheskovo sloya* (*The role of the Peterson beam and the directed waves associated with it in the propagation of a wave along a parabolic layer*). Issledovaniya po Geomagnetizmy, Aeronomii i Fizike Solntsa. Moscow: Nauka, 1973, no. 29, pp. 157–166.
36. Feinberg, E.L. *Rasprostranenie radiovoln vdol' zemnoy poverkhnosti* (*The propagation of radio waves along the earth's surface*). Moscow: Isz-vo AN SSSR, 1961. 546 pp.
37. Fok, V.A. *Problemy difraktsii i rasprostraneniya elektromagnitnykh voln* (*Diffraction and propagation problems for electromagnetic waves*). Moscow: Sovietskoe Radio, 1970. 517 pp.
38. Cherkashin, Yu.N. and Chernova, V.A. *K primeneniyu metoda parabolicheskovo uravneniya dlya rascheta volnovykh poley v neodnorodnoy ionosfere* (*On the application of the parabolic equation method for calculating wave fields in a nonhomogeneous ionosphere*). Difraktsionnye Effekty Dekametrovykh Radiovoln v Ionosfere. Moscow: Nauka, 1977, pp. 22–26.
39. Cherkashin, Yu.N. and Chernova, V.A. *Realizatsiya metoda parabolicheskovo uravneniya dlya rascheta volnovykh poley v ionosfere s elektronnoy kontsentratsiey, zadannoy v vide tret'ey stepeni* (*The realization of the parabolic method for calculating wave fields in the ionosphere with an electron density described by a third-order polynomial*). Rasprostranenie Dekametrovykh Radiovoln (The Propagation of HF Radio Waves). IZMIRAN. Moscow: 1978, pp. 5–35.
40. Chernov, Yu.A. *Vosvratno-naklonnoe zondirovanie ionosfery* (*Oblique sounding of the ionosphere*). Moscow: Svyaz', 1971. 104 pp.
41. Bremmer, H. *Terrestrial radio waves.* Theory of propagation. Amsterdam, 1949, 343 pp.
42. Budden, K.G. *Radio waves in the ionosphere.* Cambridge, 1961. 542 pp.
43. Carrara, N., De Giorgio, M.T., and Pellegrini, P.F. *Guided propagation of HF radio waves in the ionosphere.* Space Science Review vol.11 (1970), no. 4, pp. 555–592.
44. Croft, T.A. *Sky-wave backscatter: A means for observing our environment at great distances.* Rev. Geophys. and Space Phys. vol. 10 (1972), no. 1, pp. 73–155.

45. Ching, B.K. and Chiu Y.T. *A phenomenological model of global ionospheric electron density on the E, F1, and F2 regions*. J. Atmospheric and Terrestrial Physics, vol. 35 (1973), no. 9, pp. 1615–1630.
46. Gerson, N.C. *Radio wave absorption on the ionosphere.* New York: Pergamon Press, 1962. 113 pp.
47. Grossi, M.D. and Langworthy, B.M. *Geometric optics investigation of HF and VHF guided propagation in the ionospheric whispering gallery*. Radio Science, vol. 1 (1966), no. 8, pp. 877–886.
48. Fenwick, R.B. *Round-the world high-frequency propagation.* Techn. Reports no. 71/ Radiosc. Lab., Stanford Univ., 1963.

Chapter 4
The Effective Target Cross Section

4.1 INTRODUCTION

Radar systems operating in the HF band may detect objects beyond the line of the horizon, at very large distances. These targets, however, must be relatively large objects, whose linear dimensions are comparable with or larger than the operating wavelength (10–100 m).

The detection of targets with OTH radar is performed against a background of passive interference caused by backscatter from the earth's surface. The useful signal may be extracted from this interference due to its Doppler shift. It follows that OTH radar may be used only for the detection and analysis of moving targets or localized nonstationary processes in the atmosphere or on the earth's surface. In the opinion of foreign specialists, these two peculiarities of HF radar determine the class of objects and processes which may be detected and analyzed in OTH applications.

Among these processes and objects are (see [13, 15, 17, 21]): artificial ionization clouds created in the upper atmosphere by the injection of chemicals, auroral ionization, pockets of sporadic ionization in the *E* layer, moving ionization disturbances, meteor tracks, nuclear explosions in the atmosphere and on the surface, artificial earth satellite tracks, the ionized tracks of ballistic missile warheads and other space objects entering the dense layers of the atmosphere, ascending ballistic and utility rockets, aircraft, ships, and hurricanes and typhoons on the water surface.

The enumerated objects and phenomena have greatly varying radiophysical properties, and the methods by which each is observed and analyzed at long range have their own unique features. In all of these cases, however, the most important of their radar characteristics is the effective radar cross section (RCS). The specific properties of the RCS for several of the possible objects of over-the-horizon radar are described below.

4.2 THE CHARACTERISTICS OF OTH RADIO WAVE SCATTERING FROM OBJECTS

If in the propagation path there is an object whose complex permittivity $\tilde{\epsilon}_2$ differs from the permittivity of the surrounding medium $\tilde{\epsilon}_1$, then the incident electric field $\mathbf{E}_i$ will induce volume or surface currents (the former in objects with weak permittivity, and the latter in good conductors), which create a secondary scattered field $\mathbf{E}_s$. The total electric field is then given by the sum of the incident and scattered fields: $\mathbf{E}_\Sigma = \mathbf{E}_i + \mathbf{E}_s$.

For a monostatic radar, the scattering properties of objects are characterized by the effective cross section, which in the far zone is determined from the relation [5]:

$$\sigma = \lim 4\pi r^2 |\mathbf{E}_s|^2 / |\mathbf{E}_i|^2$$

where r is the distance from the scattering object to the reception point, $\mathbf{E}_s$ is the intensity of the scattered field at the reception point, and $\mathbf{E}_i$ is the intensity of the incident plane wave at the object.

For the overwhelming majority of radar targets, the radar cross section is determined by their orientation in relation to the direction of the incident radiation (with the exception of a sphere). For a monostatic radar, therefore, the most complete target description is given by a polar diagram of the backscatter, describing the dependence of the RCS on the target aspect angle [9]. Such a diagram is called the target reflectivity pattern. With a bistatic radar, the target characteristics may be analogously described by a bistatic reflectivity pattern, which shows the dependence of the RCS on the target aspect angle for a given bistatic angle ψ (the angle between the directions of propagation of the incident and scattered energy).

Depending on the ratio of the linear dimension L of the illuminated object and the wavelength λ, there are three types of scattering:

1) Rayleigh scattering ($L/\lambda \leqslant 0.3$),
2) resonant scattering ($0.3 \leqslant L/\lambda \leqslant 3$), and
3) optical scattering ($L/\lambda \geqslant 3$).

These threshold values vary somewhat, since they depend on the form of the object and the criteria being used.

For resonant scattering, there are characteristic maxima and minima in the target reflectivity pattern (RCS *versus* aspect angle), and in the corresponding bistatic diagram. It is extremely difficult to calculate these plots in this case, even for simple objects.

In the optical regime, the RCS of objects with high conductivity is determined primarily by specular reflections from those portions of the surface which are perpendicular to the direction of the incident illumination (*specular points*). The number and spatial separation of these points (on the scale of wavelengths) determine the roughness and multilobed nature of the reflectivity patterns; this irregularity may be quite significant. The RCS is therefore considered to be a random quantity in this case, and is characterized by its mean and variance, distribution function, and so on [11].

The scattered field of metallic objects is determined by the surface currents induced by the incident energy [5]. When a radio interacts with volumes of weakly absorbing plasma, the scattered field may be formed by both volume and surface currents. Surface currents arise when the electron plasma frequency f_N exceeds the operating frequency f (there is a "supercritical" or "overdense" electron concentration). In this case, the RCS may be calculated by treating the volume of plasma as a metallic object with the same shape. If the plasma frequency f_N is less than the operating frequency ("subcritical" or "underdense" electron concentration), then the weakly absorbing plasma may be treated as a lossy dielectric for the purposes of calculating the RCS.

Plasma objects may be categorized as homogeneous, regularly inhomogeneous, and statistically inhomogeneous. Independent of this classification, they may be isotropic or anisotropic. From the point of view of their interaction with electromagnetic fields, they are divided into Rayleigh and resonant scattering and optical scattering.

These various scattering mechanisms associated with plasma objects give rise to a corresponding number of methods for calculating or estimating their RCS. Even a brief presentation of these methods would be beyond the scope of this text. We refer those interested in these questions to special handbooks [7, 19].

Among the questions relating to the RCS of various objects at HF, the foreign literature has devoted the most attention to the RCS of ionospheric irregularities, ascending missiles, and aircraft [15, 16, 20].

4.3 THE RADAR CROSS SECTION OF MAGNETICALLY ORIENTED DISTURBANCES IN THE IONOSPHERIC ELECTRON DENSITY AT HIGH FREQUENCY

Under the influence of the earth's magnetic field, there frequently arise irregularities in the electron concentration, oriented along the magnetic field lines. The appearance of such disturbances is associated with the anisotropic diffusion of electrons in the presence of a magnetic field, and with the development of plasma instabilities with the movement of plasmas in crossed electric and magnetic fields.

A magnetically oriented disturbance may arise in the E, F_1, and F_2 layers of the ionosphere. They are observed most often at high and low latitudes, and occasionally in the midlatitude ionosphere. Magnetically oriented disturbances may have both natural and artificial origins. To study the mechanism of these phenomena, and their dynamics and effects on radio propagation, the oblique sounding technique is used in many bands, including the HF band. High frequency signals may be used effectively for this purpose only in the F layer, however, where the length and width of the disturbances are on the order of tens of meters and meters, respectively. The corresponding dimensions in the E layer are about ten times smaller.

Magnetically oriented inhomogeneities in the ionospheric electron density are typically not isolated occurrences, and fill large regions of space, extending for hundreds or thousands of kilometers. Therefore, even when using an extremely narrow beam, the region illuminated by the radar will contain a large number of such disturbances, all forming backscatter signals, so that the scattering is statistical in nature. In these situations the reflecting properties of the medium are characterized by a specific backscatter cross section σ_0 [4].

For the case of a gaussian spatial correlation function of the ionospheric disturbances, Booker [14] obtained the following expression for the specific backscatter cross section:

$$\sigma_0 = (2\pi)^{3/2}\,\pi^2 \left(\frac{1}{\lambda_N}\right)^4 \left(\frac{\overline{\Delta N}}{N}\right)^2 L_\perp^2 L_\parallel \times \exp\left\{-\frac{8\pi^2}{\lambda^2}\left[L_\perp^2 + (L_\parallel^2 - L_\perp^2)\sin^2\xi\right]\right\} \tag{4.1}$$

where λ_N is the wavelength corresponding to the plasma frequency in the surrounding ionosphere; $(\overline{\Delta N/N})^2$ is the mean square value of the intensity of the relative fluctuations in the ionospheric electron density; $L_\perp$ and $L_\parallel$ are the perpendicular and lengthwise correlation lengths of the electron density fluctuations; λ is the operating wave length; and ξ is the angle between the normal vector of the incident wave front and the long axis of symmetry of the ionospheric disturbance.

A peculiarity of the HF backscatter from magnetically oriented ionospheric disturbances is the so-called aspect sensitivity. Inasmuch as $L_\parallel \gg \lambda$, usually, then the value of the indicated exponent in (4.1) will be small (and hence the exponential factor itself will be large) only when the angle

ξ is close to 0, that is, when the illumination is perpendicular or almost perpendicular to the long axis of the disturbance. Such disturbances may therefore be observed with HF radar only when the geometry of the radar, disturbance, and propagation path meet very strict requirements.

4.4 THE RADAR CROSS SECTION OF METEOR TRAILS

The entry into the upper atmosphere (85–120 km) of meteors with mass on the order of 10^3–10^5 grams, moving with speed 10–75 km/s, is accompanied by the formation of ionized meteor trails about 5–20 km in length with average initial radius $R_0 \approx 1$ m [2].

In the first approximation, a meteor trail may be considered to be a circular cylinder of plasma, in which the electron concentration varies slowly along the axis. The degree of ionization is therefore characterized by the linear electron density α. In meteor trails registered by radar, the initial values of the linear electron density α_0 lies in the range 10^{11}–10^{14} cm. Under the action of the natural processes of ambipolar diffusion, recombination and the attraction of electrons to neutral molecules, the value of α decreases over time, and the radius R increases. Owing to these factors, the scattering properties of meteor trails grow weaker with time. This manifests itself as a constant decrease in the amplitude of radio reflections from the trails, or, equivalently, in a decrease in the RCS of the trails.

The existence of gradient winds in the neutral components of the atmosphere at heights of 85–120 km can lead to the deformation and breakup of the initial cylindrical form of a meteor trail. The usual process by which the amplitude of the meteor trail reflections changes is disturbed, and calculation of the radar characteristics of such meteor trails is extremely difficult. The length of time during which meteor trail reflections may be observed at HF is from a fraction of a second to several seconds or tens of seconds.

The signals reflected from meteor trails exhibit a Doppler shift caused by the movement of the trails under the action of the wind of neutral components in the atmosphere, which exists at heights of 75–120 km. This effect, in particular, is used to measure wind speed at these heights. The signals from meteor trails are always present among other signals received by OTH radars. Their characteristics contain certain information about the properties of the propagation path.

In terms of radar properties, meteor trails are classified as being either underdense or overdense. The criterion for this classification is the electron concentration in the meteor trail. If the plasma frequency f_N, corresponding to the electron concentration in the trail, exceeds the carrier frequency of the radar, then the meteor trail is overdense; otherwise, it is underdense. The initial ionization of the meteor trail is determined by the mass and initial speed of the meteorite.

Meteor trails which scatter at HF are characterized by large values of the ratio of the length of the trail L to the wavelength λ, and by small values of the ratio of the diameter of the trail to the wavelength λ. The electron distribution in a cross section of the trail may be approximated as gaussian. These assumptions allow us to consider the meteor trail to be a very narrow, infinitely long plasma cylinder, for the purposes of solving scattering problems. In this approximation, the RCS of an underdense meteor trail which is illuminated perpendicular to its axis by radio waves with collinear polarization, may be expressed with the formula [20]:

$$\sigma_{mt} = \frac{\epsilon_0^2 e^4 \alpha^2 r \lambda}{32\pi^2 m^2} \exp\left[-\frac{8\pi^2}{\lambda^2}(4Dt - R_0^2)\right]$$

where ϵ_0 is the permittivity of free space, e and m are the charge and mass of the electron, r is the distance to the radar, λ is the operating wavelength, D is the coefficient of ambipolar diffusion in m^2/s, t is the time since the meteor trail was formed, and R_0 is the initial radius of the meteor trail.

In the optical scattering regime, the finite time of the trail formation is taken into account and the formula for the RCS of an underdense meteor trail takes the form:

$$\sigma_{mt} = \frac{\epsilon_0^2 e^4 \alpha^2 v_M^2 \lambda_0^4}{4096\pi^6 m D^2} \exp\left(-\frac{8\pi^2 R_0^2}{\lambda^2}\right)$$

where v_M is the velocity of the meteor in m/s.

With a large linear electron density, the meteor trail may be considered to be a metallic cylinder with some effective radius, which increases with time. Since the electron density decreases with time, after some amount of time the meteor trail may be considered to be underdense. If the distance r from the radar to the meteor trail is many times larger than the effective radius, then the RCS of an underdense meteor trail may be written (in the Rayleigh regime) as (see [20]):

$$\sigma = \frac{r}{4}\sqrt{4Dt \ln\left(\frac{\epsilon_0 e^2 \alpha \lambda^2}{16\pi^3 Dmt}\right)} \qquad (4.2)$$

The expression under the square root sign in (4.2) corresponds to the effective radius of the meteor trail with a gaussian electron density distribution. As the time t increases, the logarithm in (4.2) decreases to zero at some time $t_f = \epsilon_0 e^2 \alpha \lambda^2/(16\pi^3 mD)$. For $t > t_f$ the calculation of the RCS becomes the formula for the underdense trail.

For the case of very strong diffusion, when the overdense ionized region forms a stretched spheroid, the main axis of which is parallel to the axis of the trail, the following expression for the RCS in the optical regime was obtained in Haukins and Winter [16]: $\sigma_{mt} = \sigma_e \lambda^4 v^2{}_m \alpha^2/(256\pi^4 D^2)$, where σ_e is the Thompson electron cross section.

4.5 THE RADAR CROSS SECTION OF ASCENDING ROCKETS

Among the various applications of over-the-horizon radar, foreign specialists consider its use for early detection of the launch of intercontinental ballistic missiles (ICBMs) to be important [15]. In the high-frequency band (3–30 MHz), the radar cross section of an ascending rocket results from the reflection of radio waves off both the body of the rocket and the adjacent exhaust stream of partially ionized products of the rocket fuel combustion [3]. The electromagnetic fields scattered from the exhaust stream and from the ICBM body are incoherent, because the boundaries, structure, and dimensions of the flame change continuously during flight. As a result, the phase of the radio waves scattered from the exhaust flame does not exhibit a constant relation to the phase of the wave reflected from the missile body. This allows the RCS of an ascending ICBM to be calculated as the sum of two terms, i.e.,

$$\sigma_{BM} = \sigma_b + \sigma_{ex} \tag{4.3}$$

where the subscripts *BM, b,* and *ex* refer to ballistic missile, body, and exhaust, respectively.

The body of an ICBM has a cylindrical shape. Therefore, σ_b for the vertical component of the field intensity may be estimated using the formula for the RCS of a metallic cylinder of finite radius [6]:

$$\sigma_b = \frac{2\pi R L^2}{\lambda} (\sin \psi) \left[\frac{\sin(2\pi L \lambda^{-1} \cos \psi)}{2\pi L \lambda^{-1} \cos \psi} \right]$$

where R is the radius of the missile body, L is the length of the rocket, λ is the wavelength, and ψ is the aspect angle between the direction of the incident wave and the velocity vector of the missile.

The scattering properties of the exhaust stream from a rocket engine result from the fact that in the process of burning the fuel mixture, some of the combustion products are partially ionized. The main contribution to the formation of free electrons in the exhaust plume is the thermal ionization of the impurities in the alkali metals. In addition, inasmuch as there is usually an excess of combustibles in the fuel mixture in comparison with oxygen, there is a significant percentage of unoxidized atoms and molecules in the exhaust. When these interact with the oxygen in the surrounding air, an afterburning process takes place, in which an additional number of free electrons are formed as a result of chemical ionization [18].

The shape, dimensions, and structure of the exhaust stream of a rocket engine depend strongly on the height and speed of the rocket [1]. Exhaust streams are therefore classified as being associated with low (to approximately 10 km), middle (from 10 to 70 km), high (70 to 180 km) and very high (greater than 180 km) altitudes. These values vary somewhat, because the shape of the flame, which is the basis of the classification, depends on the power of the rocket engine. With increasing power (thrust) in the rocket engine, the heights listed above at which the transitions from one flame type to another take place should be increased.

At low altitudes, the exhaust plume of rocket engine combustion products has the form of a free supersonic wake, whose shape is nearly cylindrical. Due to the intensely turbulent mixing of the combusion products with the external air currents, the electrophysical structure of the exhaust wake is extremely irregular, both in the axial and radial directions. The radar wavelength in the HF band will be many times larger than the scale of these random irregularities in the permittivity of the missile's exhaust plume, the dimensions of the latter being comparable to the wavelength. The electron density in a large portion of the exhaust stream is subcritical, and only in a small volume of the core of nonviscous flow near the edge of the nozzle is the electron density supercritical. In the opinion of foreign specialists, considering that at low altitudes the missile is still flying with all of its stages, its cross section may be considered to be that of the body. The resultant estimate of σ_{BM} is somewhat small, because σ_{ex} is neglected in (4.3).

At middle altitudes, the role of turbulent mixture of the combustion products with the surrounding air decreases, and the elements of the flame structures acquire a more defined outline [8]. A schematic illustration of the structure of the exhaust stream from an Atlas rocket at a height of 100 km is shown in Fig. 4.1.[1] Foreign specialists describe the components

[1] *Dreyner et al. Vykhlopnie strui raketnykh dvigateley v termosfere.* RT + K, vol. 13 (1975), no. 6, pp. 144–146.

of the exhaust plume of the ballistic missile as follows:[2]

> adjacent to the edge of the nozzle is an isoentropic core of non-viscous supersonic flow, in a substantial portion of which the electron concentration is overdense; down the flow the iosoentropic core of the stream is limited by a straight shock wave ("the Mach disk"), and on the sides by an internal shock wave;
>
> outside this core is a layer in which the combustion products mix with the air currents. The width of the layer increases with distance from the nozzle along the flow:
>
> - a far zone of turbulent interaction beyond the Mach disk;
> - an external shock wave ("leading shock wave");
> - an external flow between the leading shock wave and the contact surface of the stream of combustion products.

The dimensions of the exhaust plume at midaltitude becomes larger (and for $z > 60$ km, significantly larger) than the HF wavelength. Therefore, the ballistic missile's exhaust flame begins to make a dominant contribution to the total RCS of the missile. In addition, the missile separates from the first stage at midaltitude, and is therefore substantially smaller.

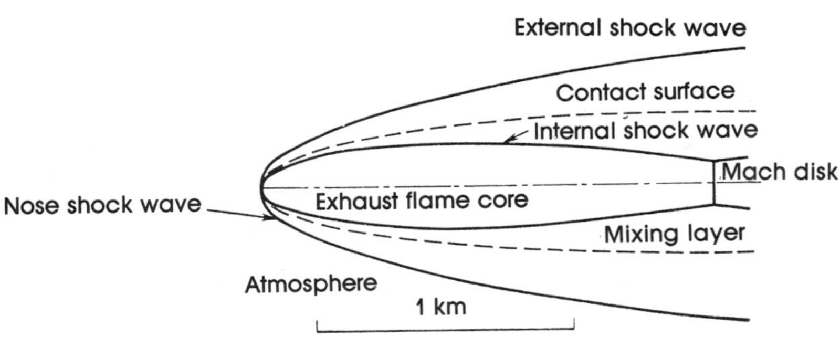

Fig. 4.1 Diagram of the exhaust flame of an Atlas rocket in the lower thermosphere at height of 100 km.

At middle altitudes, there are still afterburning processes occurring in the mixing layer of the exhaust flame [18], which are additional sources of free electrons. Therefore, at a given distance from the nozzle, the electron density in the mixing layer may be several orders of magnitude

[2] *Vlianie fakela raketnykh dvigateley na radiosvyas' s raketoy*. Obzor. VRT. issue 8(140) (1966), pp. 15–26.

higher than that along the axis of the exhaust stream. The nucleus of the flame may therefore be considered to be surrounded by a plasma cloud with an overdense electron concentration.[3]

The geometric dimensions of the main elements of the exhaust stream may be estimated with the empirical relations presented in Avduevskiy et al. [1]. The shape of the core of the stream of combustion products is well approximated by an ellipse [10]. Therefore, the RCS of the nucleus of the flame may be calculated from the formula for the RCS of a circular metallic ellipse [9]:

$$\sigma_\rho = 4\pi l^2 d^4 (l^2 \cos^2\psi + d^2 \sin^2\psi)^{-2}$$

where l is the length of the ellipse, d is the diameter of the ellipse, and ψ is the illumination aspect angle.

In addition to the nucleus of the flame, at midaltitude, a substantial contribution to the RCS is made by the far region of turbulent mixing. An estimate of this component of the RCS has been made abroad, treating it as a problem of radio scattering in statistically inhomogeneous media. An example of such an estimate is worked out in Menkes [4].

In the opinion of foreign specialists, the electrophysical structure of the exhaust flame of the rocket at high altitudes has a number of characteristic features. First, at heights of approximately 70–90 km, the duration of chemical relaxation becomes longer than the time constant associated with gaseous flow [12]. Consequently, the afterburning processes and associated production of free electrons in the mixture layer cease, and this layer no longer plays the role of a plasma cloud with an overdense electron concentration. Second, the outlines of the stream and the internal shock wave are blurred, owing to which the intensity of radio backscatter from them is sharply reduced. Besides this, the turbulent processes in the mixing layer slow down, and the turbulence in the tail region of the flame remains, apparently, undeveloped, which further reduces the backscatter intensity.

In addition to these effects, in the opinion of foreign specialists, the nose shock wave begins to play a major role in the reflection of short waves at high altitudes. The intensity of this reflection grows, since the speed of the rocket and temperature of the surrounding air increases with the height of the rocket. This leading wave diverges, which leads to an increase in the region occupied by the external flow. Partially ionized air in the upper atmosphere is compressed by a factor of 3–5 in the nose shock wave, and the shock wave has a rather sharp border (on the scale of the

[3] *Pitaevskiy, L.P. K voprosu o vozmyshcheniyakh, vyzyvaemykh v plasme bystrodvizhushchimsya telom.* Geomagnetizm i aeronomiya vol. 1, (1961) no. 2, pp. 194–208.

radio wavelength). The electron density increases in the same proportion, and the plasma frequency may exceed the radio frequency of the sounding signals at HF. In these cases, the nose wave reflects the radio signals almost as effectively as a metallic surface. The effective reflecting area may be estimated with the formula $\sigma_{\text{nose}} = a\pi\rho_1\rho_2$, where a is the Fresnel reflection coefficient, and ρ_1 and ρ_2 are the radii of curvature of the nose shock wave at the point where the wave vector of the incident radio illumination is perpendicular to the wave surface (more precisely, perpendicular to the surface of equal electron density).

At very high altitudes ($z > 180$ km), the gas flow begins a transition rrom continuous to free molecular flow. The mean free path for neutral particles is tens of meters. On the scale of the radio wavelength, therefore, the exhaust stream of combustion products from the missile is a large weakly inhomogeneous region of the ionosphere which is perturbed, the differential scattering cross section of which may be estimated with the perturbation method with the formula:

$$d\sigma = \sigma_e|\delta N(\mathbf{q})|^2 \sin^2 \psi d\Omega$$

where σ_e is the Thompson scattering cross section of the electron; $\delta N(\mathbf{q})$ is the Fourier transform of the spatial distribution of the perturbation in the electron concentration $\delta N(\mathbf{r})$:

$$\delta N(\mathbf{q}) = \int \delta N(\mathbf{r}) \exp(-i\mathbf{qr})\, d^3\mathbf{r}$$

ψ is the angle between the electric field vector of the incident wave and the scattered wave vector; and $d\Omega$ is an element of solid angle. Here $\mathbf{q} = |\mathbf{k}_s - \mathbf{k}_i|$, where $\mathbf{k}_i$ and $\mathbf{k}_s$ are the wave vectors of the incident and scattered waves.

Thus, the RCS of an ascending ICBM is a complex function of the operating frequency of the radar, the altitude of the missile, the ionospheric characteristics, the illumination aspect angle, the type of fuel mixture, and the design and thrust of the rocket engine. An example of the dependence of the RCS at a frequency of 15 MHz on the flight time of the rocket for a narrowband component of the reflected signal, borrowed from Fenster [15], is shown in the plot in Fig. 4.2. The RCS values shown here are for the case of vertically polarized illumination and were calculated with allowance for the aspect dependence of the RCS along the rocket trajectory, for the chosen propagation geometry. In Fig. 4.2, two cases are shown, both with incomplete signal hops. In the one marked $N - D$, the illuminating signal is on its downward leg from the ionosphere, and in the one marked $N + D$, the signal is on its upward leg from the earth's surface. The difference between the two results is due to the aspect dependence of the cross section.

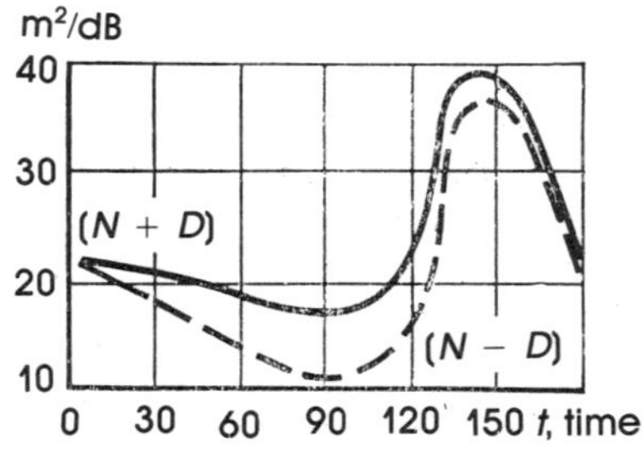

Fig. 4.2 Change in the RCS of a rocket flame as a function of flight time.

4.6 THE RADAR CROSS SECTION OF AIRCRAFT

The radar characteristics of aircraft in the HF band possess the following features: the object is completely metallic, and the scattered electromagnetic field is therefore formed by surface currents; the shape of the object is quasi-unidimensional, that is, has dimensions in two known directions (along the fuselage and along the wings); the ratio of the typical dimensions to the wavelength is not much different than unity, so that radio waves are scattered in the resonant regime.

These features make it possible to characterize the radar properties of aircraft with the help of a polarization scattering matrix [9]:

$$\begin{bmatrix} \sqrt{\sigma_{HH}}\exp(i\phi_{HH}) & \sqrt{\sigma_{HV}}\exp(i\phi_{HV}) \\ \sqrt{\sigma_{VH}}\exp(i\phi_{VH}) & \sqrt{\sigma_{VV}}\exp(i\phi_{VV}) \end{bmatrix}$$

Here the indices H and V refer to horizontal and vertical linear polarization. The first index indicates the polarization of the incident wave, and the second indicates the polarization of the scattered wave. For example, the component σ_{HV} describes the vertically polarized component of the field scattered as a result of horizontally polarized illumination.

In general, each element of scattering matrix is a function of the aspect angle and the pitch angle. The target reflectivity diagram for an aircraft, in any plane, has a segmented form. The extent to which it is divided depends primarily on the ratios l_w and l_F/λ, where l_w is the wingspan, and l_F is the length of the fuselage.

For a rough estimate of the maximum RCS for an aircraft when illuminated at zero aspect ("from the nose") at horizontal polarization, the formula for the RCS of a halfwave dipole may be used: $\sigma_{\max} = 0.86(2l_w)^2$. At nonresonant frequencies, the value $\sigma_{\max}$ decreases, approximately following the form of the resonance curve squared. In a small aspect sector ($\psi \leqslant 30°$), the reflectivity pattern for a horizontal dipole is satisfactorily approximated with the formula:

$$\sigma_{HH}(\psi) \approx \sigma_{\max} \cos^4 \psi$$

Such estimates may be rough, but for an object with as complex a configuration as an aircraft, it is practically impossible to calculate the scattering characteristics in the HF band. In practice, the RCS may be studied with electrodynamic modeling [9]. The result σ_{mod} obtained with such a model may be transformed into the RCS for actual aircraft with the formula $\sigma_{act} = \sigma_{\text{mod}} k^2$, where the modeling coefficient k is the ratio of the operating wavelength λ to the wavelength λ_m at which the model measurements were performed.

In this manner reflectivity patterns may be obtained [9]. Inasmuch as the monostatic and bistatic HF reflectivity patterns for an aircraft are multilobed, then they may be characterized by the parameters of their distributions. The median value is 10–20 dB smaller than the maximum value of the RCS. As an example, the graph of the frequency dependence of the median value of the RCS for a four-engine jet aircraft is shown in Fig. 4.3, borrowed from Fenster [15]. The RCS was measured in the frontal quadrant of aspect angles ($-45° \leqslant \psi \leqslant 45°$), and for all possible illumination polarizations.

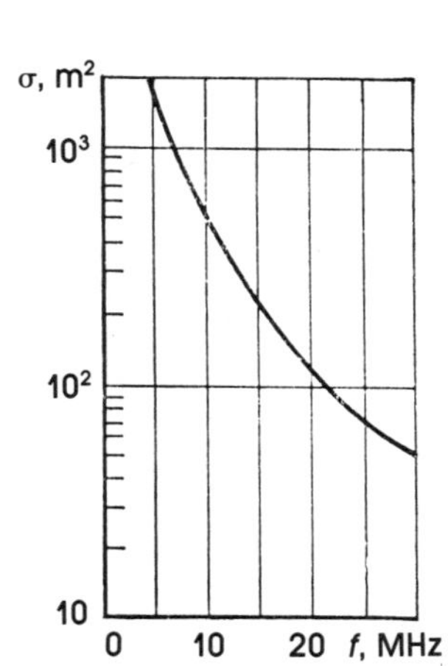

Fig. 4.3 Frequency dependence of the median RCS of an aircraft.

REFERENCES

1. Avduevskiy, V.S. *et al. Techenie v sverkhzvukovoy vzyakoy nedorasshirennoy strue (Flow in a supersonic viscous un-expanded layer).* Izv. AN SSSR. Mekhanika Zhidkosti i Gaza, (1970), no. 3, pp. 63–69.
2. Bayrachenko, I.V. *Izmereniya nachal'nykh radiusov ionizirovannykh meteornykh sledov iz parallel'nykh nablyudeniy radiometeorov na dvukh dlinakh voln (The measurement of the initial radii of ionized meteor tracks from parallel observations of radiometeors at two wavelengths).* Geomegnetizm i Aeronomiya, vol. 5 (1965), no. 3, pp. 460–464.
3. Drayper, Dzharvinen, Konli. *Analiz otrazheniya radiolokatsionnykh signalov ot turbulentnykh vykhlopnykh fakelov raket na bol'shikh vysotakh (Analysis of radar reflections from turbulent exhaust plumes of rockets at high altitudes).* Raketnaya Tekhnika i Kosmonavtika, vol. 2 (1964), no. 6, pp. 229–237.
4. Menkes. *Rasstoyanie radivoln turbulentnoy plazmoy nevysokoy plotnosti (The spacing of radio waves from a turbulent plasma with low density).* Raketnaya Tekhnika i Kosmonavtika, vol. 2 (1964), no. 6, pp. 229–237.
5. Mentzer, J.R. *Scattering and diffraction of radio waves.* Pergamon Press, 1955.
6. Mishchenko, Yu.A. *Radiolokatsionnye tseli (Radar targets).* Moscow: Voenizdat, 1966. 140 pp.
7. Radar Reflectivity Issue. Proc. of the IEEE, vol. 53 (1965), no. 8.
8. Tannehill, Anderson. *Vykhlopnyi strui raketnykh dvigateley na crednikh vycotakh (The exhaust plumes of rocket engines at medium altitudes).* Raketnaya Tekhnika i Kosmonoavtika, vol. 10 (1972), no. 1, pp. 150–157.
9. Torgovanov, V.A. and Mayzel's, E.N. *Izmerenie kharaketeristik rasseyaniya radiolokatsionnykh tseley (Measuring the characteristics of scattering from radar targets).* Moscow: Sovietskoe Radio, 1972. 232 pp.
10. Finat'ev, Yu.P. and Shcherbakov, L.A. *O vozmozhnosti approksimatsii granitsy nedorashirennoy osesimmetrichnoy strui dugoy ellipsa (On approximating the boundary of an unexpanded axially symmetric layer as an arc of an ellipse).* IFZh, vol. 17 (1969), no. 4, pp. 737–741.
11. Shtager, E.A. and Chaevskiy, E.V. *Rasseyanie voln na telakh slozhnoy formy (Scattering from bodies with complex shapes).* Moscow: Sovietskoe Radio, 1974. 240 pp.

12. Yushchenkova, N.I. *Vliyanie kinetiki elektromagnitnykh protsessov na elektrofizicheskie parametry struy nizkotemperaturnoy plazmy (The effects of the kinetics of electromagnetic processes on the electro-physical parameters of the layers of a low-temperature plasma).* Yavlenie Perenosz v Nizktemperaturnoy Plazme (The Transport Phenomenon in a Low-Temperature Plasma). Minsk: Nauka i Tekhnika, 1969, pp. 96–105.
13. Ball, Desmond S. *OTH-B radar in defence of Australia.* Electronic Today International, vol. 8 (1978), no. 2, pp. 35–40.
14. Booker, H.G. *A theory of scattering by nonisotropic irregularities with application to radar reflections from aurora.* J. Atmospheric and Terrestrial Physics, vol. 8 (1956), no. 4/5, pp. 204–221.
15. Fenster, W. *The applications design and performance of the over-the-horizon radar.* Internat. Conf. Radar-77. London: 1977, pp. 36–40.
16. Haukins, G.S. and Winter, D.F. *Radar echoes from overdense meteor trails under conditions of several diffusion.* Proc. IRE, vol. 45 (1957), no. 9, pp. 1290–1291.
17. Klass, P.J. *HF radar detects Soviet ICBM's.* Aviation Week and Space Technology, vol. 95 (1971), no. 23, pp. 38–40.
18. Pergament, H.S. and Jensen, D.E. *Influence of chemical kinetic and turbulent transport coefficients on afterburning rocket plumes.* Spacecraft and Rockets, vol. 8 (1971), no. 6, pp. 643–649.
19. Ruck, G.T., ed., *Radar Cross Section Handbook,* vols. 1,2, New York: Plenum Press, 1970. 949 pp.
20. Stuart, W.D. *Meteors. Radar Cross Section Handbook,* vol. 2, 1970, pp. 829–839.
21. Tvetin, L.H. *Ionospheric motions observed with high-frequency backscatter sounders.* J. Research NBS, vol. 65D (1961), no. 2, pp. 115–127.

Chapter 5
High-Frequency Radar Interference

5.1 INTRODUCTION

The effectiveness of a radar's detection capability bears a direct relation to the characteristics of the interference affecting the radar's operating channels. It is well known that it is more difficult to extract the desired signal from background interference in radar than it is in a communications system, because the signal is reflected from a target, as opposed to being received directly from the transmitting antenna. While this situation is characteristic of radars operating at very high frequency (VHF) and ultra high frequency (UHF), it is even more significant in HF radars. In this band, the level of external noises acting on the receiving equipment is much higher than that of the internal system noises. At the same time, obtaining a sufficiently strong target signal level is an extremely complex problem, due to the long propagation path. In addition, active interference affecting HF radars is highly irregular in frequency and has nonstationary characteristics, compared with that affecting standard radars. The interference in the radar channels is a random process with correlation parameters (nonwhite noise), which complicates the problem of extracting the signal from the noise. In light of these considerations, questions concerning the characteristics of the interference are extremely important.

Radar operation at HF is affected by both active interference and passive interference (clutter). The various forms of active interference in this band include: atmospheric interference, cosmic interference, radio frequency interference (RFI—system interference), and industrial noises. Clutter is caused by reflections from the earth's surface and from ionospheric irregularities.

On the basis of its effects on the receiving system it is possible to distinguish the following types of active interference [25]: narrow band (sinusoidal), pulsed, and fluctuating. The joint action of pulsed and fluctuating interference on the receiver is best modeled as a quasipulsed random process. The interference is characterized by its intensity, distribution, and spectral, temporal and spatial-angular characteristics. We will now examine some particular forms of interference.

5.2 INTERNAL RECEIVER NOISES

In very-high-frequency (VHF) and ultra-high-frequency (UHF) radars, the main source of interference limiting the system performance is the internal receiver noise, so long as there is no active jamming. In an over-the-horizon (OTH) radar, however, as was noted above, other forms of interference greatly exceed receiver noises, and in any well-designed OTH radar, internal noises may be neglected. The level of internal system noises in modern HF receivers is substantially lower than the minimum external interference level, even in the high end of the HF band. The issue of internal receiver noises is covered in sufficient detail in the works [11,13,19,20], and will not be considered in any more detail here.

5.3 ATMOSPHERIC INTERFERENCE

The basic source of atmospheric interference in the frequency band to 20-30 MHz is lightning between the oppositely charged masses of air, water vapor, and the earth. The ionized channel of a lightning bolt may be viewed as a huge antenna, along which flow powerful current pulses, causing the radiation of an electromagnetic pulse [11].

An oscillographic study of the forms of the currents arising from lightning bolts which strike the earth shows that these discharges are composed of two components: 1) a predischarge, moving slowly from the cloud to the earth's surface, 25–30 ms behind which follows the stepped leader, a current pulse of 100-300 A, and 2) the return stroke, a current pulse of 10–100 kA, which flows from the earth to the cloud along the path laid out by the leader. The leader radiates a continuous spectrum with a maximum in the region 30–50 kHz, which is inversely proportional to the frequency at high frequencies. The return stroke, the length of which is about 100 ms, is also characterized by a continuous spectrum with a maximum near 4-8 kHz, and inversely proportional to the square of the frequency at high frequencies [9]. Each thunder bolt should be viewed as a source of radio waves, filling a continuous frequency band, which propagates to the radio station reception site.

Atmospheric interference is caused by a large number of lightning discharges, occurring simultaneously in various regions of the entire atmosphere. Each second there are approximately 100 lightning discharges in the atmosphere [9]. Atmospheric interference is affected both by local storms, and by distant electric discharges in the atmosphere associated with worldwide centers of intense storm activity. These centers are located in the equatorial regions (equatorial Africa, equatorial America, and others) [10]. The intensity of the interference gradually decreases with increasing distance from the storm center. Propagating by ionospheric reflection, lightning discharges occurring in the equatorial regions affect the receiving apparatus even when there are no local storms.

Atmospheric interference manifests itself as a sequence of pulses, randomly distributed in amplitude and time. This phenomenon often has two components: a comparatively weak part, with a Gaussian distribution, and a much more powerful pulsed component [25]. The power spectral density of atmospheric interference in the band 5–30 MHz may reach 60-70 dB relative to kT_0 [18], where k is the Boltzmann constant and T_0 is the absolute temperature. The spectral density of the average power of nearby atmospheric interference is inversely proportional to the cube of the frequency [25]. For distant atmospheric interference, this dependence is distorted because of the frequency characteristics of the loss along the propagation path of the interference. Thunderstorms generate wideband interference in various frequency bands. The resulting frequency characteristics of the intensity of atmospheric interference changes as a function of the intensity of local storm activity, the time of day, the season, and the geographical location of the receiver.

The most detailed characteristics of atmospheric interference and the methods for calculating its strength are examined in the work [18]. The Geneva session of the CCIR (International Radio Consultative Committee) in 1963 recommended maps of the spatial-temporal distribution of atmospheric interference, obtained with the help of a worldwide network of observation stations at a series of frequencies, using standardized equipment. The maps present the spatial-temporal distribution of the average power of atmospheric noise P_{at} at the frequency 1 MHz, expressed in decibels, relative to the power of thermal noise at the antenna KT_0B_n where B_n is the effective noise bandwidth of the receiver system.

Atmospheric interference is characterized by rapid fluctuations with large dynamic range. However, the noise intensity averaged over several minutes remains almost constant during the course of any given hour, the variation rarely exceeding ± 2 dB, excluding times near sunrise or sunset, or periods of local storm activity [18].

Figures 5.1 and 5.2 show examples of maps, on which lines of constant expected atmospheric noise power levels P_{at} are drawn; the power is in decibels, at 5-dB increments, relative to KT_0B_n at the frequency 1 MHz, obtained over the entire earth's sphere over two time intervals [18]. The time intervals represent the combination of results from four-hour periods

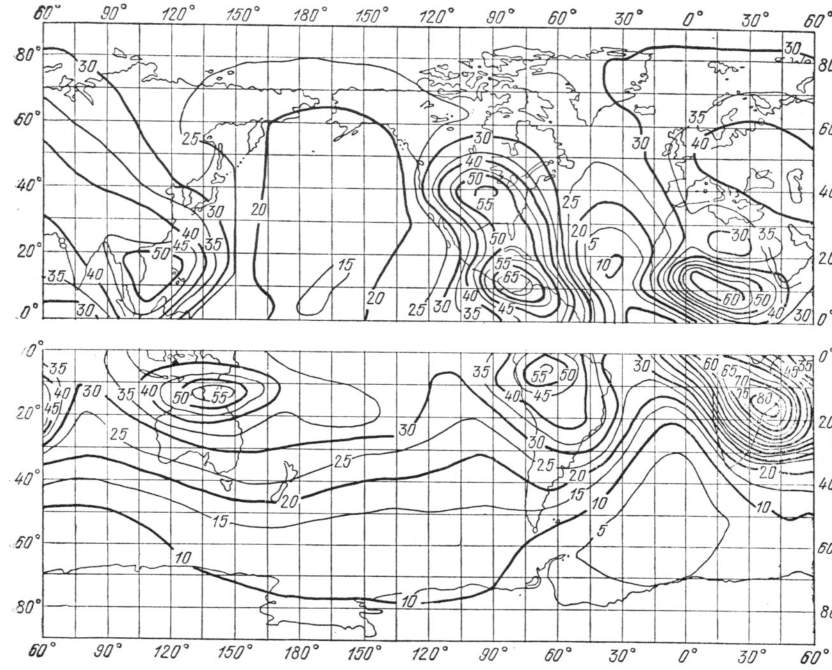

Fig. 5.1 Estimated values of the atmospheric noise power level. F_{at}. Summer, local time 8:00–12:00.

over the course of a season. The day was split into six intervals of four hours each, and the hourly parameters of the interference were averaged for each interval over one of the four seasons. Thus, 24-hour time intervals were examined for 360 hours each (a four-hour period over the course of approximately 90 days in one season). From the maps it is clear that the level of atmospheric interference depends on the geographic location of the receiver. For example, during summer days (see Fig. 5.1), the average atmospheric interference power P_{at} relative to KT_0B_n at 1 MHz may be anywhere from 5–80 dB, with the regions of strongest interference found over land. The intensity of the interference decreases with an increase in latitude. The nighttime interference levels are higher at 1 MHz than during the day, which is explained by the reduced ionospheric losses.

On the basis of materials from the record of the 322nd CCIR [18], the variation in the level of atmospheric interference in the center of the

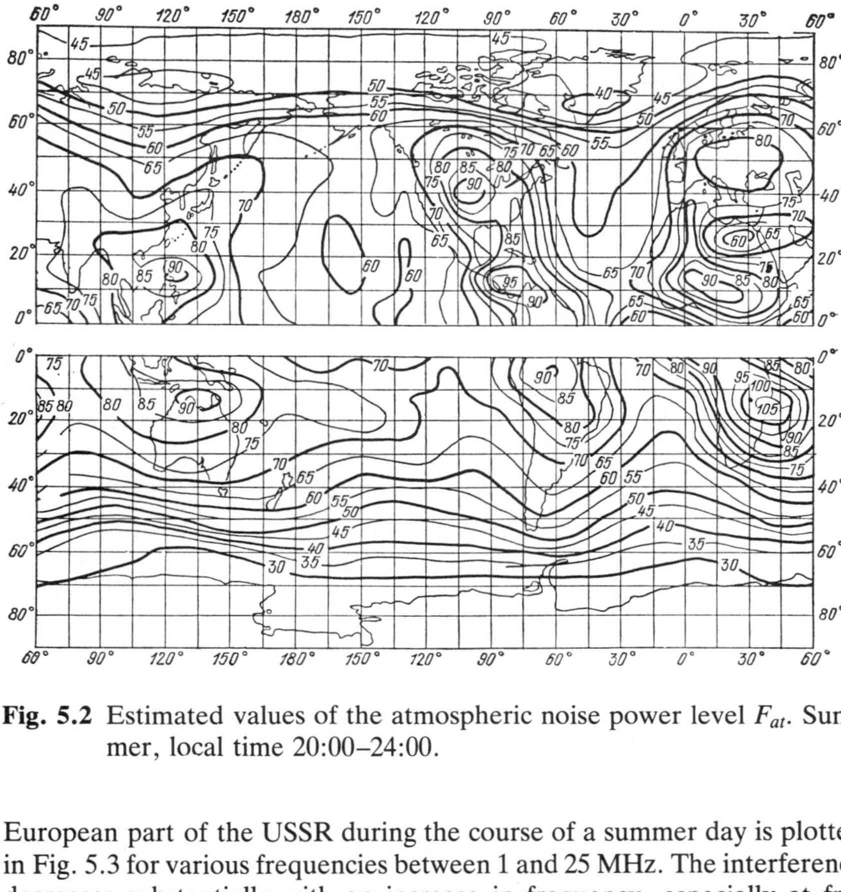

Fig. 5.2 Estimated values of the atmospheric noise power level F_{at}. Summer, local time 20:00–24:00.

European part of the USSR during the course of a summer day is plotted in Fig. 5.3 for various frequencies between 1 and 25 MHz. The interference decreases substantially with an increase in frequency, especially at frequencies exceeding 20–30 MHz. The radio waves generated by world-wide storm centers are weakly supported in the earth-ionosphere duct, and penetrate the ionosphere.

These predictions of the expected atmospheric interference levels consider with sufficient detail the systematic tendencies of their variations connected with the time of day, the season, and the frequency and location of the receiver. It should be noted that, in addition to these variations, there are various unpredictable changes in the level of atmospheric interference. For a given time of day, the interference level changes from day to day due to random variations in storm activity and propagation con-

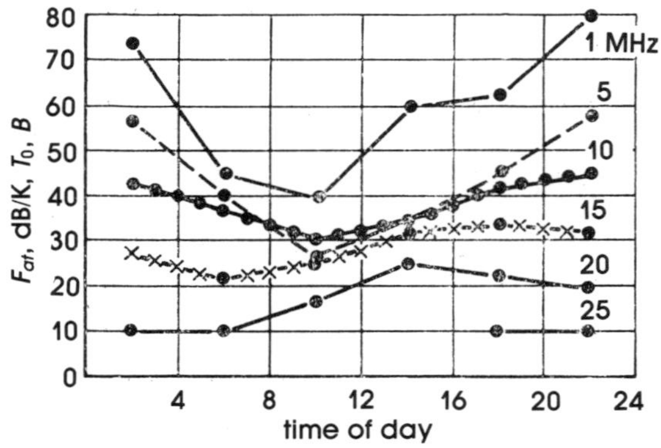

Fig. 5.3 Estimated values of atmospheric noise level F_{at} at various frequencies. Summer.

ditions. The data in the paper [18] were obtained during a period of high sun activity, but the authors indicated that during a long period of no noticeable systematic variations in atmospheric interference as a function of solar activity were observed.

Atmospheric noise is not always predominant in the band 3–30 MHz. Cosmic noise, usually in the high end of the band, frequently exceeds atmospheric interference in midlatitude and polar regions [29]. General data on the sources of electromagnetic noises and their levels are presented in Fig. 5.4 (see [29]). These curves support the observation made earlier, that atmospheric interference is much stronger at night than during the day, and drops off sharply at frequencies exceeding 20–25 MHz (see [29]).

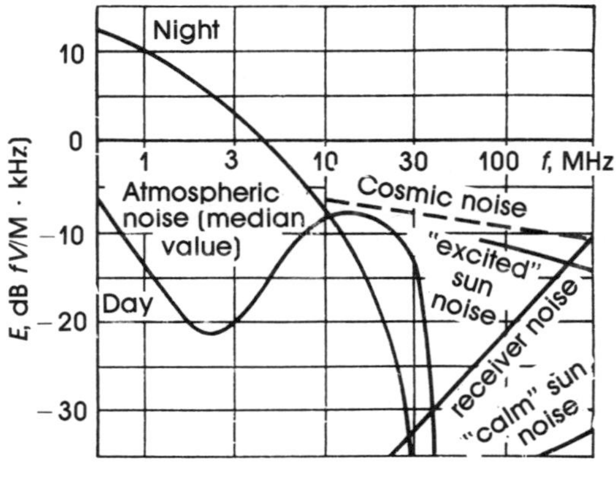

Fig. 5.4 General data on sources of electromagnetic noise and its levels.

5.4 COSMIC NOISE

Cosmic noise is caused by galactic electromagnetic radiation varying randomly in time. The main sources of radiation are radio galaxies, creating a background of radiation in the band 1–10^{16} MHz. Against this background is added the radiation from powerful discrete sources—planets and stars—with frequency spectrum 0.03–30 GHz. Thus, cosmic noise occupies a wide spectrum [21].

For the earth, one of these intense discrete sources of radiation is the sun. The intensity of radiation is characterized by the brightness temperature, which for a "calm" sun reaches 10^6 K at frequencies of 30–200 MHz (see [21]). The radiation of the sun and the planets, despite its high intensity, plays a secondary role compared with the continuous galactic radiation, because these discrete sources only have an effect when the antenna is directed towards them [7]. The continuous background of general galactic radiation is observed in all directions in the celestial sphere. It is especially intense in the plane of the celestial equator—in the directions of the constellations Scorpio and Sagittarius [17].

The earth's atmosphere passes radio waves from cosmic radiation in a wide band of frequencies—approximately 0.01–40 GHz. The lower boundary varies mainly as a function of the density of the ionosphere at the point of reception. The radiation reaching the earth's surface, as a rule, is extremely small. At frequencies below 10 MHz, cosmic interference is attenuated to a significant degree by absorption and reflection in the ionosphere. The level of cosmic interference is fairly stable. The variations which are observed in the intensity of this radiation are caused by changing ionospheric absorption. Due to the nonuniform distribution of radio sources in the celestial sphere and the rotation of the earth about its axis, the intensity of cosmic interference at a given point on the earth exhibits a marked daily variation [17].

The intensity of solar radiation depends on the state of the sun. Radio waves of various wavelengths originate in various layers of the solar atmosphere. The spectrum of solar radiation is complex and changeable. The sun is never completely calm: stormy processes in the solar atmosphere lead to the appearance of localized regions from which the solar radiation is much greater than the intensity observed for a "calm" sun [21].

The level of cosmic interference depends on the frequency. As may be seen in Fig. 5.5 [18], cosmic interference decreases with an increase in frequency. The lower bound of influence of cosmic interference changes in a seasonal and diurnal cycle, depending on the location of the receiver, and is about 15-25 MHz. Due to its fluctuating characteristics, galactic interference is close to "white" thermal noise. The distribution of cosmic interference levels is therefore nearly Gaussian in the HF operating band.

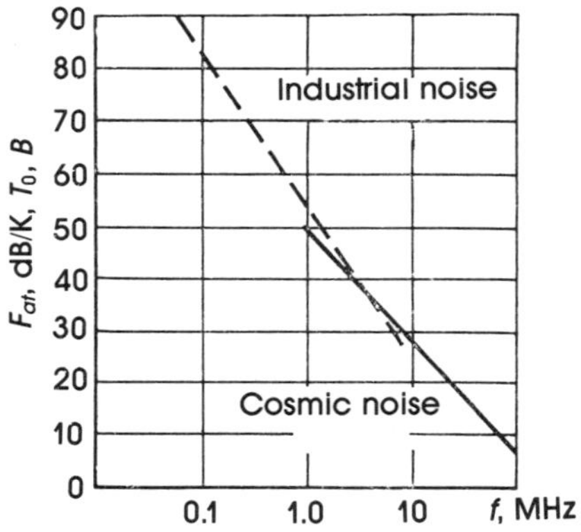

Fig. 5.5 Variation of radio interference with frequency. Summer, local time 4:00–8:00.

5.5 INDUSTRIAL INTERFERENCE

Industrial interference refers to the electromagnetic noises generated by electric and electronic devices [5]. The sources of industrial interference are commercial, medical, and scientific high-frequency equipment, electrical devices, transportation facilities, electric transmission lines, and so on. Industrial noise may affect receivers through the electrical supply networks, antenna feed devices, and indirectly (in the case of nearby interference sources).

The sources of industrial interference may be both radio frequency signal generators, and devices not intended for the generation of high frequencies, but which generate radio interference. Included in the last group at HF are various forms of transportation (automobiles, tramways, trolley buses, subways, electric trains), high-voltage transmission lines, electrical and commutation equipment, a large number of everyday electrical appliances with commutator motors, and automatic regulators, rectifiers, inverters, and so on.

Radio interference from the first group (high-frequency signal generators) is radiated both at the fundamental frequency and its harmonics. Interference from these sources has a periodic (sinusoidal) character. Radio interference from sources in the second group, of an aperiodic nature (pulsed interference), results from sudden changes in the operating current and voltage in the electrical circuits of these devices (sparks and current surges at contacts). This leads to the generation of wideband high-frequency noise. The amplitude of these currents, as a rule, decreases with increasing frequency, and depends on the switched current. Industrial interference from sources in this group is usually manifested in the band below 20 MHz as an aperiodic series of distorted rectangular pulses [12].

In practice, almost all sources of industrial interference produce continuous spectra. The voltage at individual frequencies is approximately inversely proportional to the frequency, so that the decay is approximately hyperbolic [15].

Industrial noise is usually present in large cities, where it affects reception in the band 1–100 MHz. At shorter wavelengths, this interference usually diminishes [17]. In large cities, the level and spectrum of radio noise is constantly varying, due to changes in the character and number of its sources. Therefore, it is best to locate HF receivers outside the limits of large cities [10].

Industrial noise propagates either in the form of electromagnetic ground waves, or in the form of high-frequency currents flowing in current-supporting conductors, which may themselves be interference sources. If the interference source is located close to an electrical transmission line, then interference currents will be induced in it, and may propagate for long distances. Interference currents at the lower frequencies propagate much farther along power lines than higher frequency noise. Industrial interference may also be received from distant sources through ionospheric propagation. For example, industrial noise exceeding by several decibels the level kT_0B_n at the frequency 2 MHz, was registered from a large city about 65 km from the receiver, which was completely free from local industrial noises, and was subject only to weak atmospheric noise [30]. In this latter work, it is shown that the interference level decreases with an increase in frequency both because of the characteristics of the radiated spectra and because of propagation factors.

The range to which such noises are active varies, from several meters (for many industrial high-frequency sources) to several kilometers (for electric welding and X ray apparatus). The average level of industrial noise in large cities and metropolitan areas may be estimated with the curves of Fig. 5.6([10]).

The field intensity of industrial noise diminishes with an increase in the range from the interference source. For example, the interference field intensity E, in V/m, at a distance R from industrial high-frequency devices, may be found with the formula (see [12]) $E = 21(h_a/R^2)\sqrt{P}$, where h_a is the altitude of the receiving antenna in meters, and P is the power of the generator in watts.

This formula is applicable for the conditions $0.1\lambda < R < 5\lambda$, and for $R < 0.1\lambda$ the field intensity diminishes proportionally to $1/R^3$, and for $R > 5\lambda$ proportionally to $1/R$ (where λ is the wavelength).

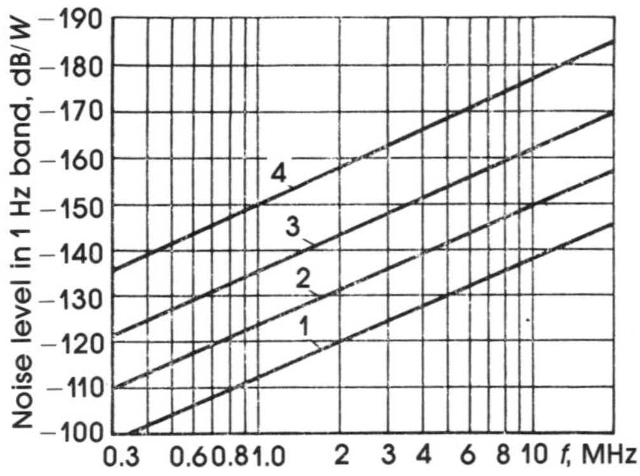

Fig. 5.6 Average level of industrial noise: 1 industrialized region; 2 residential region; 3 rural region; 4 distant, thinly settled region.

Industrial noises depend on local conditions and in a number of cases is limited by factors determining the receiving conditions. The noise level at the receiver is naturally higher during the day than at night. Industrial noise propagates mainly via transmission lines and ground waves, and therefore depends little on diurnal and seasonal variations in the ionosphere [18]. The dependence of the industrial interference level on frequency is shown in Fig. 5.5 at a "quiet" receiving point, that is, in a location chosen to minimize the industrial noise [18]. For such quiet reception (the receiver being 30–60 km from an industrial center), industrial noise will not dominate high-frequency operation.

5.6 INTERFERENCE FROM RADIO STATIONS

5.6.1 Character of the Interference

Interference created by radio stations has a narrow spectrum [25], because its energy is transmitted in a narrow frequency band. This form of interference is caused by the joint action of the carrier frequency oscillations of transmitting stations and their sidebands, with which the transmissions are modulated. In the band 3–30 MHz, narrow-band interference acts in separate narrow regions of the spectrum, which are characterized by a high noise level. The number of these narrow-band regions, and also the interference level in each of them, depends on the how much the HF band is used by transmitting stations, and on the power and range of these stations. A sample graph of the use of the HF band is shown in Fig. 5.7 (see [25]).

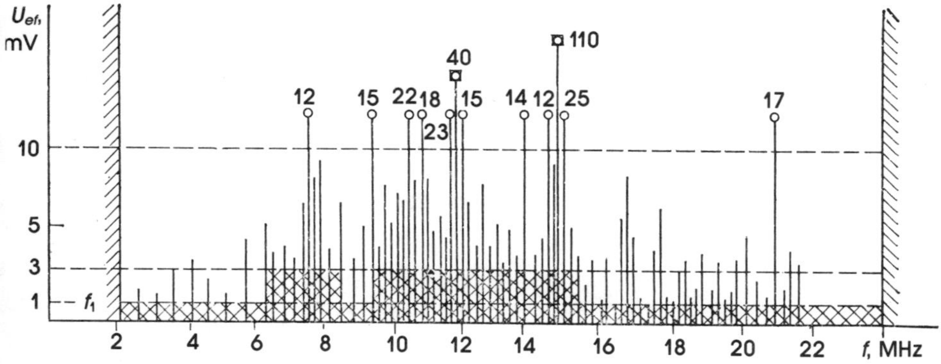

Fig. 5.7 Radio spectrum of powerful signals. The numbers indicate the signal levels exceeding 10 mV.

The signals at the input of the receiver, forming such narrow-band interference, arise from the operation of both nearby and distant radio stations. The determining factor may be the signals from distant stations, because there may be an extremely large number of these. The level of these signals changes in accordance with changes in the state of the ionosphere and attenuation along the propagation paths. In addition to continuous and relatively slow oscillations in the level of the received signals, rapid fading effects are also seen as a result of interference. Thus, if at some particular reception point, then the distribution of the signals at a panoramic receiver, such as that shown in Fig. 5.7, will continuously rise and fall. Clearly, the interference level in each portion of the band at any point in time will also depend on the location of the receiver.

From the point of view of the effects of narrow-band interference on the receiver, two questions should be considered: the general structure of the interference acting at the input of a wideband receiver, and also the interference in the operating channels. These considerations should be studied to determine how the components of the resulting interference signal should be compensated when detecting and measuring parameters of the useful signal.

5.6.2 Types of Radio Transmitters Which Generate Narrow-Band Interference

There are a large number of radio transmitters operating in the high-frequency band, serving various purposes. The main types are stationary (fixed), mobile, broadcasting, and amateur. To get an idea of the number of operating transmitters, it is sufficient to note that there are approximately 700,000 amateur radio stations, a large fraction of which operate in the high-frequency band [22].

Fixed and mobile transmitters are generally used for trunk lines of radio telephone and radio telegraph communications. Despite the measures which are taken (single-sideband operation, reduction of out-of-band and spurious radiation, and so on), the level of narrow-band noise generated by radio stations is growing continuously, and in a large portion of the HF band, significantly exceeds the level of other forms of interference.

5.6.3 The Distribution of Narrow-Band Interference

When examining the plot of narrow-band interference levels, it is necessary to consider that these interference signals are formed from independent components, which may number in the thousands in some portions of the band, and that these components all have random phases. In the case when it is not possible to examine separate components exceeding the average level, the combination of components should be considered to be a narrow band random process, which is stationary for sufficient lengths of time. This statistical model may be applied in the case when a sufficiently wide band is being considered, such as the pass band of the receiver preselector, or the most intensively used portion of the radio station band. With relatively narrow receiver bandwidths, radiation from one or another interfering transmitter will dominate. In this case, the distribution of the interference levels in the operating channel may fit a different model, such as the log-normal distribution [14].

The results of a several years of investigations have shown that for any randomly chosen frequency in the HF band, the temporal statistics of the interference level distribution in the majority of cases will be well approximated by the formula:

$$P(x) = \frac{1}{\sqrt{2\pi}\,\sigma_{x_1T}} \int_{-\infty}^{x} \exp\left[-\frac{(x - \bar{x}_T)^2}{2\sigma_{x_1T}^2}\right] dx$$

where $x = 20 \log U_i$ is the interference level, U_i is the voltage of the interference at the output of the measurement filter preceding the detector. x_T is the mean value in dB, and $\sigma_{x,T}$ is the standard deviation of the interference level during time T in dB.

5.6.4 The Time Dependence of the Interference Level at the Reception Point

This quantity depends on the state of the ionosphere and the radio wave propagation conditions. Therefore, observations of the interference are taken both during the course of the day and from day to day. In Fig. 5.8, a plot of the variation of the average interference levels over one day is shown for four frequencies [14].

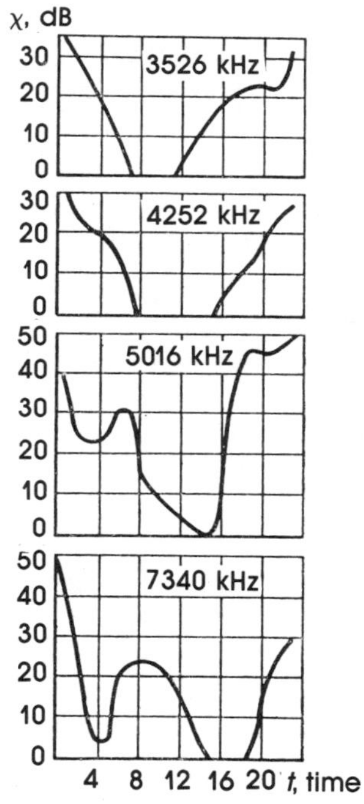

Fig. 5.8 Examples of the diurnal variation in the interference level average over an hour at various frequencies. The value x characterizes the amount by which the average interference level exceeds 1 μV at the receiver input.

It may be seen from the plots that the interference level drops sharply during the day by as much as 40–50 dB. This is explained by the fact that the absorption of radio waves changes significantly during the day. During the day, many stations are operating at higher frequencies and the interference level is low.

Week-long observations of the average interference levels measured at nine frequencies in the lower portion of the HF band, over a two-hour period (22:00–24:00) have indicated that for a given frequency, receiver site, and time of day, the interference level will vary little, usually less than 10–12 dB during a single week. From the nature of the interference, it should be expected that, in addition to diurnal variations, there are also seasonal changes in the interference level, and also variations associated with the solar cycle. The necessary experimental data, however, have not been collected.

Currently, it is possible to speak only of some mean or median values of the levels x_T during the observation time T. The quantity x_T is random, depending on the geographical location of the receiver, the frequency, the time of day, the season, the extent of solar activity, the type of receiving antenna, the receiver band, and other factors. According to data obtained in the European part of the USSR in the HF band [14], the level of x_T during the day is not greater than 30–35 dB, measured relative to the level 1 microvolt at the input to the receiver in a band of 1.2 kHz. A typical symmetrical horizontal antenna was used in these measurements; the interference level was measured at discrete times and integrated over 1.5 or 5 seconds [14].

The standard deviation $\sigma_{xi;20}$ for a twenty-minute interval at randomly chosen frequencies in the band 4–8 MHz was usually less than 5–7 dB during the day, with mean values $x_{i;20} \leqslant 20$ dB. At night, the standard deviation sometimes reached 12–14 dB, but was usually 8–10 dB. The mean interference level $x_{i;20}$ substantially exceeded the daytime values, lying in the range 20–40 dB.

5.6.5 The Distribution of Interference Levels and Channels "Free" from Interference

Despite the high level of narrow-band interference at particular frequencies, there are often times when the interference levels at frequencies in each small interval Δf are insignificant. These channels, which are "free" of interference, may then be used for operation, if the system is able to adapt to the interference conditions.

Extensive and extremely interesting material on the distribution of average interference levels $x_T(f)$ for portions of the HF band are presented in Komarovich and Sosunov [14]. These data are based on the results of several hundred thousand individual measurements, which were performed at different times of day, in different seasons, and in years of average and near-maximum solar activity. The measurements were made simultaneously at 10 frequencies, automatically switched on for six seconds, at a rate of 10, 20, or more repetitions per minute, at every frequency. These measurements were then made at the next 10 frequencies, and so on.

The variation of the mean value x_{100} and standard deviation $\sigma_{x,T}$ of the interference level in 100-kHz intervals (ΔF = 100 kHz) from 4.5 to 14 MHz is shown in Fig. 5.9. The averages were calculated from 100 elementary frequency channels of 1-kHz width, so that the distribution in percentages is the number of channels in which the interference level was less than some threshold value.

The processing of a large statistical sample for 100-kHz averages shows that in the overwhelming majority of cases, the value $x_T(f)$ satisfies a Gaussian distribution. With this distribution, if the mean value is x_{100} and the standard deviation is $\sigma_{x,T}$, 50% of the individual 1-kHz channels will exhibit an interference level less than x_{100}; in 16% the level will be no greater than $x_{100}—\sigma_{x,100}$, and in 2% the level will be no greater than $x_{100} —2\sigma_{x,100}$.

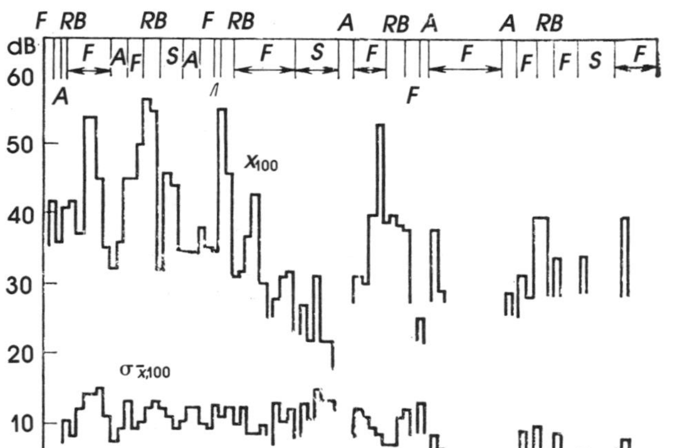

Fig. 5.9 Distribution of the mean noise level x_{100} and its standard deviation $\sigma_{x,100}$ by frequency band: F = fixed radio services; A,S = airborne and seaborne mobile services; RB = radio broadcasting.

The largest interference levels are observed in those portions of the band which are used for powerful radio broadcasting and fixed stations; a much lower interference level is observed in those portions of the band used by mobile seaborne and airborne stations. The data presented in Fig. 5.9 are for evening and nighttime. During the day, the mean interference levels will be approximately 20 dB lower. The season also has a significant effect on the interference level. The winter night interference level is approximately 12–15 dB higher than the summer night [14].

5.6.6 The Characteristics of Active Interference in the Operating Band

The level of interference in the system operating band, in general, depends on a large number of factors, such as: the geographic location of the station, the time of year and day, the operating band, and characteristics of the receiving antenna pattern. The minimum interference in the operating band will be determined by atmospheric and galactic noises, which may be predicted with the CCIR maps. The dependence on other factors is similar to that exhibited by atmospheric and galactic noise, and will therefore not be considered separately.

Frequency bands in which active interference is dominated by atmospheric and galactic noise have widths from the hundreds of hertz to several kilohertz. Such bandwidths are characteristic of the lower end of the HF band, where there is an extremely high level of narrow-band interference. The interference level, therefore, depends substantially on the pass band of the receiver. This dependence is determined by the use of the given sub-band by radioelectronic equipment and is different for the various portions of the band. As an illustration, the average interference level as a function of bandwidth is shown for various portions of the HF band in Fig. 5.10 [see Kolosov, A.A., Bogdanov, O.K., Pikalova, L.G., and Shamshchev, M.V. *Pomekhi radiopriemu v dekametrovom diapazone*—Obzor Dep. No. 3-6537, 1981].

The level of active interference in narrow-band channels, when using typical antennas, is determined mainly by the sources located within the main lobe of the antenna pattern. It is, therefore, usually impossible to predict the dependence of the interference in the operating channels as a function of azimuth. As an example, Fig. 5.11 presents the dependence of the minimum interference level on azimuth, obtained at a particular time of day. It may be seen in the figure that the minimum interference level changes by several decibels. The sharp increase in the interference level seen near 240° is explained in this case by the existence of an industrial center at this azimuth.

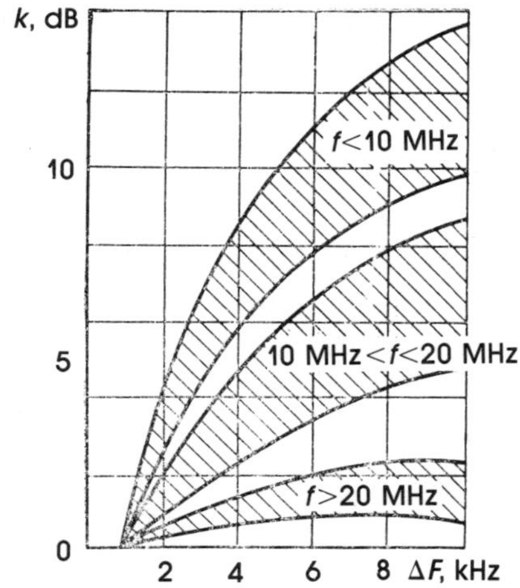

Fig. 5.10 Increase in the noise power spectral density with an increase in the system bandwidth.

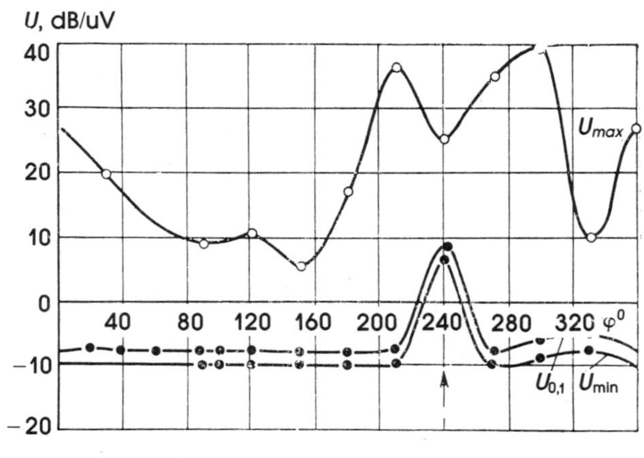

Fig. 5.11 Azimuth dependence of the interference level (the arrow indicates the direction to a large industrial center).

5.6.7 Methods for Predicting the Interference Level in Narrow-Band Channels

The minimum possible active interference level in narrow-band channels, as has already been noted, is determined by the atmospheric and galactic noise levels, which may be estimated from the CCIR predictions. Therefore, for a basic prediction of the active interference level, the maps of the CCIR 322 (see [18]) may be used, introducing certain empirical rules. An estimate of the average spectral density of active interference for a given geographical receiver location may be determined with the formula:

$$N(T, f_0, B, n) = N_{\mathrm{CCIR}}(T, f_0) + k_{vz}(f_0, + B) + k(n)$$

where T is the time of operation (month, hour), f_0 is the average operating frequency, B is the receiver bandwidth, and n is the number of separate frequency channels.

The quantities N and N_{CCIR} are measured in dB W/Hz, and k_{vz} and $k(n)$, in decibels. The empirical coefficient k_{vz} takes into account any increase in the average interference level in comparison with N_{CCIR} due to interference from other radio stations and systematic errors in the CCIR prediction for a given geographical point. The coefficient k_{vz} is determined by the use of the operating sub-band by radio-electronic equipment and depends on the receiver pass band B and the location of the operating sub-band. The value of k_{vz} depends on the system location and is determined experimentally. The coefficient $k(n)$ determines the increase in the interference level in the operating channels due to the limited number of operating frequencies.

5.7 INTERFERENCE FROM SPURIOUS AND OUT-OF-BAND RADIO TRANSMISSIONS

Radio stations transmit both primary and secondary radiation. The primary radiation is that transmitted at the frequencies within the band necessary to obtain the acceptable level of fidelity in the given signal. Most of the energy transmitted lies in this primary band.

Some portion of the transmitted energy, however, lies outside of the main band, and inasmuch as it makes no contribution to the operation of the system, may be considered interference. This includes spurious transmission and out-of-band transmission (this question is examined in more detail in Knyazev and Pchelkin [12]).

5.7.1 Spurious Radiation

This is a rather wide class of secondary radiation effects, connected with nonlinear processes in the radio transmitter, the most important of which are radiation at harmonics and subharmonics, combination radiation, parasitic radiation and intermodulation radiation. Radiation at the harmonics occurs at frequencies which are multiples of the main transmission frequency, mostly at the second and third harmonics, although powerful transmitters will radiate energy at higher-order harmonics.

The relative radiation levels at different harmonics for transmitting stations operating at frequencies below 30 MHz, taking account of the antenna gains at these frequencies, is shown in Fig. 5.12, as taken from [29]. These results are averages over a large number of stations, including high-power transmitters.

The other forms of spurious transmission usually exhibit much lower energy levels than seen at the harmonics.

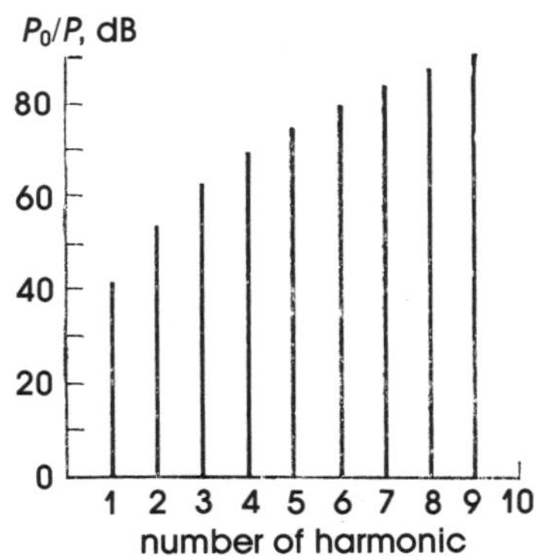

Fig. 5.12 Relative power levels at different harmonics.

5.7.2 Out-of-Band Transmission

These transmissions are in bands lying next to the main transmission band, and arise as a result of signal modulation used for communications. When considering this type of radiation, it is convenient to determine the ratio of the power radiated in the band necessary for proper operation of the system to the total radiated power, which usually covers a wider band. In [8] it is recommended that the power radiated below or above the necessary band should be no more than 0.5% of the total transmitter power. In other words, no less than 99.5% of the average radiated power should lie within the prescribed band. For a radio transmitter with average power exceeding 50 kW, an additional limitation is placed on the out-of-band radiation: it should not exceed 50 mW, independent of the station power. With this stipulation, the level at which the bandwidth is calculated is 60 dB (see [12]). The existence of spurious and out-of-band radiation in radio transmitters creates a general background interference level over a wide band, and has a noticeable effect on the general interference conditions.

5.8 THE NET (COMBINED) ACTIVE INTERFERENCE

5.8.1 Combined Interference

The net, or, as it is sometimes called, combined, interference signal [25], is caused by active interference which may be viewed as a combination of narrow-band, quasipulsed, atmospheric, and galactic noises. The data presented above on interference measurements should be considered to be data on narrow-band interference. In a large fraction of the HF band, the net noise power is usually determined mainly by the narrow-band components of radio frequency interference. When measuring the level of narrow-band interference sources in a frequency band B, it is extremely difficult to separate this form of interference from the other noise sources within the band. Therefore, it is more correct to consider the data on the mean value and standard deviation of the level of narrow-band noise sources as describing the combined interference. The exception is that portion of the band which is free of narrow-band interference, in which the dominant noise source is quasipulsed interference, which is mainly a result of atmospheric noise. According to data in Chelyshev [25], the difference in the noise level between frequency regions in which there is narrow-band interference and those where there is none is about 80–100 dB.

To estimate the extent to which atmospheric interference is pulsed, we may use the parameter V_d (see [18]), which is the ratio between the root mean square (RMS) and mean values of the interference voltage, expressed in decibels. The pulsed component of atmospheric noise is known to change with frequency; at lower frequencies and with wider receiver bandwidths, the pulsed character of the interference becomes more clear.

It was suggested in Chelyshev [25] that the parameter V_d be used to estimate the quasipulsed interference level. At the input to a receiver with a wide-band front end, quasipulsed interference appears as an extremely low-level fluctuating component, and a periodically appearing high-level pulsed component. As it passes through the narrow-band receiver, the pulsed component is filtered to a fluctuating component, sharply increasing its intensity [25].

A quantitative comparison of the average levels of atmospheric (quasipulsed) and narrow-band interference from neighboring radio station channels is presented in Khmel'nitskiy [24] (see Fig. 5.13), in the form of cumulative distributions. The measurements were made at frequencies which were used for 1500-km propagation paths at the given season and time of day. In the opinion of Khmel'nitskiy [24], the drop in the level of atmospheric and narrow-band noises would not be much different for longer paths than those illustrated in Fig. 5.13.

Only during the summer does the narrow-band source noise level approach that of atmospheric noise, with a difference in the range 3–10 dB. During the other seasons, the narrow-band noise level greatly exceeds that of atmospheric noise.

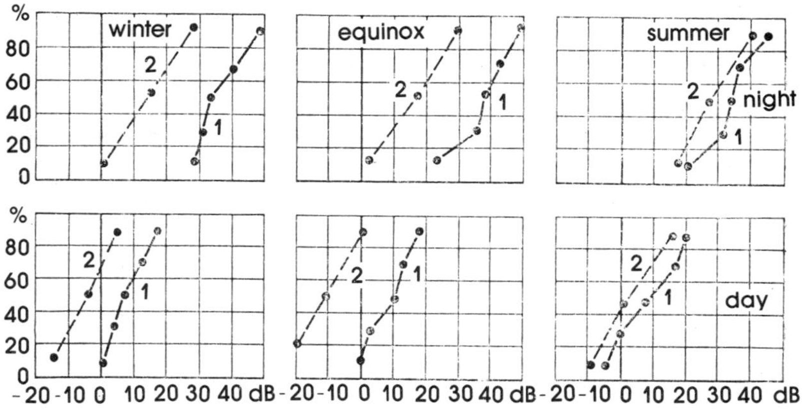

Fig. 5.13 Distribution of noise levels at frequencies used over a 1500-km path: (1) interference from neighboring channels; (2) atmospheric interference. The values along the abscissa are the interference levels relative to 1 μV, and along the ordinate is the percentage of measurements not exceeding the value on the abscissa. The receiver bandwidth was 1 kHz.

5.8.2 Dependence of the Noise Level on the Receiver Bandwidth and Type of Receiving Antenna

The noise level at the receiver input depends to a large extent on the bandwidth of the receiver. Thus, any experimental data is related to the specific bandwidth associated with the measurements, necessitating a conversion from measurements at one bandwidth to corresponding values at another.

The character of this conversion depends on the form of the interference. In those rare cases where the interference is pure fluctuation (white noise), the spectral density is uniform. In these cases, the noise power is directly proportional to the bandwidth, and the conversion results will be determined directly from the ratio of the bandwidths. With quasipulsed noise, such as atmospheric noise, the bandwidth dependence will be more complex. A graph allowing the parameter V_d to be converted from one bandwidth to another is presented in Fig. 14 (see [18]). The initial bandwidth, relative to which the conversion is shown, is 0.2 kHz.

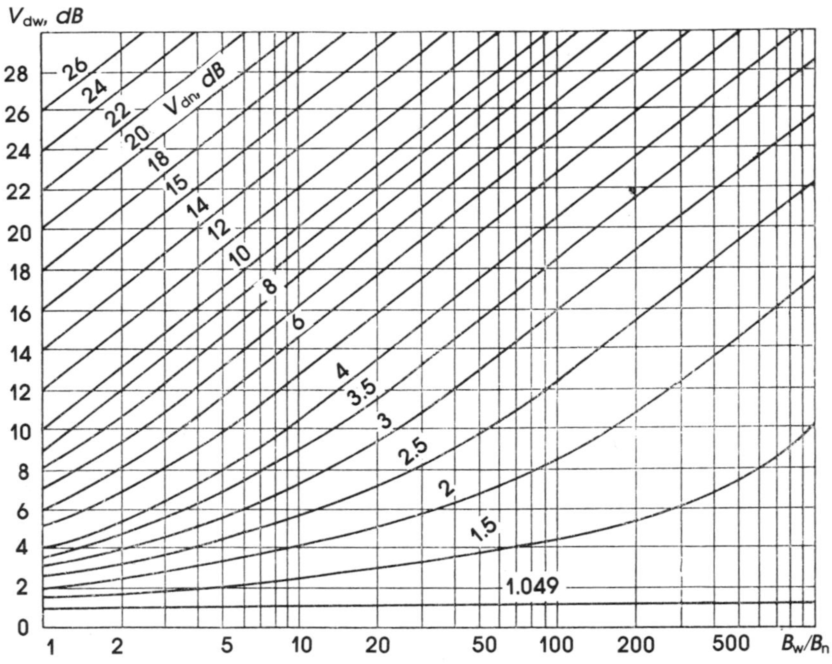

Fig. 5.14 Conversion of the parameter V_d from one bandwidth to another; c_w is the wider band, and c_u is the narrower band.

For narrow-band interference, the bandwidth dependency is determined by the character of the energy distribution of the side frequencies near the carrier frequency, which in turn depends on the form of modulation. Information on this question is extremely lacking, although it is of great interest.

The characteristics of the receiving antenna also exert an influence on the interference level at the receiver input. Measurements of signal levels depend on the pattern of the antenna with which the measurements are made. The quantitative data obtained from experimental measurements may be used directly only if the antenna in the system being analyzed is the same as that used to make the measurements. If the interference arrived uniformly from all directions, then the interference level would not depend on the form of the pattern, but the radio stations which are sources of narrow-band interference may be distributed unevenly, in various directions.

In addition, the main sources of atmospheric interference are located in particular regions with especially intense thunderstorm activity. The existence of strong sources of atmospheric noise, in tropical countries, for example, causes the interference at the receiver to be strongest from certain directions. The noise level at the receiver therefore depends on the form and orientation of the antenna pattern. For the atmospheric interference entering through the sidelobes and back lobes of the antenna, this dependence is not pronounced. For interference from radio stations, especially from powerful ones, this dependence is rather noticeable.

In light of these considerations, experimental data obtained with a particular antenna should be considered to represent only a rough indication of the results which will be obtained with a different antenna. Every type of comparative data depends to some extent on the type of receiving antenna used.

5.9 CLUTTER

5.9.1 Types of Clutter

One of the features of over-the-horizon radar, when used for oblique sounding, is the interference caused by clutter returns. The basic types of clutter energy affecting HF radars are: the earth's surface, illuminated by "hopped" signals; ionospheric irregularities; and round-the-world signals. To determine the extent to which clutter affects an HF system, it is necessary to know both its energy characteristics (the distribution of clutter power in the spatial sector observed by the OTH radar), and its spectral characteristics (Doppler shift, spectral width).

5.9.2 Hopped Signals

Hopped signals are the most powerful source of clutter power. Oblique sounding signals reflected from the earth are formed by large numbers of scattering elements. Over-the-horizon radar systems illuminate a large area of the earth's surface, so that the reflected signals which contribute to the clutter returns have a rather complex character. In addition to the fact that clutter is created by reflections from a large number of scattering elements, it is also important that the interference originates from more than just a single beam. Hopped beams illuminating the earth's surface are split by the earth's magnetic field into ordinary and extraordinary components, each of which forms its own group of reflected signals. Also, just as with other forms of interference, clutter enters the receiver not only from the direction of the main beam, but also through the sidelobes and back lobes of the antenna. In many cases, interference entering the sidelobes may dominate, as a result of its cumulative effect.

When detecting objects located relatively close to the earth's surface, signals reflected from the surface in the neighborhood of the target arrive from the same range interval as the target signals. Such reflections from objects which are not targets constitute clutter. The intensity of the clutter is directly related to the clutter cross section σ_c, which is measured in square meters: $\sigma_c = S_c\sigma_0$, where S_c is the scattering area in square meters of a surface element resolved by the radar, from which reflection occurs, and σ_0 is the clutter reflectivity, which depends on the relief features of the local surface.

5.9.3 The Clutter Reflectivity

The clutter reflectivity σ_0 is the ratio between the power scattered at some range r by a surface in a given direction, to the power which would have been scattered had the surface element scattered the incident power isotropically. The value of the clutter reflectivity may vary over a wide range depending on the local surface features. It is therefore necessary to consider some data characterizing local topography in connection with oblique HF propagation.

The actual earth's surface differs substantially from the idealization of a uniform homogeneous surface. At high frequency, the height of local relief features in many cases is commensurate with the wavelength. Even in a region of plains, the height may vary as much as 20–30 m, with a slope of 2°–3°.

In hilly terrain, the height variation is several tens of meters, with a slope of about 5°. In extremely rugged terrain, the height variation may exceed 100 m. In sandy desert regions, the height of dunes is usually no greater than 10 m, although there are dunes with a height of 30–40 m and a length of 200–300 m, the slopes reaching 30°–35°. Large height differentials are found in mountainous regions. For high mountains (higher than 1000 m), the slope may reach 30°–35°.

For the sea surface, the height of steady-state waves depends on the wind speed. Their height varies from several centimeters (slight ripples) to several meters. The slope of sea waves never exceeds 15°–20°.

The data just presented supports the fact that the dimensions of local relief features are on the same scale as a wavelength of 10–100 m. In the majority of cases, the ratio of the maximum dimension of vertical variation h to the maximum dimensions of horizontal variation l is on the order of $h/l \approx 0.1$–0.2.

When considering the scattering of radio waves from the earth's surface, it should be kept in mind that the wavelength in which we are interested is 10–100 m, while the area of the surface which may be illuminated may be tens of thousands or even hundreds of thousands of square meters [26]. In these conditions, it is sufficiently precise to use a statistical model of the surface features. For the most part, with the exception of mountainous regions, the theory of stationary random processes may be used. Nonuniformities of the earth's surface are formed over large intervals of time as the result of a number of random factors. Therefore, it is natural to expect that the distribution of heights will be Gaussian. Considering available experimental data, the Gaussian distribution may be used for the analysis of reflection from the earth's surface [26].

The variation of the clutter reflectivity σ_0 as a function of the incidence angle α, is presented in Chernov [26], on the basis of experimental results for various types of scattering surfaces (see Fig. 5.15). As may be seen from Fig. 5.15., for low incidence angles (to 5°), the value σ_0 lies between -30 and -40 dB. The data presented corresponds to a relatively uniform surface. The reflectivity in mountainous terrain is 12–14 dB higher than in flat terrain, and mixed relief, where both flat and mountainous regions are encountered, exhibits a reflectivity 6–8 dB higher. Thus, if at low incidence angles the clutter reflectivity in a flat area is -30 dB, then in a mixed region it will be approximately -23 dB.

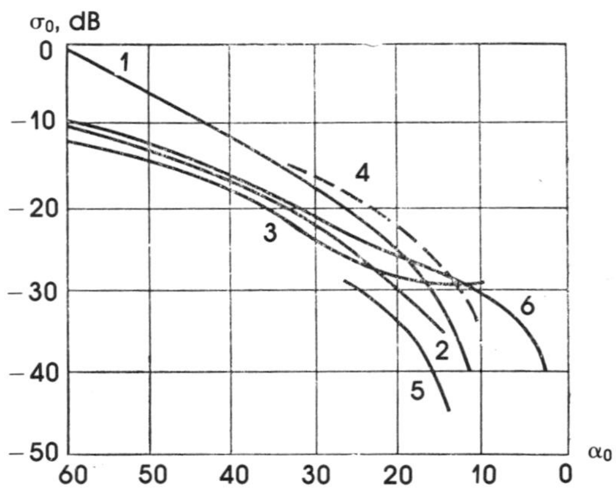

Fig. 5.15 Dependence of the clutter reflectivity on the incidence angle: (1) smooth ice or calm sea; (2) rough ice; (3) desert; (4) sea; (5) smooth land; (6) hilly plain (curve calculated for $\lambda = 30$ m).

5.9.4 Determining the Reflected Signal Level

The signal power reflected from a scattering surface may be determined using the radar equation (2.16). In this equation it is necessary to replace the target cross section σ with the effective clutter cross section σ_c, and also note that the quantity W refers to the total loss (in both directions) along the oblique propagation path, that is, the radar loss from the radar to the surface generating the clutter signals.

Thus, using (2.16):

$$P_{\text{obl}} = (P_t G_t A_r / L)\, \sigma_c (1/W_{\text{obl}}) \tag{5.1}$$

where P_{obl} is the power of the oblique signals at the receiver input in watts, L is the loss coefficient due to imperfections in the radar, σ_c is the clutter cross section, and W_{obl} is the total loss over the propagation path from the radar to the reflecting surface. The remaining quantities are the same as in (2.16). In principle, (5.1) may be used to calculate the clutter power, but in practice, this is an extremely difficult problem, due to difficulties in determining W_{obl} and σ_c (taking into account the variation of this quantity with wavelength).

Therefore (5.1) should be seen merely as an indication of the relation between the various factors determining the clutter power level.

5.9.5 The Spectral Characteristics of Obliquely Propagating Signals

The spectral characteristics of obliquely propagating signals are characterized by a Doppler shift and widening of the received signal spectrum. The Doppler shift F_D is caused by the path change along the propagation trajectory of the radio waves (see [16]): $F_D = -(f/c)dL/dt$, where f is the transmitted frequency, c is the speed of light, the phase path length $L = \int v \, dl$, v is the index of refraction, and l is the radio wave path length.

From the above expressions, it may be seen that the Doppler shift may be caused by changes both in the index of refraction, and in the geometrical path of the radio wave, resulting from changes in the height of the ionosphere, for example. Inasmuch as the ionosphere exhibits diurnal variations, the Doppler shift of obliquely propagating radio waves also undergoes daily changes. Deviations from the regular diurnal variation are observed during periods of solar flares, and also during magnetic storms [16]. Studies of the spectral characteristics of oblique transmissions show that their Doppler shifts, along midlatitude paths, are on the order of several hertz, in many cases, not exceeding 1 Hz. The sign of the Doppler shift, as well as its magnitude, changes as a function of the state of the ionosphere. In some cases, over a relatively short time interval, the Doppler shift F_b may, for example, take on values in the range of -1 to 1 Hz.

The spectrum of hopped signals is widened due to fluctuations in their amplitudes and phases, with rapid signal fluctuations having the greatest effect. The main causes of signal fluctuations at the receiver input are: interference effects; changes in the polarization plane, phase and group paths of the radio waves; changes in the angle of arrival in both the azimuth and elevation planes; and fading connected with changes in the critical frequency of the layer at which reflection occurs. These effects, which result in a widening of the hopped signal spectrum, are caused both by irregular processes in the ionosphere, and by moving ionospheric inhomogeneities of various sizes. Ionized meteor trails have a significant effect on the width of the signal spectrum, and are influenced by the dynamics of diffusion processes and wind transport. The form of the spectrum of hopped signals, appearing at the receiver input as clutter, depend on the path length, frequency, and time of day.

5.9.6 Signals Reflected from Ionospheric Irregularities

Ionospheric irregularities may be divided into three categories: 1) small, with linear dimensions up to 1 km; 2) ionized regions with linear dimensions of 1–100 km; and 3) large structures with linear dimensions larger than 100 km.

The motion of ionospheric irregularities leads to a shift in the maximum of the spectral density of the reflected signals. The velocity of chaotic motion in small irregularities lies in the range 0.5–15 m/s, which corresponds to a Doppler frequency of 0.05–1.5 Hz at a frequency of 15 MHz (see [1]). Besides this, regular movement of the structures is observed; this motion is referred to as drift or ionospheric wind [1].

The statistical characteristics of the scattered signal reflected from ionospheric irregularities is examined in [28], for operation with a pulse length of 400 ms at a frequency of 17 MHz, operating in fixed directions. In all, 400 measurements obtained from 1968 to 1972 were processed. A computer was used to calculate the correlation functions, spectral fluctuations, distribution functions, and first four moments. It was established that the distribution of signal amplitudes for hopped signals reflecting from the F layer depend on the magnetic perturbations. From data taken in 1979 (160 measurements), the distribution on a magnetically calm day was close to Gaussian in 64% of the cases, and substantially non-Gaussian in the remaining 36% of the cases. The fluctuation energy spectra for hopped signals were calculated in the band 0.02–0.1 Hz. The spectra which were obtained were linear.

5.9.7 Round-the-World Signals

Round-the-world signals, having delays of 138 ms, are signals which propagate around the earth and are received in the back lobe of the antenna. When the pulse repetition interval is less than 138 ms, round-the-world signals may generate a strong clutter response.

The overall clutter level in many cases is quite large, and significantly exceeds the target signal levels. In order to detect moving targets against a background of clutter, the same methods used in standard radar systems may be applied, namely, temporal, spatial, and frequency filtering [3, 23].

5.10 NOISE RESULTING FROM IMPERFECTIONS IN THE RECEIVER

Interference of this type is caused by the interaction of external signals from outside the main operating channel. Amongst its causes are nonlinear processes in the wide-band portion of the receiver, and the reception of signals through side channels in superheterodyne receivers [4, 6]. These phenomena appear to a lesser degree as the dynamic range of the receiver is increased.

REFERENCES

1. Alpert, Ya.L. *Rasprostranenie elektromagnitnykh voln v ionosfere (The propagation of electromagnetic waves in the ionosphere).* Moscow: Nauka, 1972. 563 pp.
2. Bubnov, I.A. *Voennaya topografiya (Military topography).* Moscow: Voenizdat, 1969. 349 pp.
3. Vishin, G.M. *Selektsiya dvizhushchikhsya tseley (Moving target detection).* Moscow: Voenizdat, 1966. 275 pp.
4. Golubev, V.N. *Chastotnaya izbiratel'nost' radiopriemnikov AM signalov (The frequency selectivity of AM receivers).* Moscow: Svyaz', 1970. 199 pp.
5. GOST 14777–76. *Radiopomekhi industrial'nye. Terminy i opredeleniya. (Industrial radio interference: terminology and definitions).*
6. Girbov, E.B. *Nelineynye yableniya v priemno-peredayushchem trakte apparatury svyazi na transistorakh (Nonlinear phenomena in transistorized communications transceivers).* Moscow: Svyaz', 1971. 243 pp.
7. Grudinskaya, G.P. *Rasprostranenie radiovoln (The propagation of radio waves).* Moscow: Vysshaya Shkola, 1975. 279 pp.
8. *The spectrum and bandwidths of radiation.* Documents of the 10th Plenary Assembly of the CCIR. Recommendation 328 (Geneva, 1963), vol. 1. Moscow: Svyaz', 1964, pp. 39–45.
9. Dolukhanov, M.P. *Rasprostranenie radiovoln (The propagation of radio waves).* Moscow: Svyaz', 1965, 400 pp.
10. Kalinin, A.I. and Cherenkova, E.L. *Rasprostranenie radiovoln i rabota radioliniy (The propagation of radio waves and the operation of radio lines).* Moscow: Svyaz', 1971. 439 pp.
11. Kalinichev, B.P. *O raspredelenii amplitud atmosfernykh pomekh (On the amplitude distribution of atmospheric interference).* Electrosvyaz', (1968), no. 2., pp. 76–77.
12. Knyazev, A.D. and Pchelkin, V.F. *Problemy obespecheniya sovmestnoy raboty radioelektronnoy apparatury (Problems in maintaining the operation of colocated radio equipment).* Moscow: Sovietskoe Radio, 1971. 200 pp.
13. Kolosov, A.A. *Rezonancnye sistemy i rezonancnye uciliteli (Resonant systems and resonant amplifiers).* Moscow: Svyaz'izdat, 1949. 559 pp.
14. Komarovich, V.F. and Sosunov V.N. *Sluchaynye radiopomiekhi i nadezhnost' KV-svyazi (Random radio interference and the reliability of KV-communications).* Moscow: Svyaz', 1977. 134 pp.
15. Lyutov, S.A. *Industrial'nye pomekhi radiopriemu i bor'ba s nimi (Industrial interference affecting receivers, and methods for suppressing it).* Moscow: Gosenergoizdat, 1951. 237 pp.

16. Namazov, S.A. and Novikov, V.D., *Khmel'nitskiy I.A. Doplerovskoe smeshchenie chastoty pri ionosfernom rasprostranenii dekametrovnykh radiovoln (The Doppler shift in HF ionospheric propagation).* Izv. Vuzov SSSR. Radiofizika, vol. 18 (1975), no. 4., pp. 473–500.
17. Zhulina, E.M., Kerblay, T.S., Kovalevskaya, E.M. *et al. Osnovy dolgosrochnovo prognozirovaniya (Fundamentals of long-term prediction)* Moscow: Nauka, 1969. 68 pp.
18. *Rasprostranenie po zemnomu sharu atmosfernykh pomekh i ikh kharakteristiki (The characteristics and propagation of atmospheric noise along the earth).* Documents of the 10th Plenary Assembly of the CCIR. Note 322. Moscow: Svyaz', 1965. 80 pp.
19. Siforov, V.I. *Radiopriemnye ustroystva (Radio receivers).* Moscow: Voenizdat, 1954. 803 pp.
20. Fal'kovich, S.E. and Muzyka, Z.N. *Chuvstvitel'nost' radiopriemnykh ustroystv s tranzistornymi ucilitelyami (The sensitivity of radio receivers with transistorized amplifiers).* Moscow: Energiya, 1970. 128 pp.
21. Visnobatiy-Kogan, G.S., Bychkov, K.V., (Pikel'nera, S.B., ed.) *et al. Fizika kosmosa (Space physics).* Malen'kaya Entsiklopediya (Little Encyclopedia. Moscow: Sov. Entsiklopediya, 1976. 655 pp.
22. Khardman, L. *Rasprostranenie chastot elektromagnitnovo spektra v usloviyakh napryazhennovo trafika (The propagation of electromagnetic frequency components in heavy radio traffic*: Obzor). Elektronika, vol. 45 (1972), no. 20, pp. 30–58.
23. Kharkevich, A.A. *Bor'ba s pomekhami.* Moscow: Nauka, 1965. 275 pp.
24. Khmel'nitskiy E.A. *Otsenka real'noy pomekhozashchishchennosti priema signalov v KV diapazone (Estimating the actual suppression of interference in the reception of KV-band signals).* Moscow: Svyaz', 1975. 232 pp.
25. Chelyshev, V.D. *Priemnye radiotsentry (Radio receiving stations).* Moscow: Svyaz', 1975. 264 pp.
26. Chernov, Yu.A. *Voxvratno-naklonnoe zondirovanie ionosfery (Oblique sounding of the ionosphere).* Moscow: Svyaz', 1971. 203 pp.
27. Cherniy, F.B. *Rasprostranenie radiovoln (The propagation of radio waves).* Moscow: Sovietskoe Radio, 1972. 463 pp.

28. Chistyakova, L.V. *Statisticheskie kharaketeristiki rasseyannovo signala pri VNZ ionosfery (The statistical characteristics of the scattered signal in oblique sounding of the ionosphere).* Aktual'nye Voprosy Rasprostraneniya Dekametrovykh Voln (Actual Questions on the Propagation of HF Waves). IZMIRAN. Moscow: 1973, pp. 130–131.
29. *Elektromagnitnaya sovmestimost' radioelektronnykh sredstv i nepred-namerennye pomekhi (The electromagnetic compatibility of radio systems in nondeliberate interference).* Sost. D.R. White. Issue 1. Obshchie voprosy EMC (General EMC issues). Mezhsistemnye pomekhi (Inter-system interference): transl. from English (Sapgira, A.I., ed.), Moscow: Sovietskoe Radio, 1977, 347 pp.
30. Pawsey, J.L., McGready L.L., and F.F. Gardner, *Ionospheric thermal radiation at radio frequencies* J. Atmospheric and Terrestrial Physics, vol. 1 (1951), no. 4, pp. 216–230.

Chapter 6
High-Frequency Anti-Interference Techniques

6.1 INTRODUCTION

As was shown in the preceding chapters, high-frequency radars must operate in extremely difficult interference conditions. Non-Soviet installations, therefore, make use of special techniques to lessen the effects of interference [10, 19]. One of these methods is the selection of operating frequencies with comparatively low levels of active interference. A scanning receiver is used for this purpose, in order to determine the usage of the frequency band continuously [19].

The wide range of possible angles of arrival of narrow-band interference from radio stations determines the effectiveness of anti-interference methods based on adaptive spatial filtering. Thus, in the US WARF OTH radar [26, 32], an adaptive receiving antenna system with digital correlation processing is used, providing for the adaptive formation of nulls in the antenna pattern in those directions from which active interference is being received.

Comparatively wideband signals are used in the known non-Soviet OTH radars [32]. Meanwhile, a substantial number of interfering radio station signals (radio telegraph signals, for example) are narrow-band sources. Methods of countering such interference are widely known [5, 6, 17].

In the construction of foreign OTH radars, it turned out that the clutter levels greatly exceeded the useful signal levels, which necessitated the application of effective methods for reducing the clutter levels [11, 19]. The earliest of these methods were used in one of the first US OTH radars, of the MADRE type. They were further developed in conjunction with digital processing in later systems [26, 32].

On the basis of foreign materials, some methods of countering active interference which have been used in foreign OTH radar systems will be examined in this chapter. These include: adapting to the interference conditions, adaptive spatial filtering, and minimizing the effects of narrow-band interference.

6.2 SELECTING THE CHANNEL WITH THE MINIMUM LEVEL OF ACTIVE INTERFERENCE

As was shown in Ch. 5, the dominating source of interference in the HF band is radio system interference, caused by the multiple users of this portion of the radio spectrum. In addition to diurnal and seasonal variations, this type of interference is characterized by significant nonuniformity even in small portions of the frequency band, and temporal nonstationarity of the individual narrow-band components. As a result, despite the extreme saturation of the HF band, there are time-frequency areas which may be used by specially designed radars [19] [see Komarovich, V.F. and Sosynov, V.N. *Sluchaynie pomekhi i nadezhnost' KV svyazi* (*Random interference and the reliability of KV communications*). Moscow: Svyaz', 1977. 134 pp.] A typical spectrum of external noise in a portion of the HF band is shown in Fig. 6.1, as measured with a filter with a 5 kHz bandwidth. The level of atmospheric and cosmic interference is about −140 dBW. The largest peaks in the interference correspond to frequencies at which powerful radio stations are operating.

As may be seen in Fig. 6.1, there are portions of the band which are suitable for radar in terms of propagation conditions and which are practically free of radio interference, with widths which are sufficient for comparatively narrow-band operation (5–10 kHz). Operating the radar in these channels allows the system to come close to its potential performance in actual interference conditions, the level of interference being determined only by atmospheric and cosmic noises.

Radio system interference is nonstationary, in that its frequency distribution varies randomly in time. The frequency-time areas with low interference levels in which the signal energy should be placed may be selected with frequency adaptivity, which necessitates real-time analysis of the frequency band, selection of the channels with minimal or acceptable interference levels, and radar tuning [19]. Adaptive frequency selection, as opposed to formal frequency allocation, is considered the most promising means of countering radio interference in the HF band [20].

Unlike communications systems, with their inherent spatial separation of users, radar transmission and reception may be performed in a localized region, which substantially improves the prospects for practical realization of frequency adaptivity [19]. Adaptive frequency selection is seen to increase the capabilities of radars significantly, both in maintaining operation in channels with acceptable interference levels, and in improving their sensitivity. Both of these characteristics may be used to quantify estimates of the effectiveness of frequency adaptivity.

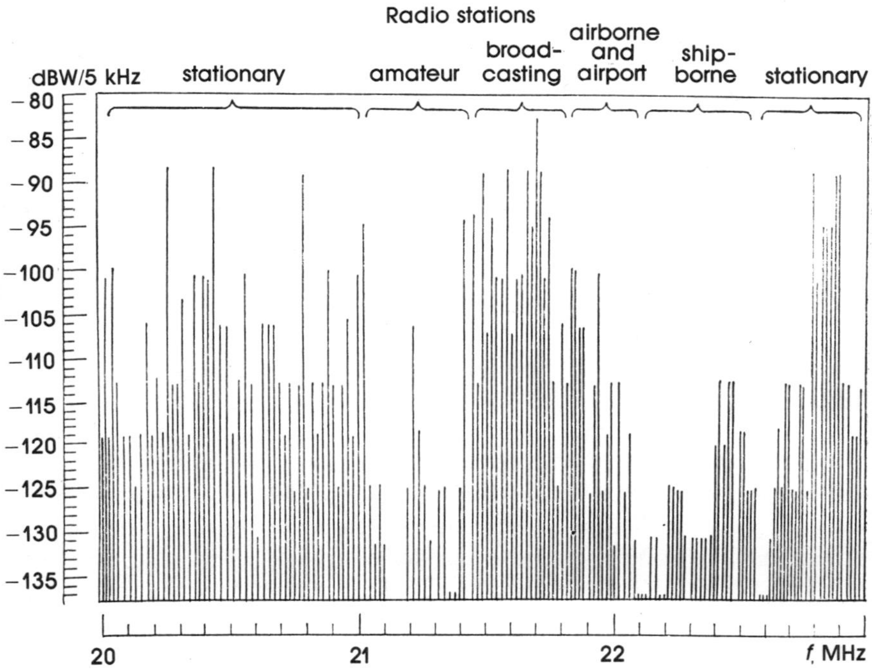

Fig. 6.1 Noise and interference spectrum in part of the HF band.

If a particular set of fixed frequencies is not specified for radar operation, then, knowing the spectral width of the transmitted signal B_s, the combination of sub-bands allowed for operation may be viewed as multiple independent channels. The channels may be considered independent if they are separated by the bandwidth $B_s > 3$–5 kHz. The probability H of maintaining satisfactory performance (relative to interference) is then given by the probability of being able to choose, from among n independent frequency channels, one with acceptable characteristics for radar operation, i.e.:

$$H = 1 - (1 - P)^n \tag{6.1}$$

where P is the probability that an arbitrarily chosen ith channel is "free" in that portion of the band which may be used for propagation. In other words, H is the probability that there is at least one free channel out of the n possible channels.

To achieve potential performance and meet signal processing demands, it is necessary that the interference in the operating channels not exceed some threshold level $U_{i.thr}$, and that the length of time the channel is free is not less than the time constant associated with the signal processing T. The group of radio channels may then be placed in the form of a number of elements with discrete transitions from one state to another. The probability that a channel with the necessary energy and time characteristics will exist is then

$$P = P_U P_\tau \tag{6.2}$$

where $P_U = P_U\,(U_1 \leqslant U_{i.thr})$ is the probability that there is a channel in which the interference lies below the threshold $U_{i.thr}$, and $P_\tau = P_\tau\,(\tau \geqslant T)$ with $U_1 \leqslant U_{i.thr}$ is the conditional probability that the time τ during which the channel with the acceptable interference level exists will not be less than the processing time T. It is clear that $P_U\,(U_i \leqslant U_{i.thr})$ is determined by the probability distribution of interference levels in the ensemble of radio frequency channels. The second factor in (6.2) determines, as said in Cox and Smith [8], the reliability of using a channel for the interval T relative to the threshold level $U_{i.thr}$. The value of the probability $P_\tau\,(\tau \geqslant T)$ with $U \leqslant U_{i.thr}$ is equal to the value of the function $\mathcal{F}_\tau(T)$, characterizing the distribution of the lifetimes of channels relative to the threshold $U_{i.thr}$. The time between threshold crossings during a short usage of a radio frequency (RF) channel is satisfactorily described by a Poisson distribution. This allows us to use the exponential distribution to describe the temporal characteristics of the channels [8]:

$$\mathcal{F}_\tau(T) = \exp(-\tau/\tau_0) \tag{6.3}$$

The parameter τ_0 fully determines the exponential distribution, for which both the mean value and standard deviation are equal to τ_0. The quantity τ_0 strongly depends on the chosen threshold, that is, $\tau_0 = \tau_0(U_{i.thr})$. This dependence is satisfactorily described by the expression:

$$\tau_0 = \tau_{0\overline{U}}\, P_U/(1 - P_U) \tag{6.4}$$

where $\tau_{0\overline{U}}$ is the mean time during which the median interference level $\overline{U}_i$ is not exceeded, and P_U is the value of the cumulative distribution function over the ensemble of radio frequency channels for $U_i = U_{i.thr}$.

Now, using (6.2)–(6.4) in (6.1) and taking account of actual predicted interference conditions, it is possible to estimate the probability of maintaining satisfactory radar operation in actual interference conditions. If

(6.1) is transformed into the form:

$$P = 1 - \sqrt[n]{1 - H} \tag{6.5}$$

then, using a model interference intensity distribution, it is possible to estimate the expected radar sensitivity using frequency adaptivity. The analysis is substantially simplified when the interference model gives an easily inverted analytical expression, i.e., has a fairly simple rule for determining the argument U_i from value of the cumulative distribution function P_U. The Freshe exponential-hyperbolic distribution, for example, possesses this characteristic, exhibiting sufficient agreement with experimental results for small signals [15], i.e.:

$$P_U(U_i) = P_0{}^{(U_i/U_i)^\nu} \tag{6.6}$$

where P_U and U_{i0} are the coordinates of the points fixing the position of the distribution such that $P_U(U_i) = P_0$, and ν is the index determining the spread of the distribution. As the index ν is increased, the variance decreases, and in the limit $\nu \rightarrow \infty$, the cumulative distribution becomes a step function (the probability density function is a delta function) [15]. We will use the model of (6.6) to estimate the effectiveness of frequency adaptivity. As was shown above, a measure of this effectiveness may be the increase in sensitivity for a given value of the probability H (the probability that there is a "noise-free" channel). It is clear that if there is no frequency adaptivity, then the interference level in the operating channel will be determined by the random quantity Θ, which has the distribution $P_U(U_i)$. The given value of H will then be maintained so long as the threshold level $U_{i.thr}$, determining the sensitivity, corresponds to the value of the distribution function $P_U(U_{i.thr}) = H$ (recall that $P_U(U_{i.thr})$ is the probability that the interference in one or more of the channels is below the threshold value $U_{i.thr}$, and that H is the necessary probability of a channel being free). Then, assuming that the duration of channel usage is small in comparison with the "lifetime" of the channels (the time during which the interference is sufficiently low), it may be shown that the expected increase in actual sensitivity (that is, reduction in the energy threshold) due to the use of frequency adaptivity will be, in decibels:

$$\Delta = 20 \ln \frac{U_{i.thr}}{U'_{i.thr}} = \frac{20}{\nu} \ln \frac{\ln(1 - \sqrt[n]{1 - H})}{\ln H} \tag{6.7}$$

where $U_{i.thr}$ and $U'_{i.thr}$ are the threshold levels when operating with and without frequency adaptivity, respectively.

Analysis of (6.7) supports the natural assumption that with an increase in the variance of the interference amplitude, the effectiveness of frequency adaptivity grows (this is evident from the inverse proportional relationship between the sensitivity increase Δ and the index of the exponential-hyperbolic distribution ν). In the allowable frequency sub-bands, the value of ν for actual interference conditions is assumed to be ≈ 2 (see [15]). Using this value of ν, plots of the expected increase in sensitivity Δ *versus* the number of possible operating channels for various values of H are shown in Fig. 6.2. It is clear from the figure that in actual conditions, tens or even hundreds of channels may be used for high-frequency OTH radar operation with frequency adaptivity, the number depending on the width of the transmitted signal spectrum and the widths of the allowable bands [19]. The effectiveness of frequency adaptivity may then reach 20 dB or more, as estimated from the increase in sensitivity.

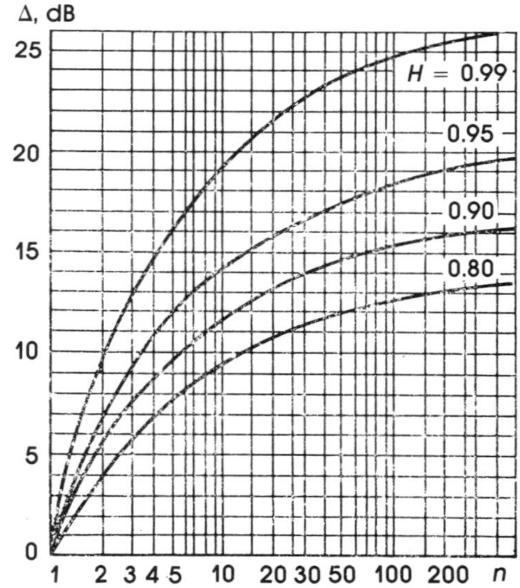

Fig. 6.2 Expected gain in sensitivity, using frequency adaptivity.

6.3 ADAPTIVE SPATIAL FILTERING

The use of spatial filtering with constant parameters in HF equipment is complicated by the lack of *a priori* knowledge of the location, number and spectral characteristics of interference sources, and also by the variability of the levels and structures of their associated fields [26, 32]. In these situations, adaptive filtering is extremely effective, allowing the antenna pattern and time-domain filtering parameters to be altered to match the current interference conditions.

As is known from non-Soviet publications [26, 32], adaptive receiver antenna arrays have been used in high-frequency radars. We will examine briefly the basic operating principles of adaptive antenna systems, which are discussed in a number of books [2, 14, 24, 25].

In receiver antenna arrays, adaptive beam steering is accomplished by regulating the amplitude and phase of the signal and interference voltages U_{si} and U_{ii}, taken from the output of the individual array elements (or groups of elements); this is equivalent to multiplying these voltages by complex weighting coefficients w_i. In the simplest case—when receiving narrow-band signals (in those systems where the product of the bandwidth and the maximum difference in delays between the various antenna points $\Delta\tau_d$ is significantly less than unity)—adaptive processing consists of a weighted summation of the combined signals $x_i = U_{si} + U_{ii}$ (Fig. 6.3), that is,

$$U_{\text{out}} = y = \Sigma\, w_i^{*T} \,(U_{si} + U_{ii}) \tag{6.8}$$

where the notation $*T$ indicates the complex conjugate. Equation (6.8) may be placed in the matrix form:

$$\mathbf{y} = \mathbf{W}^{*T}\mathbf{X} \tag{6.9}$$

where

$$\mathbf{W}^* = \begin{vmatrix} w_1 \\ w_2 \\ \cdots \\ w_n \end{vmatrix} = \text{column vector of weights}$$

and

$$\mathbf{X} = \begin{vmatrix} x_1 \\ x_2 \\ \cdots \\ x_n \end{vmatrix} = \text{column vector of input signals}$$

and * denotes the Hermitian conjugate.

With the condition that the number of receiver channels exceeds the number of sources, it is possible, with the correct selection of weighting

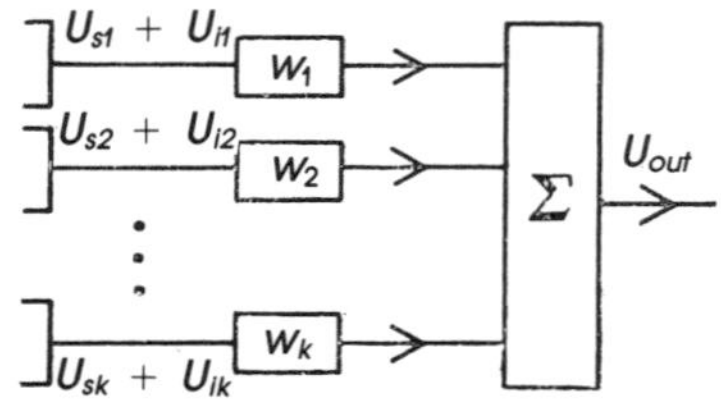

Fig. 6.3 Block diagram of adaptive signal processor.

coefficients, to achieve coherent summation of the useful signal arriving from some determined direction, and to cancel interference arriving from other directions. Thus, the main beam of the receiving antenna is steered to the direction from which the useful signal is arriving, and the pattern is nulled in the directions of strong interference.

The role of interelement summation is greater when the differences between the individual signal amplitudes U_{si} are smaller. If the signal is received essentially by one narrow-beam antenna, and the remaining elements are, in effect, weakly directed antennas for receiving interference, then interelement summation plays no significant role.

If the condition $B\Delta\tau_d \ll 1$ is not satisfied, then the amplitude-phase relations between the effective signal paths to the individual elements will be different for the different frequency components of the signal (and interference). To maintain effective weighted processing of all of the spectral components, tapped delay lines are included in the individual feeds. The voltages taken from these taps are weighted and summed.

The weighting coefficients, which optimize the system in accordance with some given criterion, are formed with the help of processors or analog devices with correlation feedback, using some particular algorithm [2, 22, 24, 25].

The most appropriate criteria, from the point of view of effectiveness and simplicity of construction, are the maximum signal-to-interference ratio criterion, and the criteria of obtaining the minimum mean square error and the minimum antenna system output noise power.

When using the first criterion, the optimum weighting vector is (see [24])

$$\mathbf{w}_{\text{opt}} = \tilde{k}\Phi^{-1}\mathbf{s} \tag{6.10}$$

where Φ^{-1} is the inverse of the correlation matrix Φ, $\mathbf{s}$ is the vector of the useful signal voltages, and $\tilde{k}$ is a nonzero complex number.

The covariance matrix is not known beforehand, and changes during operation. Therefore, with adaptive operation, the changing interference covariance matrix is estimated continuously or at discrete times.

To use the least mean square error (LMSE) criterion [16], we must minimize the quantity:

$$E[\epsilon^2] = E[(\mathbf{y} - \mathbf{r})^2] \tag{6.11}$$

that is, the mean value of the square of the deviation of the output voltage **y** from the reference signal **r**, which is generated within the antenna system, and equal to the expected output signal. Adaptive regulation of the weighting coefficients to meet the LMSE criterion, or the minimum output power criterion, is realized using the gradient method and the principle of correlation feedback. Algorithms based on the minimum output power criterion do not require the generation of a reference signal, which significantly simplifies their realization [22]. It should be noted that the weights derived from the LMSE criterion are extremely close to the optimum weights given by (6.10).

There are various issues particular to the application of adaptive antenna arrays in high-frequency systems, mainly associated with the structure of the interference fields. In the HF band, the structure of the field has, as a rule, a fluctuating character, which is caused by such features of its propagation as its multiple beams and scattering at ionospheric irregularities. As a result, the amplitude-phase relations between the voltages and individual antenna elements do not remain constant, and fluctuations are supported (both rapid and slow).

In the HF band, the correlation time of the rapid fluctuations is usually anywhere from a fraction of a second to several seconds [20]. In these conditions, it is necessary that the time required to calculate the adaptive array weighting coefficients so as to cancel the interference, and then to set these weights, should be much smaller than 1 s [26, 32].

The rapid fluctuations in the interference field structure make it necessary to increase the demand for fast-acting adaptive processing in the HF band. These requirements are usually not too difficult to meet, however, especially since the narrow-band character of the signals and interference affecting HF band performance lends itself to digital processing.

Thus, adaptive spatial interference cancelation is fully realizable, using both analog and digital processing. In digital systems, the quantities discussed above (interference, signals, weights) are all discrete-time functions [26, 32]. We will examine the basic principles of adaptive interference cancelation with the example of adaptive antenna systems utilizing a "correlation loop," based on the method of steepest descents [2, 18, 22, 24, 25], and providing wide-band operation.

6.4 CORRELATION LOOP ADAPTIVE ANTENNA SYSTEMS

We will first examine the operating principles of adaptive antenna systems using correlation loops, in which the operations of combining the useful signals and canceling the interference are separate [10, 22, 25]. In these systems, an unsteered narrow-beam channel is created by the nonadaptive coherent summation of the useful signal components. In addition, in a k-channel adaptive antenna system, $k - 1$ broad-beam cancelation channels are formed, spanning the sidelobes of the main beam. The voltages from the cancelation channels, when added to the main channel (with the corresponding weights), effect automatic coherent cancelation of interference received in the sidelobes of the main antenna. This is equivalent to forming a pattern with nulls in the directions of interference sources. Such an antenna pattern is shown in Fig. 6.4.

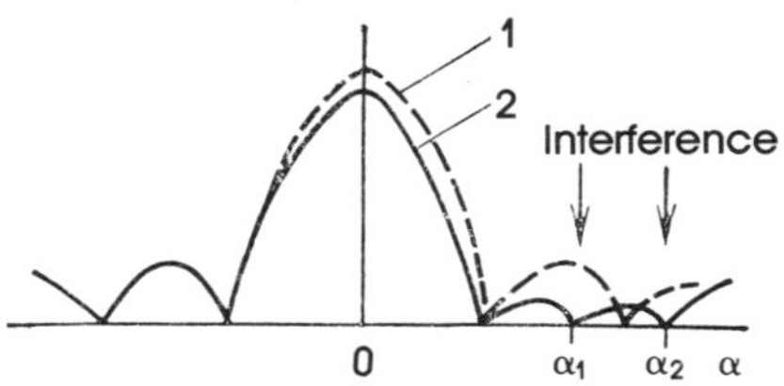

Fig. 6.4 Example of the form of an adaptive antenna pattern, without interference (curve 1), and with interference located at angles ϑ_1 and ϑ_2 (curve 2).

The voltages controlling the cancelation channel weights are formed with a correlation loop, from the output to the input of the processor. This is referred to as a correlation loop, because the control voltages are formed with the help of correlators, the inputs of which are fed with the output voltages and the cancelation voltages.

Thus, the examined systems are combinations of main and auxiliary antennas, and coherent autocancelation devices with feedback loops (correlation autocancelers) [10, 22].

We will now examine the operating principles of such adaptive cancelers and the potential capabilities of the coherent cancelation of correlated interference which underlies them. Let a simple single-channel canceler perform summation of the main and cancelation inputs, U_{i0} and U_{ik}, respectively (Fig. 6.5). The cancelation voltages are weighted with a balanced amplifier, whose gain is regulated at the weighting value w, which for a single-channel device is a scalar quantity. As was shown, the weights are generated so as to minimize the output noise power.

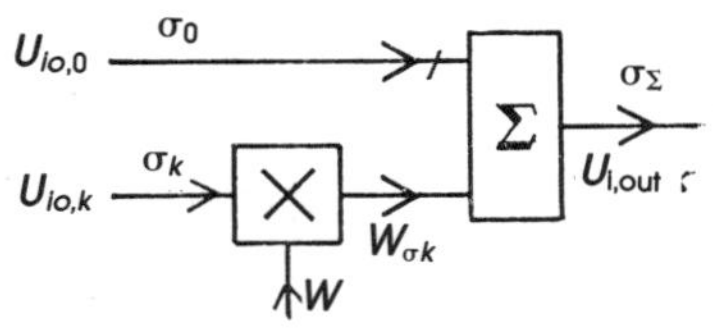

Fig. 6.5 Block diagram of simplified cancelation device.

Assuming that the interference is stationary, we obtain for the variance of the output noise voltage [22]:

$$\sigma_\Sigma^2 = \sigma_0^2 + w^2\sigma_c^2 + 2\rho_i w\sigma_c\sigma_0 \tag{6.12}$$

where ρ_i is the correlation coefficient of the instantaneous voltages U_{i0} and U_{ic} in the main and cancelation channels $(-1 \leqslant \rho_i \leqslant 1)$. We will vary the weight w and determine the value w_{opt} which minimizes σ_Σ^2. For this condition:

$$\left.\frac{\partial\sigma_\Sigma^2}{\partial w}\right|_{w=w_{\text{opt}}} = 2w_{\text{opt}}\sigma_c^2 + 2\rho_i\sigma_0\sigma_c = 0$$

whence,

$$w_{\text{opt}} = -\rho_i\sigma_0/\sigma_c \tag{6.13}$$

With this value, the residual noise variance is

$$\sigma_{\Sigma\min}^2 = \sigma_0^2(1 - \rho_i^2) \tag{6.14}$$

The effectiveness of interference cancelation may be estimated with the ratio:

$$k_{\text{opt}} = \sigma_0^2/\sigma_{\Sigma\min}^2 = 1/(1 - \rho_i^2) \tag{6.15}$$

which may be called the interference suppression coefficient. Thus, the primary condition for effective interference cancelation is a high level of correlation between the interference voltages (the value ρ_i^2 should be close to unity). The value ρ_i, however, depends on a number of factors, and mainly on the value of the phase shift ϕ between U_{i0} and U_{ic}. In actual conditions, the angle ϕ may assume very different values, inasmuch as it depends on the difference in the paths by which the interference reaches the main and cancelation antennas. This path difference depends on such random factors as the spatial location of the interference source with respect to the antennas.

We will now consider the possibility of designing devices to perform automatic cancelation, independent of the amplitude-phase relations between the voltages at the inputs. One of the simplest of such devices is the single-channel quadrature canceler (see Fig. 6.6) [10]. In this system the cancelation channel consists of two quadrature subchannels. In each subchannel there is a regulated amplifier (RA), and a correlator (combination of a multiplier and integrator), the output voltage of which regulates the amplifier gain. The two channels are placed in quadrature with a 90°-phase shifter.

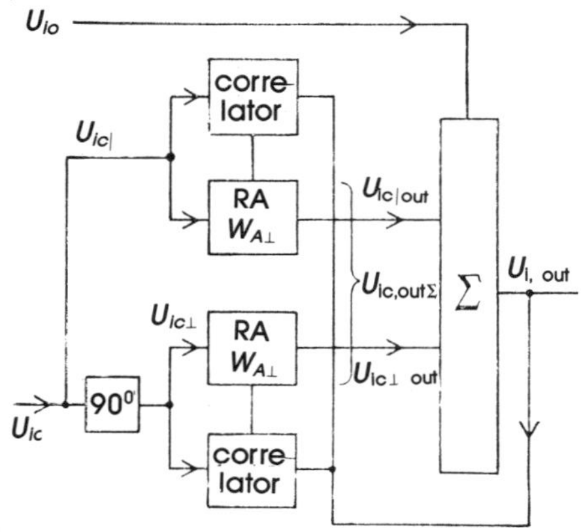

Fig. 6.6 Block diagram of a single-channel quadrature canceler (RA-regulated amplifier).

For clarity, we will begin examining the operating principles of this canceler with the simple case when $U_{i0}(t)$ and $U_{ic}(t)$ are periodic voltages of a single frequency, differing from one another in amplitude and phase. A vector diagram of the voltages in the canceler for this case is shown in Fig. 6.7. The vector $\mathbf{U}_{i0}$ may be placed in the form of the sum of two components: $U_{i0\|}$, which is in phase with U_{ic} (or 180° out of phase), and the corresponding quadrature component $U_{i0\perp}$. In order to cancel the interference, it is sufficient to cancel each of these components. This is accomplished with the help of the corresponding quadrature channels, at the outputs of which, in steady state and with full correlation, are formed the cancellation voltages:

$$U_{ic|\text{out}} = W_{A|}U_{ic|}, \quad U_{ic\perp\text{out}} = W_{A\perp}U_{ic\perp} \tag{6.16}$$

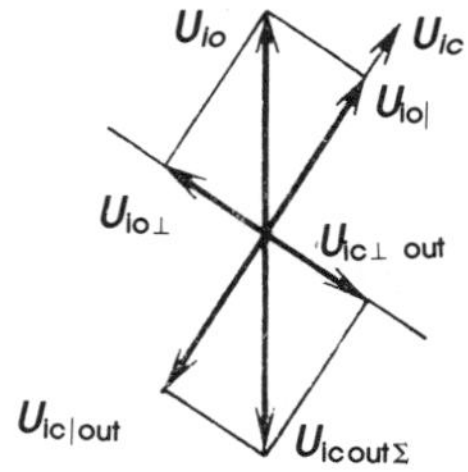

Fig. 6.7 Vector diagram of voltages in canceler.

Accordingly, the resulting vector is the sum $U_{ic.\,\mathrm{out}\,\Sigma}$ of the canceler subchannel voltages, and is equal to and in opposite phase with the voltage U_{i0}.

In accordance with [10], we will consider the operation of a single-channel canceler for the simple case of stationary narrow-band ($\Delta f_i \ll f_{op}$) interference, and show that its operation automatically sets the gains $W_{A|}$ and $W_{A\perp}$ in the regulated amplifier, thus canceling the interference. We will assume that the system is in steady state, and that the canceler is stable, which is determined by the parameters of its tuning and the parameters of the input signal [10]. We will also assume that the effects of the useful signal may be neglected. The interference voltage at the output of the adder is

$$U_\Sigma(t) = U_{ic}(t) + W_{A|}U_{ic|}(t) + W_{A\perp}U_{ic\perp}(t) \tag{6.17}$$

The gains in the regulated amplifiers are proportional to the voltages formed at the correlator outputs, i.e.:

$$W_{A|} = -\alpha_| E[U_{ic|}(t)\,U_\Sigma(t)] \tag{6.18}$$

$$W_{A\perp} = -\alpha_\perp E[U_{ic\perp}(t)\,U_\Sigma(t)] \tag{6.19}$$

where E is the time average, and $\alpha_|$ and $\alpha_\perp$ are proportionality constants. Using the value of $U_\Sigma(t)$ from (6.17) in (6.18), we obtain

$$\begin{aligned} W_{A|} = {} & -\alpha_| E[U_{ic|}(t)\,U_{i0}(t)] - \alpha_| W_{A|} E[U_{ic|}^2(t)] \\ & - \alpha_| W_{A\perp} E[U_{ic|}(t)\,U_{ic\perp}(t)] \end{aligned} \tag{6.20}$$

Inasmuch as $U_{ic|}(t)$ and $U_{ic\perp}(t)$ are uncorrelated at any given instant, the last term of (6.20) is equal to zero. Then, considering that

$$E[U_{ic|}(t)\,U_{i0}(t)] = \rho_i\sigma_0\sigma_c \quad \text{and} \quad E[U_{ic}^2(t)] = \sigma_c^2$$

we obtain

$$W_{A|} = -\alpha_|\,\rho_{i|}\sigma_0\sigma_c/(1 + \alpha_|\sigma_c^2) \tag{6.21}$$

where $\rho_{i|}$ is the correlation coefficient between the instantaneous voltages in the main channel and the first (in-phase) cancelation subchannel.

With strong feedback, when it is possible to assume that $\alpha_|\sigma_c^2 \gg 1$, (6.21) takes the form:

$$W_{A|} \approx -\rho_{i|}\sigma_0/\sigma_c \tag{6.22}$$

Substituting the expression for $U_\Sigma(t)$ in (6.19), and performing the analogous operations for the second subchannel, we obtain for the gain of this second cancellation channel:

$$W_{A\perp} \approx -\rho_{i\perp}\sigma_0/\sigma_c \tag{6.23}$$

where $\rho_{i\perp}$ is the correlation coefficient of the instantaneous interference voltages in the main channel and quadrature cancelation channel.

In accordance with (6.13), the values obtained for the cancelation channel gains, (6.22) and (6.23), maintain the optimal cancelation of each of the interference components. The variance of the interference voltage at the output of the canceler will then be

$$\sigma_{\Sigma\min}^2 = \sigma_0^2(1 - |\tilde{\rho}_i|^2) \tag{6.24}$$

where $\tilde{\rho}_i$ is the correlation coefficient of the complex interference amplitudes in the main and cancelation inputs. The square of its modulus $|\tilde{\rho}_i|^2 = \rho_{i|}^2 + \rho_{i\perp}^2$. In accordance with (6.24), the interference suppression coefficient for the quadrature canceler will be $k_{is} = 1/(1 - |\tilde{\rho}_i|^2)$. To obtain the values of $|\tilde{\rho}_i|$ close to unity, necessary for effective cancelation, the amplitude-phase responses of the receiver channels must be extremely well matched. The use of digital filters makes this much easier to accomplish.

In typical HF interference conditions, when the interference fields are fluctuating, the value of ρ_i also depends on the directive properties and spatial location of the individual antennas [4]. This is taken into consideration in foreign adaptive antenna systems, especially in the selection of the patterns of the cancelation antennas and their location relative to the main antenna.

We will now consider the case when in addition to interference, both the main and auxiliary antennas receive useful signals $U_{s0}(t)$ and $U_{sc}(t)$ (see [10]). In these conditions, the voltage at the output of the canceler may, in place of (6.17), be placed in the form:

$$\begin{aligned} U_{\Sigma}(t) &= U_{s0}(t) + U_{i0}(t) + W_{A|}U_{sc}(t) + W_{A|}U_{sc|}(t) \\ &\quad + W_{A\perp}U_{sc\perp}(t) + W_{A\perp}U_{ic\perp}(t) \end{aligned}$$

In place of (6.18) and (6.19), we obtain for the regulated amplifier gains:

$$\begin{aligned} W_{A|} &= -\alpha_{|}E[(U_{sc}(t) + U_{ic|}(t))U_{\Sigma}(t)] \\ W_{A\perp} &= -\alpha_{\perp}E[(U_{sc\perp}(t) + U_{ic\perp}(t))U_{\Sigma}(t)] \end{aligned}$$

For simplicity and clarity, we assume that $U_{s0}(t)$ and $U_{sc}(t)$ are harmonic oscillations, with phase difference ϕ and amplitudes U_{ms0} and U_{msc}; we also assume that $U_{msc}/U_{ms0} = b < 1$, and $|\rho_i| = 1$.

With these conditions and strong correlation feedback, expressions have been worked out in [10], allowing the canceler output signal-to-interference ratio γ_c to be determined, along with the gain η in the signal-to-interference ratio due to the use of automatic cancelations:

$$\gamma_c = (P_s/P_i)_{\text{out}} = 1:(P_s/P_i)_c \quad \eta = \gamma_c:(P_s/P_i)_{\text{in}}$$

Here $(P_s/P_i)_{\text{out}}$ is the ratio of the powers of the useful signal and the interference at the canceler output; $(P_s/P_i)_c$ is the ratio of the signal and interference powers at the canceler input; and $(P_s/P_i)_{\text{in}}$ is the ratio of the signal and interference powers at the main input.

Thus, with a useful signal at both inputs of the adaptive canceler, and with $|\tilde{\rho}_i|$, there is a simple inverse proportional relationship between the signal-to-interference ratio at the output and input of the canceler. Evidently, it is necessary to maintain a minimum signal level and as high an interference level as possible in the cancelation channel. The reduction in η and γ_c which occurs when the useful signal is present in both the main and cancelation channels has two causes. First there is a reduction (due to partial self-cancelation) of the output signal level coming from the main beam of the main antenna; in other words, not only is interference canceled in the sidelobes, but some of the signal itself is canceled in the main beam. Secondly, the interference is incompletely canceled in the sidelobes. This last effect is explained by the fact that the useful signal causes the weights $W_{A|}$ and $W_{A\perp}$ which are formed in the adaptive network to differ significantly from their optimum values, as determined by (6.22) and (6.23).

Consequently, there arises the problem of protecting the main beam of the main antenna pattern from cancelation in an adaptive array.

In many cases, in standard ground radar stations, for example, high duty factors are used and the targets are of small extent. The target pulses are, as a rule, of a comparatively low level, and are short in comparison with the adjustment time of the adaptive antenna system. In these conditions, the adaptive system barely reacts to the target signal. It is a different situation with some foreign over-the-horizon radars, where long pulses are used, and the adaptation time must be kept relatively short due to rapidly changing external conditions [26, 32] [see Arkad'ev, I.D. *Vestnik protivovozdushnoy oborony (Over-the-horizon radar)* 1980, no. 9]. In this case it is necessary to take special measures to ensure that the signal is not canceled within the main beam. In the opinion of foreign specialists, this may be accomplished using time, frequency and spatial differences between the signal and interference [25], and also various algorithms [12, 18].

When using timing considerations, the weights (for example, $W_{A|}$ and $W_{A\perp}$) may be formed by gating the interference when the reflected target pulses are absent. The resulting weights are "frozen," that is, held steady for a certain length of time (the repetition interval, for example).

Frequency differences may be used in those cases when the interference has spectral components which do not overlap the signal spectrum. These components may be extracted with the help of corresponding frequency filters, and used to control the adaptive array. In this case, as when using timing differences, the weights which are generated are almost optimal, and the problem of incomplete interference cancelation is generally eliminated. This does not, however, remove the difficulty of signal components penetrating the cancelation channel and causing signal amplitude variations (reduction or increase, depending on the phase relation between U_{s0} and U_{sc}).

As was shown in Applebaum and Chapman [25], the applicability of timing and frequency methods to elimination of the signal from the adaptive network depends on the type of radar and the operating conditions. We will, therefore, devote the greatest attention to the problem of protecting the main beam from signal cancelation, utilizing the spatial differences between the signal and interference. In adaptive antenna systems using correlation loops with separate primary and cancelation antennas, this problem may be solved by creating notches in the cancelation antenna patterns in the directions of useful signal sources. We will consider as examples two variations of this solution for adaptive arrays, the elements of which are radiators in a linear uniformly spaced array [25].

These methods include: preliminary spatial filtering, preventing reception of the signal by the cancelation antennas; use of a pilot signal; and limiting the weighting coefficients so as to avoid distorting the main beam of the primary antenna. This last method is examined in Sec. 6.5.

We will consider briefly various devices which effect the first two methods. First, we will examine a system which uses comparatively weakly directional cancelation antennas, the patterns of which are notched in the direction of the primary antenna's main beam. In the system whose structure is illustrated in Fig. 6.8 (see [25]), the voltages forming the primary and cancelation antenna patterns are taken from the outputs of the variable phase shifters which perform the scanning. The first pattern is obtained by taking a weighted sum of neighboring element voltages, and the other patterns are obtained by element-to-element subtraction of the voltages. The cancellation patterns have nulls in the direction of the main beam of the primary antenna. Various combinations of the individual difference patterns are available, providing several patterns (from 1 to $n - 1$), with the necessary nulls (in both direction and shape).

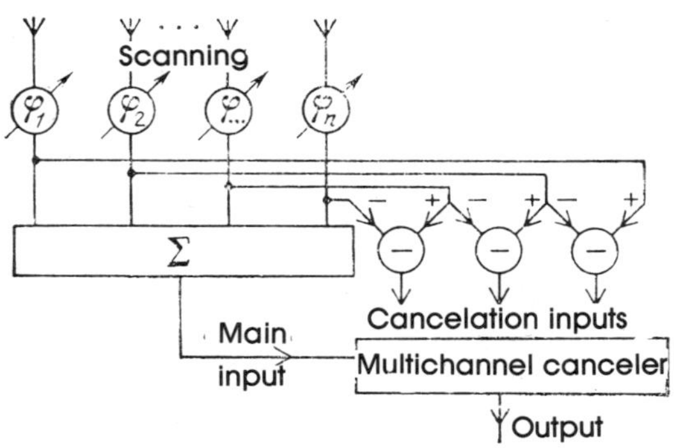

Fig. 6.8 Block diagram of an adaptive antenna system using auxiliary (cancelation) antennas.

Figure 6.9 illustrates an antenna system in which a beam-forming matrix is used with an n-element antenna to form n orthogonal beams [25]. Depending on the angle of arrival of the desired signal, any one of these beams may be the primary beam, and the others will be the auxiliary beams. Due to the orthogonality of the beams, in any direction where one beam has its maximum, the other beams will have nulls. This prevents the desired signal from being received by the auxiliary antennas.

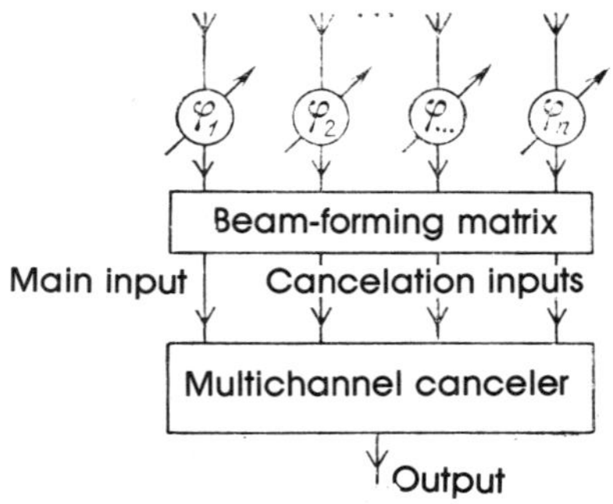

Fig. 6.9 Block diagram of an adpative antenna system in which *n* orthogonal beams are formed.

When considering various types of auxiliary antennas for adaptive cancelation, it should be remembered that besides effective spatial filtration of the desired signal, they should also provide a much higher interference level than the main antenna in the sector being canceled [10].

Pilot signals are also used in foreign systems, in which the desired signal level is maximized with in-phase summation of all the individual element signals, and simultaneous adaptive formation of nulls in the directions of interference (Fig. 6.10) [16, 25]. Here the antennas connected to the inputs of a multichannel canceler possess identical characteristics, and there is no need for their patterns to exhibit low response in the direction of the main beam. The pilot signal is a continuous (usually sinusoidal) oscillation, which lies in the receiver band, but which may be filtered out later. When there are phase shifters used for scanning, the pilot signals U_{p1}, U_{p2}, . . . , U_{pn} are fed to all receiver channels of the multichannel canceler, in phase. If there are no phase shifters, the pilot signals are injected with the phase shifts necessary to scan the beam. The pilot signal U_{p0} is fed to the "reference" input, with a level chosen to exceed the interference level significantly. The adder output signal is subtracted from the reference signal. In the process of canceling the voltages of the pilot signal U_{p0} and interference, complex weights are formed adaptively, providing in-phase addition of the signals arriving from the direction of the main beam, and forming notches in the pattern in directions of spatially concentrated interference. The advantage of this approach is the highly effective compensation of amplitude and phase errors with the injection of the reference signals in direct proximity to the receiving element aperture. Among its drawbacks are a reduction in the dynamic range and the need to filter out the reference signal in the output voltage.

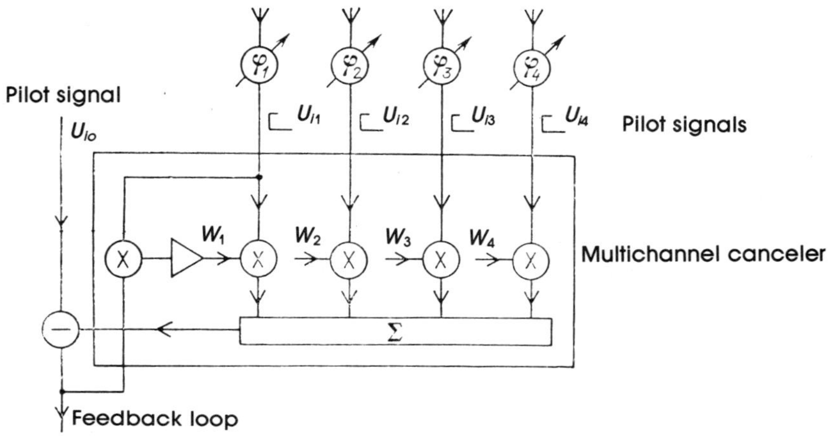

Fig. 6.10 Block diagram of an adaptive antenna system using a pilot signal [25].

These adaptive antenna systems may be implemented with both analog and digital processing. Digital processors are used most widely in foreign systems, however, due to their advantages—highly balanced receiver channels, wide dynamic range and elimination of crosstalk [7, 16, 18].

6.5 SOME ALGORITHMS FOR ADAPTIVE SPATIAL FILTERING (FROM FOREIGN LITERATURE)

We will begin our survey of spatial filtering algorithms with LMSE algorithms. These algorithms require no *a priori* information about the structure of the interference or direction of its arrival.

The adaptive signal processor generates the complex weights and performs the summation (Fig. 6.11(a)) [2]. The signal x_i from each element passes through an amplifier whose gain and phase are regulated. Then all of the signals are added, forming the output array signal $\mathbf{y}(j)$, which is compared with the reference signal $\mathbf{r}(j)$ (the given adaptive system response). The difference between the reference signal and the actual output signal gives an error signal $\epsilon(j) = \mathbf{r}(j) - \mathbf{y}(j) = \mathbf{r}(j) - \mathbf{W}^T\mathbf{X}(j)$, where j is the time of the jth sample, and $\mathbf{W}^T$ is the transposed weighting vector.

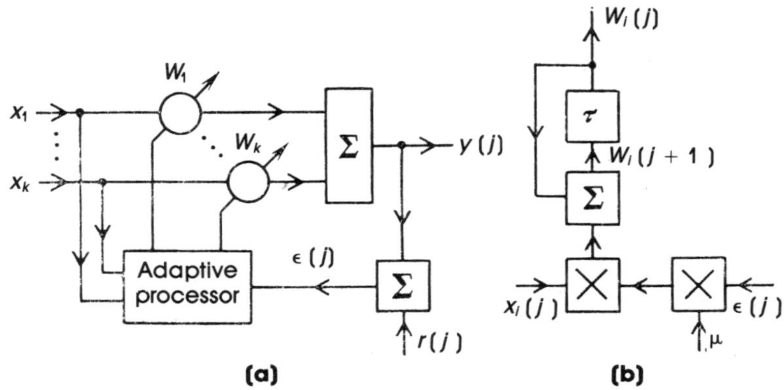

Fig. 6.11 Implementation of LMSE algorithm.
(a) block diagram of adaptive antenna system
(b) block diagram showing generation of weights

The error signal acts to control the complex weights through the feedback loop. The expected value of the square error, will be (considering that $\mathbf{W}^T\mathbf{X}(j) = \mathbf{W}\mathbf{X}^T(j)$)

$$E[\epsilon^2(j)] = E([\mathbf{r}(j) - \mathbf{y}(j)]^2)$$
$$= E[\mathbf{r}^2(j) + \mathbf{W}^T\mathbf{X}(j)\mathbf{X}^T(j)\mathbf{W} - 2\mathbf{W}\mathbf{X}^T(j)\mathbf{r}(j)]$$

or

$$E[\epsilon^2(j)] = E[\mathbf{r}^2(j)] + \mathbf{W}^T\mathbf{R}(x, x)\mathbf{W} - 2\mathbf{W}^T\mathbf{R}(x, r) \tag{6.25}$$

where

$$\mathbf{R}(x, x) = E[\mathbf{X}(j)\ \mathbf{X}^T(j)]$$
$$= E\begin{bmatrix} x_{0j}x_{0j} & x_{0j}x_{1j} & x_{0j}x_{2j} & \cdots & x_{0j}x_{nj} \\ x_{1j}x_{0j} & x_{1j}x_{1j} & x_{1j}x_{2j} & \cdots & x_{1j}x_{nj} \\ \cdots & \cdots & \cdots & \cdots & \cdots \\ \cdots & \cdots & \cdots & \cdots & \cdots \\ x_{nj}x_{0j} & \cdots & \cdots & \cdots & x_{nj}x_{nj} \end{bmatrix}$$

and

$$\mathbf{R}(x, r) = E[\mathbf{X}(j)r(j)] = E\begin{bmatrix} x_{0j}r_j \\ x_{1j}r_j \\ \cdots \\ x_{nj}r_j \end{bmatrix}$$

$\mathbf{R}(x, x)$ is the correlation matrix of the signals at the element inputs, and $\mathbf{R}(x, r)$ is the column vector of correlation coefficients between the input signals and the given response at the output of the adaptive antenna system.

Thus, the mean square error is a second-order function of the weights. The minimum error corresponds to the point where the gradient of the mean square error function is zero. Differentiating (6.25) by W, we obtain the gradient of the error function:

$$\nabla E[\epsilon^2(j)] = 2[\mathbf{R}(x, x)\mathbf{W} - \mathbf{R}(x, r)] \tag{6.26}$$

Setting this to zero, we obtain the value of the optimum weighting vector, $\mathbf{W}_{\text{opt}} = \mathbf{R}^{-1}(x, x)\mathbf{R}(x, r)$.

Algorithms for adaptive filtering usually make use of recursive methods. This includes, in particular, the method of steepest descents, which is one of the gradient descent methods. Using the method of steepest descents, the weighting vector is changed in the direction of the estimate of the negative gradient $\hat{\nabla}(j)$ of the function $E[\epsilon^2(j)]$, relative to the weighting vector $\mathbf{W}$. In light of (6.26), we have

$$\begin{aligned} \mathbf{W}(j + 1) &= \mathbf{W}(j) - \mu\hat{\nabla}(j) \\ &= \mathbf{W}(j) - 2\mu[\mathbf{R}(x, x)\mathbf{W}(j) - \mathbf{R}(x, r)] \end{aligned} \tag{6.27}$$

where j and $j + 1$ are discrete times for two successive iterations, and μ is a scalar constant determining the rate of convergence and stability of the estimate. The algorithm is not usable in this form, because the matrices $\mathbf{R}(x, x)$ and $\mathbf{R}(x, r)$ are unknown.

Another form of the gradient descent method, not requiring knowledge of the matrices $\mathbf{R}(x, x)$ and $\mathbf{R}(x, r)$, was proposed by Widrow and Hoff [14, 16]. It is different in that in place of the average values of the matrices $\mathbf{R}(x, x)$ and $\mathbf{R}(x, r)$, their sample estimates are used:

$$\mathbf{R}(x, x) = E[\mathbf{X}(j)\mathbf{X}^T(j)], \quad \mathbf{R}(x, r) = E[\mathbf{r}(j)\mathbf{X}(j)]$$

The estimate of the gradient is performed on the jth sample $\epsilon^2(j)$:

$$\hat{\nabla}(j) = \nabla[\epsilon^2(j)] = 2\epsilon(j)\,\nabla\,[\epsilon(j)] = 2\epsilon(j)\,\nabla\,[\mathbf{r}(j) - \mathbf{y}(j)]$$
$$= 2\epsilon(j)\,\nabla\,[\mathbf{r}(j) - \mathbf{W}^T(j)\mathbf{X}(j)] = -2\epsilon(j)\mathbf{X}(j)$$

Then, using (6.27), we obtain the following recursive rule for determining the weighting coefficients for the Widrow-Hoff algorithm:

$$\mathbf{W}(j + 1) = \mathbf{W}(j) + 2\mu\epsilon(j)\mathbf{X}(j) \tag{6.28}$$

i.e., the increment in the weighting vector over the iteration time is proportional to the product of the input signal vector and the error vector.

The scalar constant μ is chosen so as to ensure the convergence of the algorithm. In practice, μ is chosen from the condition:

$$0 < \mu \ll 1/P_{\text{in}} \tag{6.29}$$

where P is the total input signal power.

The process of establishing the weights is described by the sum of exponentials with time constants $\tau_i = 1/4\mu\lambda_i$, $i = 1, 2, \ldots, k$, where λ_i is the ith eigenvalue of the matrix $\mathbf{R}(x, x)$, and k is the number of adaptive antenna elements.

Inasmuch as the eigenvalues λ_i are proportional to the input signal power, the process of adaptive weighting is faster with greater signal power and larger values of μ. If the eingenvalues of the matrix $\mathbf{R}(x, x)$ differ widely, however, the rate of convergence will be low. This situation is observed when strong and weak interference sources are acting simultaneously, or when several sources with nearly equal intensities, and lying in a narrow angular sector, are acting at the same time. The rate of convergence will then be limited by the lowest interference power level (i.e., by the lowest value λ_i). At the same time, strong interference power will not allow the convergence to be accelerated by increasing the parameter μ, because this will violate the convergence condition (6.29). These considerations will generally be valid for all algorithms using a gradient method to generate the weighting coefficients.

The design of a device to generate the weights automatically, using the algorithm described by (6.28), is shown in Fig. 6.11(b). The realization of this algorithm is hampered by the requirement for knowledge of the target reference signal at each iteration. A pilot signal may be used to overcome this difficulty. This scheme, however, entails filtering out the pilot signal at the output of the antenna system, and also reduces the dynamic range. Furthermore, the existence of the reference signal causes the weighting vector to deviate from its optimum value.

In a number of radar systems, the angle of arrival and spectral characteristics of the useful signal may be estimated beforehand or calculated with sufficient accuracy, so that the cross-correlation vector between **X** and **r** (the **P**-vector) may be assumed to be known. In this case, the Griffiths modified LMSE, or **P**-vector, algorithm may be used [7]. It is a combination of the algorithms described by (6.27) and (6.28), and is written in the following form:

$$\mathbf{W}(j + 1) = \mathbf{W}(j) + \mu[\mathbf{R}(x, r) - \mathbf{y}(j)\mathbf{X}(j)]$$

This modified algorithm differs from the algorithm (6.28) in that the sample estimate of the column vector $\mathbf{R}(x, r)$ is replaced with its average value, which, as was indicated above, is known *a priori.* The other elements, $\mathbf{y}(j)$ and $\mathbf{X}(j)$, may be obtained through measurement. Another of its important advantages is the low number of arithmetic operations and low amount of memory required for each recursion; they are proportional to the number of filter elements. The number of iterations required is strongly dependent on the interference conditions. The results of a simulation of the operation of such a system are presented as an example in [7], for a four-element antenna system, with four delay lines in each element, using a signal and two spatially concentrated interference sources. Values of the mean square error were found to be close to optimum only for an extremely large number of iterations, nearly 20,000. At the same time, as was also shown, satisfactory convergence may be obtained with a much smaller number of iterations in some conditions.

An important problem is the protection of the main beam, that is, the maintenance of high directivity in the direction from which the desired signal is arriving, while suppressing interference in the sidelobes by the required amount. Some solutions to this problem were considered in Sec. 6.4. One solution is the use of a limited LMSE algorithm, proposed by Frost [18], which is an improved LMSE algorithm. The criterion in this solution is that the total antenna output power be a minimum, with the condition that the desired amplitude-frequency response be maintained in the expected direction of arrival of the desired signal.

We will consider this algorithm briefly in accordance with Frost [18], as applied to a wide–band signal processor with real-valued weights. The corresponding block diagram is shown in Fig. 6.12. Without loss of generality, we will assume that the desired signal arrives from the direction normal to the array aperture, which consists of K rows and L columns of elements. In this case, the voltages at the taps of each vertical column of antenna elements, which have identical signal delays, are in phase, and

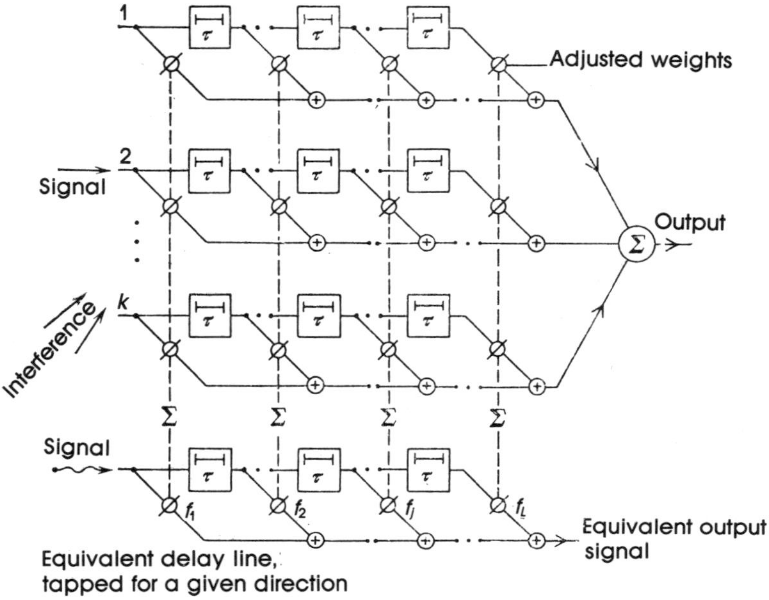

Fig. 6.12 Block diagram of implementation of limited LMSE algorithm.

the array processor is equivalent to a single tapped delay line (DL), in which each weight f_i is equal to the sum of the weights in the corresponding column. The necessary frequency response of the antenna system is obtained in the direction of arrival of the signal through selection of L of the summation coefficients. The remaining KL-L degrees of freedom in selecting the weights may be used to minimize the output power from interference sources which lie in directions other than that from which the signal is arriving.

The estimate of the adaptive antenna system output at the jth sample is

$$\mathbf{Y}(j) = \mathbf{W}^T\mathbf{X}(j) = \mathbf{X}^T(j)\mathbf{W}$$

The corresponding average output power is

$$E[Y^2(j)] = E[\mathbf{W}^T\mathbf{X}(j)\mathbf{X}^T(j)\mathbf{W}] = \mathbf{W}^T\mathbf{R}(x, x)\mathbf{W} \tag{6.30}$$

The following limitation is placed on the sum of the weights in the ith column:

$$\mathbf{C}_i^T\mathbf{W} = f(i), \quad i = 1, 2, \ldots, L, \tag{6.31}$$

where $\mathbf{C}_i$ is the KL-dimensional column vector, in which the elements corresponding to a column of the antenna array are equal to one, and the others are zero, i.e.:

$$\mathbf{C}_i = \left|\begin{matrix} 0 \\ \cdots \\ 0 \\ \cdots \\ 1 \\ \cdots \\ 1 \\ \cdots \\ 0 \\ \cdots \\ 0 \end{matrix}\right| \begin{matrix} \left.\vphantom{\begin{matrix}0\\ \cdots\\ 0\end{matrix}}\right\} K \\ \\ \left.\vphantom{\begin{matrix}1\\ \cdots\\ 1\end{matrix}}\right\} i\text{th group of } K \text{ elements} \\ \\ \left.\vphantom{\begin{matrix}0\\ \cdots\\ 0\end{matrix}}\right\} K \end{matrix}$$

Introducing the limitation matrix formed from the vectors $\mathbf{C}_i$:

$$\mathbf{C} = \left|\begin{matrix} c_{11} & c_{12} & \cdots & c_{1L} \\ c_{21} & c_{22} & \cdots & c_{2L} \\ \cdots & \cdots & \cdots & \cdots \\ c_{(K-1)1} & c_{(K-1)2} & \cdots & c_{(K-1)L} \\ c_{K1} & c_{K2} & \cdots & c_{KL} \end{matrix}\right|$$

and the L-dimensional vector $\mathbf{F}$ of the weights in the equivalent tapped delay line, where

$$\mathbf{F} = \left|\begin{matrix} f_1 \\ f_2 \\ \cdots \\ f_L \end{matrix}\right|$$

and the condition (6.31) may be placed in the form:

$$\mathbf{C}^T\mathbf{W} = \mathbf{F} \tag{6.32}$$

Inasmuch as the receiving properties of the adaptive antenna system in a given direction are fixed by L constraints, minimization of the interference power from other directions is accomplished by minimizing (6.30) while fulfilling the condition of (6.31). The optimum weighting vector may be found using the method of Lagrange multipliers (see [18]):

$$\mathbf{W}_{\text{opt}} = \mathbf{R}^{-1}(x, x)\mathbf{C}[\mathbf{C}^T\mathbf{R}^{-1}(x, x)\mathbf{C}]^{-1}\mathbf{F} \tag{6.33}$$

Direct calculation and inversion of the matrix $\mathbf{R}(x, x)$ requires a large amount of memory and many calculations. As was shown in Frost [18], the optimization is significantly simplified using the method of steepest descents, and approximating the matrix $\mathbf{R}(x, x)$ as the scalar product of the vector $\mathbf{X}(j)$ with itself. The weight vector is initially determined so as to satisfy (6.33); with each recursion, it moves in the direction opposite to that of the gradient. Choosing the Lagrange multiplier such that $\mathbf{W}(j + 1)$ satisfies the condition (6.33), and approximating $\mathbf{R}(x, x)$ after the jth iteration as the product $\mathbf{X}(j)\mathbf{X}^T(j)$, the following stochastic algorithm for determining the limited least mean square error is obtained in Frost [18]:

$$\mathbf{W}(0) = \mathbf{F}_1,$$
$$\mathbf{W}(j + 1) = \mathbf{P}[\mathbf{W}(j) - \mu\mathbf{Y}(j)\mathbf{X}(j)] + \mathbf{F}_1$$

where $\mathbf{F}_1 = \mathbf{C}(\mathbf{C}^T\mathbf{C})^{-1}\mathbf{F}$, $\mathbf{P} = \mathbf{I} - \mathbf{C}(\mathbf{C}^T\mathbf{C})^{-1}\mathbf{C}^T$, and $\mathbf{I}$ is the identity matrix.

As is mentioned in Frost [18], among the advantages of the algorithm is its self-correcting property, which allows it to be used for an unlimited duration, implemented with a computer, with no deviation from the applied limits resulting from accumulating errors. A.A. Pistol′kors has proposed a version of the limited LMSE algorithm which uses complex weights, and does not require a delay line [12].

The Griffiths-Frost algorithm was used in the adaptive antenna system of the U.S. WARF over-the-horizon radar facility. A uniformly spaced array was used, with 256 elements and a total aperture of 2.56 km [*Note*: A more detailed description of the radar is provided in Griffiths [26], in Washburn and Sweeney [32], and Ch. 9 of this book]. The array is divided into $K = 8$ subarrays, with 32 elements in each. Chebyshev weighting is applied to the signals from the subarrays, which are then passed to eight receiver channels, which perform amplification, conversion of the signals to video frequency, and adaptive processing of the array signals using tapped delay elements. In each of the eight channels there are $L - 1$ delay elements. A linear frequency-modulated (LFM) signal is used for transmission, the modulation period being $\frac{1}{60}$ s. With the help of a local (reference) LFM modulator and a mixer, the signal spectrum is converted to video frequency. Further analog-digital transformation is performed with frequency bins of 1920 Hz, which corresponds to 32 samples during the linear frequency modulation period.

Foreign specialists have studied modified and limited LMSE algorithms [26, 32]. When calculating $\mathbf{R}(x, r)$, the pilot signal is taken to be $r(j) = \cos \omega_0 j$. Inasmuch as $\mathbf{R}(x, r)$ is the time correlation vector of the quantities $r(j)$ and the components $x(j)$, which all vary harmonically, then the elements of the correlation vector are also harmonic oscillations. Using (6.29), the value of the parameter μ is taken to be $\mu = \beta/(L\Sigma_1^8 P_{xi})$, where P_{xi} is the average power at the output of the ith subarray. The quantity β, which maintains normal operation of the adaptive antenna system, lies in the range 0.02–0.5. As was shown in Griffiths [26], and Washburn and Sweeney [32], its value should not exceed two if the algorithm is to remain stable. In the study considered in Griffiths [26], β was set to 0.1.

In investigating the limited LMSE algorithm, an additional constraint was applied—that the frequency response should be uniform in the direction of the main beam. This resulted in the requirement that the sum of the weights should be zero, except for one (corresponding, for example, to $l = L/2$), for which it should be equal to one.

This condition is written as:

$$\sum_{i=1}^{8} W_{il} = 0, \quad l = 0, 1, \ldots, L, \quad l \neq L/2$$
$$\sum_{i=1}^{8} W_{i,L/2} = 1 \qquad (6.34)$$

In accordance with this limitation, the matrices $\mathbf{C}$ and $\mathbf{F}$ in (6.32) take the form:

$$\mathbf{C} = \overset{L}{\begin{bmatrix} 1 & 0 & \ldots & 0 \\ 1 & 0 & \ldots & 0 \\ \cdots & \cdots & \cdots & \cdots \\ 1 & 0 & \ldots & 0 \\ \cdots & \cdots & \cdots & \cdots \\ 0 & 1 & \ldots & 0 \\ 0 & 1 & \ldots & 0 \\ \cdots & \cdots & \cdots & \cdots \\ 0 & 0 & \ldots & 1 \\ 0 & 0 & \ldots & 1 \\ \cdots & \cdots & \cdots & \cdots \\ 0 & 0 & \ldots & 1 \end{bmatrix}} \begin{matrix} \left.\vphantom{\begin{matrix}1\\1\\1\\1\end{matrix}}\right\} 8 \\ \left.\vphantom{\begin{matrix}1\\1\end{matrix}}\right\} \\ \left.\vphantom{\begin{matrix}1\\1\\1\\1\end{matrix}}\right\} 8 \end{matrix} \; 8(L+1), \quad \mathbf{F} = \begin{bmatrix} 0 \\ \ldots \\ 0 \\ 1 \\ 0 \\ \ldots \\ 0 \end{bmatrix}$$

For these conditions, the components of the weight vector may be calculated as follows ($K = 8$):

$$w_1(j+1) = w_1(j) - \mu y(j)x_1(j) - \frac{1}{8}\sum_{i=1}^{8}[w_i(j) - \mu y(j)x_i(j)] + \frac{f_1}{8}$$

$$\cdot$$
$$\cdot$$
$$\cdot$$

$$w_8(j+1) = w_8(j) - \mu y(j)x_8(j) - \frac{1}{8}\sum_{i=1}^{8}[w_i(j) - \mu y(j)x_i(j) + \frac{f_1}{8}$$

$$\cdot$$
$$\cdot$$
$$\cdot$$

$$w_{2k}(j+1) = w_{16}(j+1) = w_{16}(j) - \mu y(j)x_{16}(j) - \frac{1}{8}\sum_{i=9}^{16}[w_i(j) - \mu y(j)x_i(j)] + \frac{f_2}{8}$$

$$\cdot$$
$$\cdot$$
$$\cdot$$

$$w_{Lk}(j+1) = w_{8L}(j+1) = w_{8L}(j) - \mu y(j)x_{8L}(j) - \frac{1}{8}\sum_{i=8(L-1)+1}^{8L}[w_l(j) - \mu y(j)x_i(j)] + \frac{f_L}{8}$$

Here, in light of (6.34), we have $f = 0$ for $l \neq L/2$, and $f = 1$ for $l = L/2$. The total number of computations for this algorithm for the 40 weights (eight subarrays with five taps in each), was 95 real multiplications and 240 real additions. For a modified LMSE algorithm, the number of multiplications was 81 and the number of additions was 120. In investigating the **P**-vector algorithm, a study was made of the transition processes when receiving a simulator signal in a background of interference. All of the weights were simultaneously set to zero, after which the process of establishing the output signal-to-noise ratio was observed. The adaptation time was 40 ms, which corresponds to 80 adaptation cycles. These results are quite different from those presented in Frost [18], which is apparently explained by the significant differences in the interference conditions.

The adaptation time constant was found through experiment to be about 0.1 s, which, in the opinion of Griffiths [26], was sufficient to separate the fluctuations taking place in the structure of the fields. In Griffiths [26], and Washburn and Sweeney [32], data obtained in detecting an actual target were processed using the **P**-vector and limited LMSE algorithms. With two-dimensional Fourier transforms of the weighting coefficients, families of curves were calculated for the frequency dependence of the pilot signal at the output of the adaptive antenna system for 16 azimuths (spanning 0.25°) in the band of frequencies 0–920 Hz (Figs. 6.13(a, b)). The expected signal corresponds to the azimuth with index eight and frequency 460 Hz. It is clear that for the limited LMSE algorithm (as opposed to the **P**-vector algorithm), the frequency response of the receiver in the direction of the expected signal is practically uniform. In conclusion, we will consider briefly the issue of increasing the rate of convergence of the processing algorithm, that is, decreasing the time necessary to establish the optimal weights [7, 13, 18, 27].

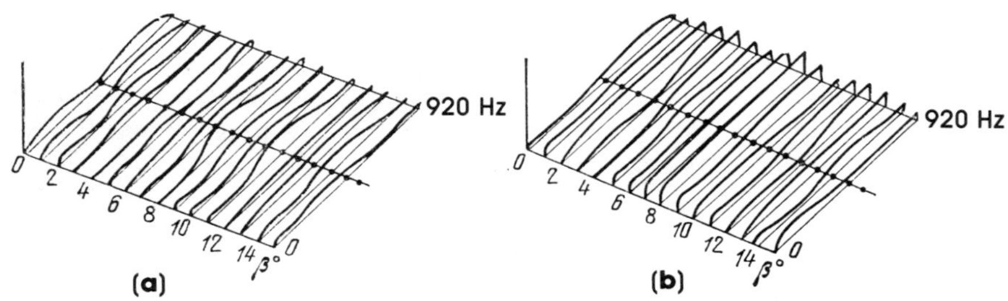

Fig. 6.13 Angle-frequency dependence of adaptive antenna system pilot signal, using (a) **P**-vector algorithm, and (b) limited LMSE algorithm.

The rate of convergence depends to a great extent on the nature of the interference, the number of channels and the algorithm used. The rate of convergence is greatest when the interference sources are in different directions, and create voltages which vary little at the input to the canceller. The rate of convergence is lowest when the interference sources are close together and have different strengths; this is especially true if there is a large number of receiver channels. The time required to set the weights is approximately proportional to their number. As foreign studies have shown, practically all known digital algorithms using both maximum signal-to-noise and least mean-square error criteria have relatively slow rates of convergence. Therefore, in typical high-frequency interference conditions, these algorithms are effective only for systems with comparatively small numbers of weighting coefficients (on the order of 10–100) [32]. As was shown in Reed, Mallet, and Brennan [27], significantly higher rates of convergence may be obtained with direct inversion of a sample (estimated) covariance matrix. This also assumes direct use of the algorithm for calculating the weights given in (6.10). The covariance matrix Φ is replaced by its estimate $\hat{\Phi}$ from N samples of the input signal (interference): $\hat{\Phi} = (1/N)\Sigma_1^N \mathbf{X}_1\mathbf{X}_i^*$, where * denotes complex conjugate.

If the acceptable loss in the signal-to-noise ratio due to the use of the estimated covariance matrix $\hat{\Phi}$ in place of the actual matrix is set at 3 dB, then the required number of samples N used to estimate the matrix should be at least twice the number of components in the input signal [27]. This is substantially smaller (by several orders of magnitude) than in other known adaptive algorithms. An important advantage of this method is the fact that the rate of convergence does not depend on the spatial distribution of the interference sources. Its shortcomings include the degradation in the estimate of the covariance matrix caused by the presence of the desired signal, and also the large number of calculations required. The algorithm is simplified somewhat if a recursive method is used to invert the covariance matrix [27]. In Ch. 8 we will consider the issue of optimizing space-time signal processing in a given finite time interval, against criteria expressed in terms of the detection characteristics in interference with unknown parameters.

6.6 PROTECTION AGAINST NARROW-BAND INTERFERENCE

As was shown in Ch. 5, the high-frequency band contains many radio transmissions which constitute narrow-band interference, characterized by its narrow spectral width. Therefore, one of the methods of improving the performance of HF systems is the use of devices designed to suppress narrow-band interference. As has been shown by their use in non-Soviet systems, however, the application of these devices in the HF band is attended by various problems. One of these is the fact that in the absence of *a priori* information about the spectral width of the narrow-band interference and about its center frequency, the device must be adaptive. In non-Soviet radars which use time selection (range gating, for instance [10, 19]), the receiver is switched on for certain intervals T_s. Continuous narrow-band interference is thus subject to modulation and becomes pulsed, which has a significant influence on the system's anti-interference performance.

In addition, as both foreign and domestic researchers have shown, in addition to narrow-band interference, a high-frequency receiver is also subject to fluctuating external active interference (atmospheric or cosmic), the intensity of which may not always be neglected (in comparison with the intensity of the narrow-band interference).

With these considerations in mind, we will assess the possibilities for suppressing narrow-band interference, and the design principles of anti-interference devices, for the case when the interference has a narrow spectrum, the width of which is significantly narrower than the width of the useful signal spectrum.

An analysis is presented in Goykhman and Tsybulin [6], treating the optimum methods for detecting wide-band discrete signals against a background of a combination of fluctuating ("quasiwhite") noise and narrow-band interference with known parameters. It is assumed that the spectrum of the interference is known, and that optimum signal processing is performed using, first, a "whitening" filter, which makes the spectrum of the noise-interference combination uniform, and then a filter matched to the signal as transformed by the whitening filter. The increase in signal-to-interference ratio, realized as a result of using *a priori* knowledge about the interference spectrum, is estimated with the help of the coefficient [6] $\Gamma = \gamma^2_{0\max}/\gamma^2_0$, where $\gamma^2_{0\max}$ is the signal-to-interference power ratio at the output of the optimum receiver, and γ^2_0 is the signal-to-interference ratio obtained by processing the "raw" signal and interference with a matched filter (that is, without a whitening filter).

The general expression for the coefficient Γ, with no gating in the receiver, takes the form [6]:

$$\Gamma = \frac{2(1 - \eta + \eta\delta)}{\delta B_s} \int_0^{B_s/2} \frac{df}{\eta + (1 - \eta)Y_i^2(f)}$$

where η is the ratio of the power spectral density of the noise to the maximum power spectral density of the narrow-band interference, δ is the ratio of the effective signal bandwidth B_n to the effective narrow-band interference bandwidth B_1, and $Y_1^2(f)$ is the normalized power spectrum of the interference at the output of the receiver.

The gating process has practically no influence on the structure of the useful wide-band signal and noise, inasmuch as their correlation times are much smaller than the length of the gate. The interference spectrum, on the other hand, undergoes substantial spreading as a result of the gating, causing a corresponding change in $Y_i(f)$, and, consequently, in Γ. It is easiest and most illustrative to consider the effects of gating on anti-interference performance for the case when the receiver input is presented with quasiwhite noise and either harmonic interference or narrow-band noise interference with a rectangular spectrum.

We will denote the gate length as T_s. Considering that for harmonic interference:

$$Y_{ih}^2(f) = \sin^2(\pi f T_s)/(\pi f T_s)^2 \quad \text{and} \quad B_{ih} = 1/T_s$$

and setting $B_s T_s = B_1$, we obtain

$$\Gamma_h = \frac{2(1 - \eta_h + \eta_h B_1)}{\pi B_1^2} \int_0^{\pi B_1/2} \frac{dx}{\eta_h + (1 - \eta_h)(\sin^2 x)/x^2}$$

where η_h is the value of η after gating the harmonic interference.

An expression for the coefficient Γ for the case when the interference is noise-like with rectangular spectrum, acting in combination with quasi-white noise, takes the form [5]:

$$\Gamma_i = \frac{1 - \eta_i(1 - \delta_1)}{\sigma_1^2} \int_0^{\delta_1} \frac{dx}{\eta_i + (1 - \eta_i)I(x)} \tag{6.35}$$

where

$$I(x) = \frac{\displaystyle\int_0^{\pi a} \frac{\sin x}{x} \cos bx \, dx - \frac{1}{\pi a} \int_0^{\pi a} \sin x \cos bx \, dx}{\displaystyle\int_0^{\pi a} \frac{\sin x}{x} \, dx - \frac{1}{\pi a} \int_0^{\pi a} \sin x \, dx}$$

where δ_1 and η_i are the values of δ and η after gating has altered the envelope of the interference spectrum, the coefficient $a = B_i T_s$ characterizes the amount of interference "accumulated" during the gating interval, and $b = \Delta f / B_i$ is the current normalized tuning deviation.

Figure 6.14 presents plots of B_i as a function of δ for various values of η_i and a. The figure shows a relatively weak dependence of Γ_i on δ, the ratio of the signal and interference bandwidths. This is explained by two effects associated with the increase in δ: on the one hand, the narrow-band interference becomes more intense, inasmuch as its spectrum becomes narrower (and, consequently, more distinct from the wideband signal), and on the other hand, as δ grows, the fraction of narrow-band interference in the interference-noise mixture decreases, so that suppressing the interference has a lesser effect. For the same reason, increases in Γ_i are small for very large values of η_i (the dashed lines in Fig. 6.14); it is important only to maintain a satisfactory level for the quantity a.

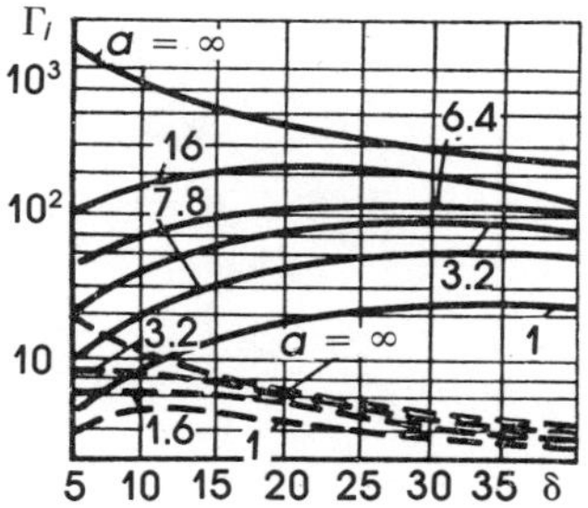

Fig. 6.14 Γ_i as a function of k for various values of η_i and a (solid lines correspond to $\eta = 10^{-4}$, and dashed lines to $\eta = 10^{-2}$).

To estimate the effectiveness of anti-interference methods based on the use of *a priori* knowledge of the interference spectrum in one or another actual interference situation, we will consider the dependence of Γ on the ratio Q of the input powers of the quasiwhite noise P_n and narrow-band interference P_i. The quantity Q has a single-valued relation to η through the relations $\eta_h = Q/B_1$ and $\eta = Q/\delta_1$.

Goykhman [5] presents equations which may be used to estimate the increase in anti-interference performance due to the use of *a priori* knowledge about the interference spectrum, as a function of the input interference-to-noise power ratio, including the effects of gating. As an example, in Fig. 6.15 are plotted data, calculated on a computer, showing the dependence of Γ_h and Γ_i on the value of $q = Q^{-1}$, for various values of B_1 (for Γ_h) and a (for Γ_i). The curves for $\Gamma_i = \phi(q)$ were calculated for the case of narrow-band interference whose spectral width is 5% of the spectral width of the useful signal ($\delta = 20$), and for various values of a. Calculations using the equations developed in this section, and examination of curves in Goykhman [5] and Goykhman and Tsybulin [6], indicate that in these conditions, with gating, it is expedient to use the whitening filter method of suppressing narrow-band interference only when the interference levels exceed the quasiwhite noise levels by a rather large amount (15–30 dB). When $a \geqslant 10$, it is possible to suppress narrow-band interference by 15–25 dB. In addition, as non-Soviet specialists have noted [31], rejection of the portion of the signal spectrum occupied by the narrow-band interference leads to a degradation in signal reception and a corresponding reduction in signal-to-noise ratio. Thus, Fig. 6.16 shows the results obtained in Sussman and Ferrari [31] for the loss in output signal amplitude due to interference suppression, for a phase shift keying (PSK) signal, as a function of $g = B_{\text{rej}}/B_s$, the ratio of the rejected bandwidth (at the 10-dB level) to the signal bandwidth. The curves were calculated using Chebyshev 2nd- and 4th-order filters as rejection filters, with the interference in the center

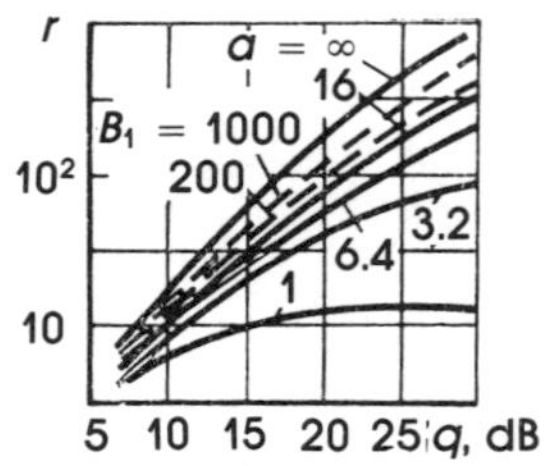

Fig. 6.15 Γ_h (dashed lines) and Γ_i (solid lines) as functions of $q = Q^{-1}$ for various values of B_1 (for Γ_h) and a (for Γ_i).

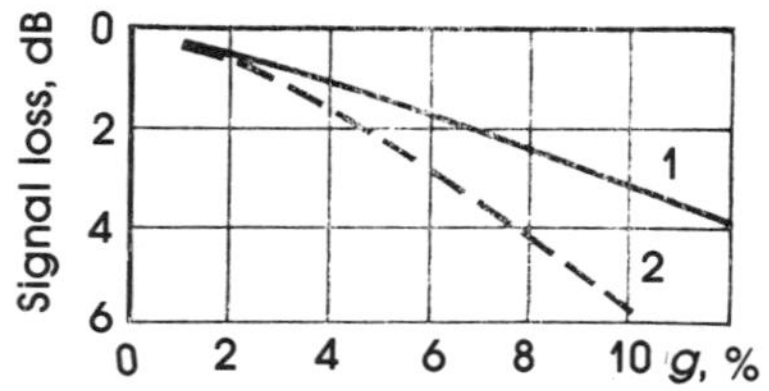

Fig. 6.16 The loss in useful signal amplitude as a function of the ratio $g = B_{rej}/B_s$.
1—2nd order Chebyshev filter
2—4th order Chebyshev filter

of the signal band (worst case). With rejection of 10% of the signal bandwidth, the loss reaches 6 dB. Significantly lower losses are obtained when filters with linear phase-frequency responses are used.

6.7 METHODS FOR SUPPRESSING NARROW-BAND INTERFERENCE

The main methods used to suppress narrow-band interference in radio systems are adaptive filtering and the use of adaptive rejection filters [see Kirillov, N.E. *Pomekhoustoychivaya peredacha soobshcheniy po lineynym kanalam so sluchayno izmenyaushchimisya parametrami* (*Anti-interference information transmission using linear channels with randomly varying parameters*). Moscow: Svyas', 1971. 256 pp.].

6.7.1 Adaptive Filtration

Adaptive filtration consists of "training" the filter on the basis of interference, and using the results of these measurements in subsequent signal reception [see Fal'ko, A.I. *Raznesenniy priem s obucheniem v kanalakh s sosredotochennymi i fluktuatsionnymi pomekhami* (*Spaced reception combining channels with narrow-band and fluctuating interference*). Radiotekhnika i Elektronika, (1975), no. 10, pp. 20–70]. In its simplest form, this consists of rejecting those portions of the signal spectrum occupied by interference. Devices used to realize this method provide high stability relative to narrow-band interference.

Near-optimal adaptive filtering of narrow-band interference may be realized as follows. First, information is obtained on the spectral structure of the interference from training samples, preferably taken in the absence of the desired signal (which is not essential if the interference level greatly exceeds the signal level). Then, the information thus obtained is used to optimize the signal processing adaptively (by adjusting a digital filter, for example). The time necessary to effect this optimization should be substantially shorter than the period over which the interference parameters vary.

6.7.2 Adaptive Band-Rejection Filters

Adaptive band-rejection filters for suppressing narrow-band (periodic) interference are designed on the basis of correlation cancelers [17]. In these filters (see Fig. 6.17), part of the wideband signal and narrow-band interference is diverted from the primary input (1) through a delay line to the auxiliary (cancelation) input (2). The delay time $\Delta\tau$ is chosen so that the wideband signal components will be uncorrelated at the primary and auxiliary inputs, and thus will not be canceled. In addition, the delay $\Delta\tau$ should be significantly shorter than the correlation time of the narrow-band interference, so that the interference voltages at the two inputs will be strongly correlated, and the interference thus effectively canceled.

The effectiveness of band-rejection filters in suppressing narrow-band interference depends on the ratio of the interference and signal bandwidths, B_i and B_s. Figure 6.18 shows a plot of the quantity k_B, characterizing the improvement in the signal-to-interference ratio at the filter output relative to its input value, as a function of B_i/B_s. In these calculations, it

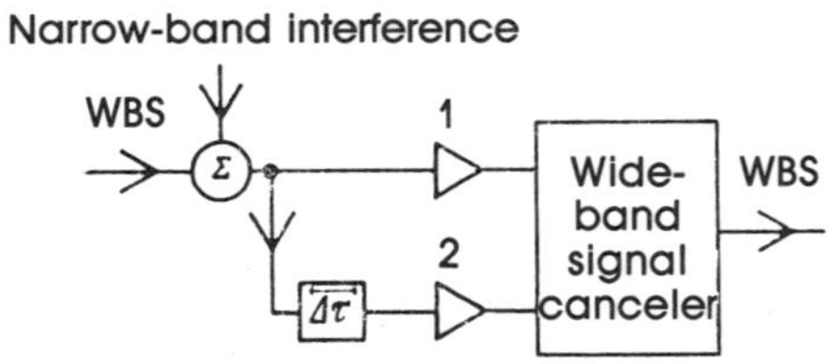

Fig. 6.17 Block diagram of an adaptive rejection filter.

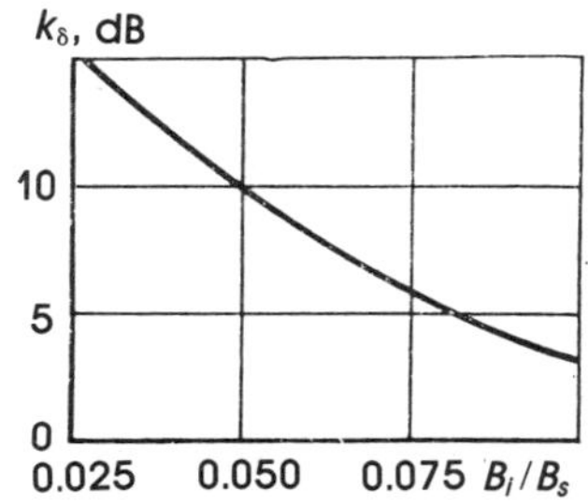

Fig. 6.18 Gain in signal-to-interference ratio as a function of the ratio of the interference and signal bandwidths, B_i/B_s.

was assumed that the narrow-band interference has a Gaussian spectral envelope, the signal spectrum is uniform within the receiver passband, and that $\Delta\tau = B_s^{-1}$.

These results bear witness to the sharp reduction in the effectiveness of adaptive band-rejection filters when the interference bandwidth increases. Another drawback to this form of adaptive filtering is the fact that the signal is split into two components at the output of the filter, due to the fact that the cancelation channel is delayed by $\Delta\tau$ relative to the main channel. The effectiveness may be increased by reducing the delay $\Delta\tau$. However, with $\Delta\tau < B_s^{-1}$, the output signal begins to exhibit a marked dependence upon the position of the interference spectrum within the receiver passband, and the range resolution is worsened due to the splitting of the signal.

6.8 THE INFLUENCE OF THE TRANSMITTED WAVEFORM ON THE ANTI-INTERFERENCE PERFORMANCE AND ACCURACY OF THE RADAR

As is known from general radar theory [9, 21, 23, 28], the accuracy of a radar and its ability to perform in noise and interference is determined to a large extent by the waveform of the transmitted signal. To assess the target range and velocity resolution performance of a particular waveform, the *ambiguity function,* or two-dimensional autocorrelation function, is often used. This function determines the output of an optimum receiver when the time delay and frequency of the input signal differ from the expected values τ and F.

The normalized two-dimensional autocorrelation function of a signal with the complex amplitude $Y(t) = U(t) \exp (i2\pi f_0 t)$, for a matched filter receiver, will take the form:

$$\varphi(\tau, F) = \frac{\left| \int_{-\infty}^{\infty} U(t)\, U^*(t - \tau) \exp (i2\pi Ft)\, dt \right|}{\left| \int_{-\infty}^{\infty} (U(t))^2\, dt \right|}$$

From the form of the ambiguity function, it is possible to estimate for a given radar waveform the ability to resolve targets in range (by the quantity τ) and velocity (by F). The optimality of the signal is determined by the extent to which its characteristics match the conditions established by the targets and interference sources [28]. It is unwise to use signals which possess high sidelobe levels in the operating range of time delays and Doppler shifts. Such sidelobes give rise to intense interference from distinct targets. The separation of weak signals from different targets is difficult in such interference, which degrades the performance of the radar in target-dense situations. The performance is further worsened by clutter received through these signal sidelobes. This is especially significant for over-the-horizon radars, which are subject to high clutter levels over a wide range interval [19, 26, 30].

In view of the intense clutter distributions over a wide spatial region, an acceptable signal is a short pulse with a Gaussian envelope [9, 28]. This signal offers relatively high range resolution, and sidelobes which decrease monotonically with τ. The first foreign (non-Soviet) OTH radars used such short sounding pulses (approximately 100 ms in the MADRE radar) [11]. However, to detect targets beyond the horizon and measure their parameters in conditions of external interference and noise, it is necessary to maintain high illumination energy. This necessitates the use of extremely high pulse power when transmitting short pulses, which involves serious technical difficulties. In these cases, it is necessary to increase the pulse duration [19]. In order to maintain satisfactory range resolution with these longer pulses, non-Soviet radar systems use intrapulse modulation, most often phase shift keying, or linear frequency modulation (LFM), which increases the signal bandwidth. The product of the pulse duration T and the signal bandwidth Δf (the time-bandwidth product B_s) is much greater than unity for these signals. Through pulse compression in the receiver, the duration of the reflected signals may be reduced by a factor of B_s, that is, to the value $1/\Delta f$.

Issues concerning the generation of complex signals, and their properties, are covered in detail in a number of works [9, 21, 22, 23, 29]. We will therefore restrict ourselves to a brief examination of the properties of PSK and LFM signals, directing our attention to their ambiguity functions.

In phase shift keying, a Barker code is often used, in which the number of bits in a pulse satisfies $N \leq 13$. In the case of Barker code signal at zero Doppler frequency, almost all of the energy is concentrated within the main peak of the ambiguity function; the sidelobe levels are extremely low. Thus, for $N = 13$, the amplitude of the sidelobes will be approximately $\dfrac{1}{N} = 0.07$. If the receiver is mistuned ($F_D \neq 0$), the amplitude of the sidelobes grows. An M-ary sequence is also sometimes used with PSK signals [9]. In this case, the average amplitude of the sidelobes is $0.8/\sqrt{N}$ to $1.0\sqrt{N}$, with peaks reaching the value $3/\sqrt{N}$.

The use of PSK signals is effective when they have large time-bandwidth products. The generation and processing of these signals does not present any special difficulties, especially when using an M-ary sequence. However, as has been shown by non-Soviet experience, the maximum signal bandwidth of an OTH pulse radar is limited by the effects of dispersion (primarily phase dispersion) as the wave propagates through the ionosphere. Such dispersion results in both poorer resolution and coordinate measurement, and in a lower signal-to-interference ratio, i.e., in poorer anti-interference performance. This complicates the use of large time-bandwidth product PSK signals in high-frequency radars.

The use of linear frequency modulation signals provides monotonically decreasing range sidelobes, with no spikes. As is known from radar theory, a significant reduction in the ambiguity function sidelobe levels may be obtained with weighted signal processing. This processing may be performed in the frequency domain, using filters whose frequency responses fall off at the edges of the signal spectrum, or in the time domain, by applying a correction to the frequency variation or the pulse envelope [9]. Thus, when using a frequency-correcting Hamming filter, with a response approximated by the expression $|k(f)| = 0.08 + 0.92 (\cos^2 \pi f)/\Delta f$, the maximum sidelobe level relative to the main peak will not exceed -42.8 dB. This results in the main peak being widened by a factor of about 1.5, which in turn causes a slight reduction in the signal-to-noise ratio (about 1.5 dB) [9]. When the LFM pulses are received by a matched filter, the compressed pulses are shifted in time. The signal processing must be made more complex to eliminate this problem.

A practically "clean" region around the central peak of the ambiguity function may be obtained if coherent pulses with random frequency-time structures are repeated periodically, the spectral width of the pulses substantially exceeding the pulse repetition frequency [28, 29]. Such signals provide excellent resolution of densely-grouped multiple targets, but the coordinate measurements are ambiguous. It may be necessary to avoid this difficulty in some cases by using two or three such pulse sequences simultaneously, with different repetition frequencies [29]. Thus, the structure of the transmitted signal has a significant effect on a whole set of important radar characteristics, and also on its anti-interference capability; the choice of a particular waveform is determined by the objectives of the radar and the associated requirements.

In concluding this chapter, it should be mentioned that, in addition to unintentional interference in the HF band, over-the-horizon radars may be subject to active jamming from special transmitters. These jammers may have various time and frequency characteristics. The characteristics of the most common types of jammers are discussed in Vakin and Shustov [3]. Non-Soviet specialists believe that the effective use of barrage jamming is complicated by the wide bandwidth of over-the-horizon radar signals. It is much simpler to effect spot jamming in these conditions [11]. In countering such jamming, the non-Soviet researchers believe that the anti-interference techniques described in this section are applicable, especially the use of adaptive frequency tuning.

REFERENCES

1. Garbuzov, Yu.V., Motora, A.A., and Dudnik, B.S. *et al. Adaptivnaya kompensatsiya pomekh, korrelirovannykh v shirokom vremennom intervale (Adaptive cancellation of interference correlated over a long time interval)*. Radiotekhnika: mezhvedom. nauchno-tekhnich. sb. Khar'kov, 1978, issue 47, pp. 64–68.
2. Burakov, V.A., Zorin, L.A., and Ratynskiy, M.V. *et al. Adaptivnaya obrabotka signalov v antennykh reshetkakh (Adaptive processing of signals in antenna arrays)*: Obzor (Survey). Zarubezhnaya Radioelektronika, (1976), no. 8, pp. 35–59.
3. Vakin, S.A., and Shustov, L.N. *Osnovy radioprotivodeystviya i radiotekhnicheskoy rasvedki (Fundamentals of radio countermeasures and radio intelligence)*. Moscow: Sovietskoe Radio, 1969. 443 pp.
4. Goykhman, E.Sh. and Zitetskiy, A.K. *K voprosu o korrelyatsii signalov v raznesennykh antennakh (The correlation of signals in spatially separated antennas)*. Radiotekhnika i Elektronika, (1978), no. 2., pp. 437–439.
5. Goykhman, E.Sh. *O potentsial'noy pomekhoustoychivosti sistem priema impulsnykh shirokopolosnykh signalov na fone sosredotochennykh po spektru pomekh i kvasibelovo shuma pri vremennom strobirovanii (On the potential anti-interference performance of a receiver of pulsed wide–band signals in narrow-band interference and quasi-white noise with time gating)*. Radiotekhnika, (1977), no. 8, pp. 92–94.
6. Goykhman, E.Sh. and Tsybulin V.A. *O potentsial'noy pomekhoustoychivosti sistem priema impul 'snykh shirokopolosnykh signalov na fone sovokupnosti sosredotochennykh po spektru pomekh i kvasibelovo shuma (On the potential anti-interference performance of a receiver of pulsed wide–band signals in a combination of narrow-band interference and quasi-white noise)*. Radiotekhnika, vol. 26 (1971), no. 12, pp. 24–28.
7. Griffiths, L. *A simple adaptive algorithm for real-time processing in antenna arrays*. Proc. IEEE, (1969), no. 10, p. 1696.
8. Cox, D. and Smith, V. *Teoriya vosstanovleniya (The theory of excitation)*: transl. from English. Moscow: Sovietskoe Radio, 1967. 299 pp.
9. Cook, C. and Bernfeld, M. *Radar Signals*. London and New York: Academic Press, 1967.
10. Maximov, M.V. *Zashchita ot radiopomekh*. Moscow: Sovietskoe Radio, 1976. 495 pp. *(Radar Anti-Jamming Techniques*. Boston and London: Artech House, 1979).

11. Mishchenko, Yu.A. *Zagorizontnaya radiolokatsiya (Over-the-horizon radar)*. Moscow: Voenizdat, 1972. 96 pp.
12. Pistol 'kors, A.A. *Zashchita glavnovo maksimuma v adaptivnykh antennykh reshetkakh (Protecting the main beam in adaptive antenna arrays)*. Radiotekhnika, (1980), no. 12, pp. 8–19.
13. Pistol'kors, A.A. *O raschete perekhodnykh protsessov v adaptivnoy antennoy reshetke (Calculating transport processes in adaptive antenna arrays)*. Antenny (Antennas). Moscow: Svyaz', 1980, Sec. 28, pp. 13–26.
14. Pistol'kors, A.A. and Litvinov, O.S. *Vvedenie v teoriyu adaptivnykh antenn. Statsionarniy rezhim. (Introduction to the theory of adaptive antennas, for stationary inputs)*. Radiotekhnika, (1979), no. 5, pp. 7–16.
15. Platonov, D.M. and Surkova, N.A. *Analiticheskie approksimatsii empiricheskikh zakonov raspredeleniya intensivnosti pomekh v KV-diapazone (Analytic approximations of empirical distributions of interference strength in the KV-band)*. Geomagnetizm i Aeronomiya, vol. 15 (1975), no. 6, pp. 1016–1020.
16. Widrow, B., Mantey, P., Griffiths, L., and Goode B. *Adaptive antenna systems*. Proc. IEEE, (1969), no. 10, p. 2143.
17. Widrow, B. *et al. Adaptive noise cancelling: Principles and applications*. Proc. IEEE, (1975), no. 12, p. 1692.
18. Frost, O. *An algorithm for linearly constrained array processing*. Proc. IEEE, (1975), no. 8, p. 926.
19. Headrick, D.M. and Skolnik, M.I. *Over-the-horizon radar in the HF band*. Proc. IEEE, (1972), no. 8, pp. 5–16.
20. Khmel'nitskiy, E.A. *Otsenka real'noy pomekhosashchishchennosti priema signalov v KV diapazone (Estimating the actual anti-interference performance when receiving KU-band signals)*. Moscow: Svyaz', 1975. 232 pp.
21. Shirman, Ya.D. *Rasreshenie i szhatie signalov (Resolution and compression of signals)*. Moscow: Sovietskoe Radio, 1974. 360 pp.
22. Shirman, Ya.D. and Manzhos, V.N. *Teoriya i tekhnika obrabotki radiolokatsionnoy informatsii na fone pomekh (The theory and techniques for processing radar information in interference)*. Moscow: Radio i Svyaz', 1981. 416 pp.
23. Pestryakova, V.B., Ed. *Shumopodobnye signaly v sistemakh peredachi informatsii (Spread spectrum signals in information transmitting systems)*. Moscow: Sovietskoe Radio, 1973. 423 pp.
24. Applebaum, S.P. *Adaptive arrays*. Syracuse University Research Corp. Rep. SPL, 1974, June, p. 769.

25. Applebaum, S.P. and Chapman, D.J. *Adaptive arrays with main beam constraints*. IEEE Trans., vol. AP-24 (1976), no. 5, pp. 650–662.
26. Griffiths, L. *Time domain adaptive beamforming of HF backscatter radar signals*. IEEE Trans., vol. AP-24 (1976), no. 5, pp. 707–720.
27. Reed, J., Mallet, J., and Brennan L. *Rapid convergence rate in adaptive arrays*. IEEE Trans., vol. AES-10 (1974), no. 6, pp. 853–863.
28. Rihaczek, A.W. *Radar signal design for target resolution*. Proc. IEEE, vol. 53 (1965), no. 2, pp. 116–128.
29. Rihaczek, A.W. *Radar waveform selection—a simplified approach*. IEEE Trans., vol. AES-7 (1971), no. 6, pp. 1078–1086.
30. Shearman, F.T.R., Bagwell, D.J., and Sandham, W.A. *Progress in remote sensing of sea-state and oceanic waves by HF radar*. Radar-77 Internat. Conf., London, 1977 p. 41–45.
31. Sussman, S.M. and Ferrari, E.J. *The effect of notch filters on the correlation properties of a PN signal*. IEEE Trans., vol. AES-10 (1974), no. 3, pp. 385–390.
32. Washburn, W. and Sweeney, L.E. *An on-line adaptive beamforming capability for HF backscatter radar*. IEEE Trans., vol. AP-24 (1976), no. 5, pp. 721–731.

Chapter 7
Signal Detection and Parameter Estimation

7.1 INTRODUCTION

In this chapter, we will examine algorithms for optimum detection of reflected targets signals (prior to any "track" processing), and estimation of their useful parameters. The design of optimum algorithms for detecting a single signal, in interference whose intensity is not known *a priori,* is accomplished with general simple models of signal and interference, using optimization criteria examined in Sec. 7.2. In Sec. 7.3–7.6, we will examine the features of the corresponding models and optimization criteria for various specific types of signals and interference, for multiple option (more than a single possible signal) and multiple target (multiple simultaneous signals) detection, and also for estimating signal parameters. Detection characteristics will be considered in Sec. 7.7. Questions concerning secondary (track) processing will be examined briefly in Sec. 7.8.

7.2 CRITERIA FOR OPTIMUM RADAR DETECTION

7.2.1 The Characteristics of Signals and Interference which are Significant in Developing Signal Detection Algorithms

Let the transmitter radiate a periodic coherent sequence of pulses, with both amplitude and phase modulation (in general). The signal reflected from the target is subject to random amplitude and phase fluctuations. The spectral width of these fluctuations may be significantly less than, comparable with, or greater than the pulse repetition frequency f_p, depending on the characteristics of the reflecting object and the extent of the radar's unambiguous range. In the first case, the received signal will be practically coherent over intervals of tens of repetition periods, lasting from fractions to tens of seconds. In the second case, the signal amplitude-phase fluctuations will be independent, even in successive pulses. It is assumed in this case, however, that these fluctuations do not disturb the coherence of the signal over intervals of length τ, the length of the complex modulation applied to the transmitted signal. The average received signal power will not be known *a priori,* and may vary within wide limits as a function of propagation conditions along the over-the-horizon path. As was indicated in Ch. 5, target signals are received in a background of active interference (cosmic, atmospheric and radio frequency interference), and also clutter (earth reflections). We will note briefly those features of the various forms of interference which are used as the basis of their mathematical models.

Wide-band active interference, for those cases when the system is operating at frequencies with minimum relative levels of such interference, may be modeled as Gaussian white noise with variable power, determined by the fading of distant radio signals. The interval over which such interference is quasistationary at one frequency may be from fractions to tens of seconds, or even longer. Varying the operating frequency may lead to irregular changes in the power of an interference component.

Clutter (earth reflections propagating by oblique paths) differs from active interference in that its components, arriving from some range and some angle, have the same modulation as the transmitted signal, and, in addition, clutter returns fluctuate in time. The spectrum of these fluctuations, at midlatitudes, is such that its energy is concentrated in a narrow band no wider than 1 Hz. With rapid changes in the operating frequency, the coherence of the narrow-band clutter components is limited, naturally, only by the intervals over which operation is maintained at one frequency.

The main feature particular to interference in over-the-horizon radars, and to radio communications, is its significant temporal nonstationarity, and the dependence of the interference power on the operating frequency. Accordingly, detection algorithms should be designed so as to maintain a constant or limited probability of false alarm, independent of those parameters of the interference not known *a priori.* On the other hand, in these conditions it is necessary to maximize the probability of detection of the useful signals, along with the accuracy of the radar, making maximum use of those sample values with comparatively high signal-to-interference values. It is then necessary to consider the limitations on the number of samples of the input information. Analogous problems frequently arise for ordinary radars as well. These questions are widely covered in the literature, for example in the works [4, 16, 19, 21], and others.

In this chapter we will examine some methods, from both US and Soviet literature, for designing optimal and quasioptimal algorithms for coherent-noncoherent processing of slowly and rapidly fluctuating signals, in interference whose power levels and temporal variations in a given resolution cell are not known *a priori.* The structure of the associated decision rules will be considered, and their characteristics will be estimated.

7.2.2 Models of Signals and Interference

A statistical model of a reflected target signal and interference, for a discrete input sample X, may be represented as follows.

Let the observation period T_0, in which a useful signal with relatively constant time delay and Doppler shift may be present, consist of K intervals T_k, each of which includes m intervals T_{jk}, equal to the pulse repetition period T_p, of the transmitted signal (see Fig. 7.1). Discrete sampling is performed at each T_{jk}, consisting of the reading of complex amplitudes X_{ijk} of the signal at the output of the receiver's intermediate frequency (IF) amplifier, where $i = 1, \ldots, n, j = 1, \ldots, m, k = 1, \ldots, K$. Here, n is the number of samples in T_{jk}, m is the number of intervals T_{jk} in T_k, and K is the number of intervals T_k in T_0. The complete input sample $\mathbf{X} = \{X_{ijk}\}$ corresponds to an interval of primary signal processing, the goal of which is to perform preliminary signal detection in each range-velocity resolution cell. The quantity $N = mn$ is the number of samples in the interval T_k.

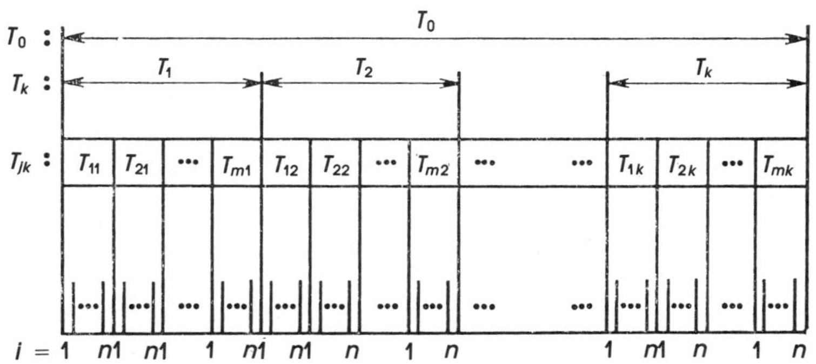

Fig. 7.1 Division of the observation interval T_0, and the relations of the sample indices x_{ijk}.

[*Note*: It is very important to understand the time divisions the authors have established here, inasmuch as they are used extensively throughout the rest of the chapter; therefore, at the risk of redundancy, their meanings are repeated here. The single subscript k identifies a group of pulses, or repetition periods. The double subscript jk identifies the jth pulse or repetition period within the kth group. The triple subscript ijk references the ith sample taken on the jth pulse or repetition period in the kth group. For low bandwidth signals, each of the K groups of m samples corresponds to a coherent processing period; for high bandwidth signals, the division into groups of m pulses is somewhat arbitrary, since the signal is no longer coherent from pulse to pulse.—Tr.]

Let $\mathbf{X} = \mathbf{S} + \mathbf{Z}$, where $\mathbf{S} = \{S_{ijk}\}$ is the input sample due to the useful reflected target signal, and $\mathbf{Z} = \{Z_{ijk}\}$ is the sample of additive interference, so that we have the superposition $X_{ijk} = S_{ijk} + Z_{ijk}$. To simplify the design of optimum algorithms and the analysis of their characteristics, both the signal and interference will be assumed to be random Gaussian, possibly nonstationary, processes, with zero mean values and unknown intensities.

Further, we will assume that over the interval T_0, the signal may in general be placed in the form $S_{ijk} = U_{jk}a_ib_j$, where U_{jk} are complex Gaussian quantities with average values $\mathbf{E}[U_{jk}] = 0$ and $\mathbf{E}[|U_{jk}|^2] = \nu_{jk}$, determining the signal strength over the intervals T_{jk}; $\mathbf{a} = \{a_i\}$, $i = 1, \ldots, n$, is the vector determining intraperiod (in particular, intrapulse) amplitude-phase modulation of the expected received signal in T_{jk}; $\mathbf{b} = \{b_j\}$ is the vector determining the pulse-to-pulse phase modulation of the slowly fluctuating signal, caused by a Doppler shift. The time delay and Doppler frequencies associated with the useful signal, on which **a** and **b** depend, are assumed to be known. Later, these limitations, and the assumption of a single target, will be removed. It is convenient to consider the vectors **a** and **b** to be normalized, so that $\mathbf{a}^*\mathbf{a} = \mathbf{b}^*\mathbf{b} = 1$, where * denotes the Hermitian conjugate (transposition and complex conjugation).

Based on the signal properties described above, we will divide all possible forms of signals into two types. A slowly fluctuating signal will be assumed to be completely coherent within the intervals T_{jk} (repetition periods) and T_k (groups of repetition periods), and independent in different intervals T_k, i.e. $U_{jk} = U_k$. A rapidly fluctuating signal will be fully coherent over each interval T_{jk}, and independent in different intervals T_{jk}; in other words, a rapidly fluctuating signal is one that is not coherent from pulse to pulse.

Interference is classified as being one of three types, on the basis of its nonstationarity: stationary, nonstationary, and very nonstationary. In all cases Z_{ijk} are independent complex Gaussian random quantities with $\mathbf{E}[Z_{ijk}] = 0$. In the case of stationary interference, $\mathbf{E}[|Z_{ijk}|^2] = \epsilon$, where ϵ is the intensity of the interference, which is unknown *a priori*. For nonstationary interference, $\mathbf{E}[|Z_{ijk}|^2] = \epsilon_k$, where ϵ_k is the interference power in the interval T_k, unknown *a priori* and different for different k. In the case of very nonstationary interference, $\mathbf{E}[|Z_{ijk}|^2] = \epsilon_{jk}$, where ϵ_{jk} is the interference power in each repetition period T_{jk}, unknown *a priori* and different for different combinations of j and k. In other words, the intensity of stationary interference does not change during the observation period T_0, the intensity of non-stationary interference changes from one group of repetition periods to the next, and very nonstationary interference changes from one repetition period to the next. An analogous model for nonstationary interference is used in Kulikov and Trifonov [15].

To simplify the description of the model of clutter with an unknown fluctuation spectrum, the time sequence of samples with duration T_0 is first transformed into the frequency domain. It is assumed that in this domain the interference, with an unknown fluctuation spectrum within the limits of the repetition frequency, is transformed, with the help of discrete narrow-band digital filters with practically nonoverlapping frequency responses, or with corresponding weighted Fouier transforms, into an uncorrelated sequence of clutter samples with different intensities unknown *a priori,* analogous to the case of very nonstationary interference. In other words, clutter frequency samples are analogous to very nonstationary interference time samples. These statistical classifications of the various forms of signals and interference will be repeated briefly and made more concrete in the following sections.

7.2.3 Criteria for Optimum Detection of Signals in a Background of Interference with Incompletely Known Statistical Parameters

During the observation time T_0, let the discrete sampling $\mathbf{X} = \{X_{ijk}\}$ of complex amplitudes be taken, at the output of the receiver's IF amplifier, for example. Since the complex amplitudes X_{ijk} may be placed in the form of pairs of real quadrature components, each sample may be viewed as a point in a $2NK$-dimensional real Euclidean space E_{2NK}, i.e., $\mathbf{X} \in E_{2NK}$. [*Note:* The number $2nK$ was used several times in the original for the dimensionality of this space, but inasmuch as there are nmK samples taken during the observation time, the value would appear to be $2nmK$, or $2NK$, where $N = nm$, and that is the value used in this translation –Tr.]

Let the probability density for $X, p(\mathbf{X}|\theta)$, be known *a priori*, accurate to a finite set of unknown parameters $\theta \in \Omega$. That is, $p(\mathbf{X}|\theta)$ is the probability of obtaining the sample $\mathbf{X}$ given the values of the unknown parameters set by θ. The combination of parameters θ may include various parameters of the useful detected signals, and also parameters of the interference which are unknown *a priori*. Let $\Omega = \Omega_0 \cup \Omega_1$, where Ω_0 and Ω_1 are subsets of Ω, corresponding to the absence or presence of the useful signal in the sample $\mathbf{X}$. It is sometimes convenient to assign to Ω_1 those quantities θ which correspond to the presence of a signal whose level is such that the signal-to-interference power ratio is sufficient for assured signal detection with an optimum detector.

An arbitrary threshold decision rule for binary signal detection may be described with the help of the function $\psi(\mathbf{X})$, which for any $\mathbf{X} \in E_{2NK}$ takes the value zero or one, corresponding to the decision on whether or not the signal is present. In practice, however, the binary detection decision rule may be described in the form of the condition $f(\mathbf{X}) \geqslant C$ where $f(\mathbf{X})$ is some function of the random input data, and C is the detection threshold.

The case when this condition is satisfied corresponds to $\psi(\mathbf{X}) = 1$. The quality of detection using $\psi(\mathbf{X})$ is usually characterized by the probability of correct detection $D(\theta)$ with $\theta \in \Omega_1$, and the probability of false alarm $F(\theta)$ with $\theta \in \Omega_0$. In other words, the probability of detection is the probability that the decision rule indicates the presence of a signal when the signal is actually present ($\theta \in \Omega_1$), and the probability of false alarm is the probability that the detection rule indicates the presence of a signal when the signal is actually absent ($\theta \in \Omega_0$).

A natural requirement for the decision rule, when there are unknown interference parameters, is that the probability of false alarm be limited to some acceptable value F_0, i.e.:

$$F(\theta) \leq F_0 \quad \text{for } \theta \in \Omega_0 \tag{7.1}$$

On the other hand, the optimum decision rule should also, in some sense, maximize the probability of detection $D(\theta)$, at least in some important controlled region where the parameters of the signal and interference are unknown. The most commonly known condition of this type is the requirement for maximizing the minimum value of $D(\theta)$ in Ω_1, i.e.:

$$\min_{\theta \in \Omega_1} D(\theta) = \max_{\psi} \tag{7.2}$$

where the maximum is sought amongst all possible $\psi(\mathbf{X})$ satisfying the condition (7.1). In a more strict construction, max and min should be replaced by sup and inf, which will apply to any region of Ω_1 and for any given set of limited functions $\psi(\mathbf{X})$.

Let $D(\theta, \psi)$ and $F(\theta, \psi)$ be the probabilities of detection and false alarm for the decision rule ψ. Then the decision rule ψ^0, satisfying the condition (7.1), will be called the maximin decision rule for signal detection, if for all ψ satisfying (7.1):

$$\min_{\theta \in \Omega_1} D(\theta, \psi^0) \geq \min_{\theta \in \Omega_1} D(\theta, \psi) \tag{7.3}$$

It should be noted that these problems are particular cases of the problem of verifying complex statistical hypotheses, one of the branches of mathematical statistics [17], and the corresponding optimum decision rules are tests with the largest guaranteed power (probability of detection) in Ω_1. In the detection problems being considered, the regions Ω_0 and Ω_1 should be determined with a parameter (or parameters) characterizing the signal-to-interference ratio in the input sample $\mathbf{X}$ (or in subsets of this sample). For a single parameter $\gamma \equiv \gamma(\theta)$, the hypothesis regions Ω_0 and Ω_1 may be described in the form:

$$H_0: \gamma = 0 \quad \text{and} \quad H_1: \gamma \geqslant \gamma_0 \tag{7.4}$$

where γ_0 is some given threshold of the signal-to-interference ratio.

The usefulness of describing the controlled regions with the help of (7.4) consists in the simple physical concept underlying the conditions; the fact that the information required is available to the decision processor; the invariance of the parameter $\gamma = \gamma(\theta)$, the regions Ω_0 and Ω_1 and the problem as a whole relative to the group G of scale transformations of the sample space and the corresponding (induced) group $\overline{G}$ of transformations in the space of unknown parameters; and, finally, the usually monotonic dependence of $D(\theta)$ on γ, which simplifies the solution of the minimax problem. With the existence of K samples with different γ_k, $k = 1, \ldots, K$, the corresponding decision rule is

$$H_0: \gamma_k = 0, \quad k = 1, \ldots, K, \quad \text{and} \quad H_1: \gamma_k \geqslant \gamma_{0k} \tag{7.5}$$

In terms of γ, the maximin criterion (7.3) may be replaced with the equivalent minimax criterion, which, together with (7.1), minimizes the threshold signal-to-interference ratio(s) γ_0 or γ_{0k}, while providing a guaranteed probability of detection D_0, i.e.:

$$\max_k \gamma_{0k} = \min_\psi \quad \text{for } D(\theta) \geqslant D_0 \tag{7.6}$$

General methods for finding minimax decision rules for binary detection of signals in background interference with unknown parameters, using the above optimality criteria, and using the principles of similarity and invariance, are presented, for example, in Korado [8, 9]. The general structure of minimax decision rules corresponds to the structure of the Neyman-Pearson rule for the least favorable *a priori* distribution of the random parameters in the problem, and takes the form [17]:

$$\frac{\int_{\Omega_1} p(\mathbf{X}|\theta) p_1'(\theta)\, d\theta}{\int_{\Omega_0} p(\mathbf{X}|\theta) p_0'(\theta)\, d\theta} \geqslant C \tag{7.7}$$

where $p_1'(\theta)$ and $p_0'(\theta)$ are the least favorable pair of *a priori* distributions for the parameters θ in Ω_1 and Ω_0, and C is a constant determined by the condition $\max F(\theta) = F_0$. In the given case, the least favorable are those distributions for which the largest average probability of detection, determined by the optimum rule (7.7), takes its minimum value. The most convenient indication of a least favorable distribution is a minimum value for $D(\theta)$ and a maximum value for $F(\theta)$, in the regions where the distribution is concentrated for the "signal present" and "signal absent" hypotheses, respectively.

Thus, when detecting a signal in a background of stationary interference, where the unknown parameter of the interference is its intensity ϵ, and the problem is invariant relative to the group of scale transformations G of the input sample $\mathbf{X} \rightarrow \rho\mathbf{X}$, the least favorable distribution for ϵ takes the form $d\epsilon/\epsilon$. Here, for $\Omega_1 : \gamma \geqslant \gamma_0$, the least favorable distribution of γ is concentrated at the point γ_0.

In the given case the invariance of the detection problem lies in the fact that a scale transformation of a sample does not remove the probability distribution $\mathbf{X}$, with the existence and absence of the signal, from the family $p(\mathbf{X}|\theta)$ with $\theta \in \Omega_1$ or $\theta \in \Omega_0$ respectively, and only changes the values of the unknown parameters θ, leaving them in the regions Ω_1 and Ω_0. Here we have in mind scale transformations, the set of which with $0 < \rho < \infty$ forms a commutative group G with multiplication by unity, which is the identity transformation with $\rho = 1$. Thus, the minimax decision rule is found through direct integration in the numerator and denominator of (7.7), and takes the following form:

$$\frac{\int_0^\infty p(\mathbf{X}|\gamma_0, \epsilon)\, d\epsilon/\epsilon}{\int_0^\infty p(\mathbf{X}|0, \epsilon)\, d\epsilon/\epsilon} \geqslant C \tag{7.8}$$

When there are K independent "subsamples" $\mathbf{X}_k$ with different unknown parameters γ_k and ϵ_k, the least favorable distributions of ϵ_k and γ_k are independent and analogous to those presented above, with γ_k concentrated at the points $\gamma_k = \gamma_{0k}$. Accordingly, the decision rule takes the form:

$$\prod_{k=1}^{K} \frac{\int_0^\infty p(\mathbf{X}_k|\gamma_{0k}, \epsilon_k)\, d\epsilon_k/\epsilon_k}{\int_0^\infty p(\mathbf{X}_k|0, \epsilon_k)\, d\epsilon_k/\epsilon_k} \geqslant C \tag{7.9}$$

Of interest in some cases is the decision rule satisfying a partial modification of the minimax optimality criterion, when the controlled region Ω_1 is limited either by the region of weak (threshold) signals, with $\gamma \rightarrow 0$, for example, or the region of strong signals with $\gamma \rightarrow \infty$. In this case, it is possible to find so-called locally and asymptotically minimax decision rules, equivalent to limits of the minimax decision rules with the parameters γ_0 or γ_{0k} approaching zero or infinity. These decision rules are usually simpler than the minimax, while being sufficiently effective, becoming more effective with a larger number n of input samples in those regions where γ approaches zero or infinity. In the limit $n \to \infty$, these decision rules become equivalent to the minimax decision rule, and are asymptotically ($n \to \infty$) uniformly the decision rules with the highest power (probability of detection) in all regions Ω_{01}: $\gamma_k > 0$, $k = 1, \ldots, K$.

These detection problems may also be solved using the maximum likelihood decision rule, the general form of which is given by the inequality (see [17]):

$$\frac{\sup\limits_{\theta \in \Omega_1} p(\mathbf{X}|\theta)}{\sup\limits_{\theta \in \Omega_0} p(\mathbf{X}|\theta)} \geqslant C \tag{7.10}$$

where C is a constant, chosen from the condition (7.1), limiting the probability of false alarm. It would be more precise to refer to the rule of (7.10) as the maximum likelihood ratio test, but we will use the shortened name for this decision rule. For problems which are invariant relative to the group of transformations G, the maximum likelihood decision rule (7.10) is also G-invariant. At least for the cases considered here, maximum likelihood decision rules are simple, and in some particular cases are the same as the minimax decision rule. These rules are frequently very effective in the entire region Ω_1, which is explained by the fact that the structure of the maximum likelihood ratio corresponds to the structure of the Neyman-Pearson decision rule with the highest probability of detection, with known fixed values of θ for the two hypotheses. These values of θ are taken to be their maximum likelihood estimates, which are asymptotically optimum for $k \to \infty$ (see [19]). The maximum likelihood decision rule approximately satisfies the criterion for minimizing the maximum energy loss η (in decibels) in comparison with optimum signal detection, with values of θ known *a priori* for both hypotheses:

$$\max_{\theta \in \Omega_1} \eta(\psi, \theta) = \min_{\psi} \tag{7.11}$$

It should be noted that the decision rules satisfying the optimality criteria presented above are quite different from Bayesian decision rules, which, by classical theory, are optimum when the interference parameters are known *a priori.* The importance of this difference lies in the fact that

Bayesian decision rules will not in general satisfy the basic requirement for limiting the probability of false alarm (7.1) for all possible values of the interference parameters, and are therefore, in principle, not applicable to detection problems with new conditions, not known *a priori.*

In Sec. 7.3–7.5, we will consider the optimum detection of signals (using the optimality criteria presented above), in a background of uniform, cosmic and atmospheric interference with unknown intensity, in a background of nonstationary radio frequency interference, with unknown temporal variations, and in a background of nonstationary clutter with unknown fluctuation spectrum. The effectiveness of the optimum decision rules will also be estimated. The effectiveness of optimum decision rules when the interference parameters are unknown *a priori* is theoretically lower than the effectiveness of the corresponding optimum decision rules without this uncertainty. The corresponding energy losses, resulting in a higher signal-to-interference requirement, are an unavoidable consequence of the uncertainty associated with the interference parameters.

7.3 DETECTING SIGNALS IN A BACKGROUND OF STATIONARY INTERFERENCE WITH UNKNOWN INTENSITY

7.3.1 The Signal and Interference Model

We will proceed using the following formulation (see Sec. 7.2). Let the slowly fluctuating signal be given by $\mathbf{S} = \{S_{ijk}\}$, where $S_{ijk} = U_k S_{ij}$; U_k are independent Gaussian complex random quantities, with $\mathbf{E}[U_k] = 0$, $\mathbf{E}[|U_k|^2] = \nu$; the signal intensity, unknown *a priori*; $S_{ij} = a_i b_j$ is the form of the signal, which is known with an accuracy up to the time delay and Doppler frequency; $\mathbf{a} = \{a_i\}$ is the form of the intrapulse signal modulation; $\mathbf{b} = \{b_j\}$ is the form of the pulse-to-pulse signal modulation; $\mathbf{E}[\ ]$ denotes the expected value. [*Note.* The Russian original uses the phrases *intra-period* and *interperiod,* referring to the repetition period. Because these phrases are awkward, and may lead to confusion through their similarity, they have been replaced with *intrapulse* and *pulse-to-pulse* in this translation. For CW operation, these should be interpreted as *intraperiod* and *period-to-period*. —Tr.]

A rapidly fluctuating signal has the form $S_{ijk} = U_{jk} a_i$, where U_{jk} are independent Gaussian complex random quantities with $\mathbf{E}[U_{jk}] = 0$ and $\mathbf{E}[|U_{jk}|^2] = \nu$. The interference is $\mathbf{Z} = \{Z_{ijk}\}$, where Z_{ijk} are independent Gaussian complex random variables with $\mathbf{E}[Z_{ijk}] = 0$, and $\mathbf{E}[|Z_{ijk}|^2] = \epsilon$, the latter being the interference intensity, unknown *a priori*. The signal-to-interference ratio is $\gamma = \nu/\epsilon$.

7.3.2 Detecting a Slowly Fluctuating Signal in a Background of Interference With Unknown Intensity

In accordance with [10], we introduce the preliminary orthogonal transformation of the input sample $\mathbf{X} = \mathbf{S} + \mathbf{Z}$, of the form $\mathbf{V} = B\mathbf{X}$, where $\mathbf{X}$ is the column vector of input samples, $\mathbf{V}$ is the column vector of transformed inputs, B is the linear unitary transformation whose first row is proportional to $\mathbf{S}^*$, and * denotes the Hermitian conjugate (transposition and complex conjugation).

Considering the specifics of the signal and interference, this transformation may be placed in the form of the sequential transformations B_1 and B_2; $B = B_2B_1$, where B_1 is the matrix of simultaneous intrapulse unitary transformations, such that the first row of B_1 is proportional to $\mathbf{a}^*$; B_2 is the matrix of simultaneous pulse-to-pulse unitary transformations, such that the first row of B_2 is proportional to $\mathbf{b}^*$. Then B_1 transforms the signal to the first sample values of each T_{mk}, and B_2 takes the signal from these values to the first sample value of each T_k.

The interference distribution does not change through these transformations. As a result, the Hermitian correlation matrix of the signal M_s is diagonal, with K nonzero elements on the main diagonal, equal to $\nu\mathbf{a}^*\mathbf{a}\mathbf{b}^*\mathbf{b} = \nu$, since $\mathbf{a}^*\mathbf{a} = \mathbf{b}^*\mathbf{b} = 1$.

Thus, the probability density function for V, in the presence of a signal, takes the form:

$$p(V|\gamma, \epsilon) = (\pi\epsilon)^{-nmK}(1 + \gamma)^{-K}$$
$$\exp\left[-\epsilon^{-1}(1 + \gamma)^{-1}\sum_{k=1}^{K}|V_{11k}|^2 - \epsilon^{-1}\sum_{k=1}^{K}\sum_{i-2}^{n}\sum_{j-2}^{m}|\mathbf{V}_{ijk}|^2\right]$$

Placing $P(V|\gamma, \epsilon)$ and $P(V|0, \epsilon)$ in (7.8), integrating the numerator and denominator over $d\epsilon/\epsilon$, and simplifying the result, leads to an equivalent minimax decision rule:

$$\sum_{k=1}^{K}|V_{11k}|^2 \geqslant C\sum_{i,j,k}^{n,m,K}|V_{ijk}|^2 \tag{7.12}$$

Because

$$V_{11k} = \sum_{j=1}^{m} b_j^* \sum_{i=1}^{n} a_i^* X_{ijk}$$

and

$$\sum_{i,j,k}^{n,m,K} |V_{ijk}|^2 = \sum_{i,j,k}^{n,m,K} |X_{ijk}|^2$$

then, in terms of the input sample X, the optimum algorithm decides on the presence of a signal of the form **S** when the following condition is met:

$$\sum_{k}^{K} \left| \sum_{j=1}^{m} b_j^* \sum_{i=1}^{n} a_i^* X_{ijk} \right|^2 \geqslant C \sum_{i,j,k}^{n,m,K} |X_{ijk}|^2 \tag{7.13}$$

where the constant C is chosen from the condition that a given probability of false alarm F_0 be maintained. For scalars, * denotes complex conjugation.

The structure of the optimum decision rule based on (7.13) is shown in Fig. 7.2. Here, IPP is intrapulse processing, with the formula:

$$Y_{jk} = \sum_{i}^{n} a_i^* X_{ijk}$$

effecting the coherent autocorrelation summation with the expected signal in each repetition period T_p; PPP is the pulse-to-pulse processing, which effects the coherent pulse-to-pulse summation of the expected signal with the formula

$$V_k = V_{11k} = \sum_{j}^{m} b_j^* Y_{jk}$$

D is a square-law detector, forming the quantity $\{V_k|^2$; is noncoherent integration with the formula $\sum_k |V_k|^2$; EII is an estimate of the interference intensity (scaled by NK), from the formula

$$t = NK\hat{\epsilon} = \sum_{i,j,k}^{n,m,K} |X_{ijk}|^2$$

C is a multiplier equal to the threshold constant C; THR is comparison with the threshold. It is not difficult to see that the structure of the rule (7.13) is a special case of the structure of an optimum detector of random signals [10].

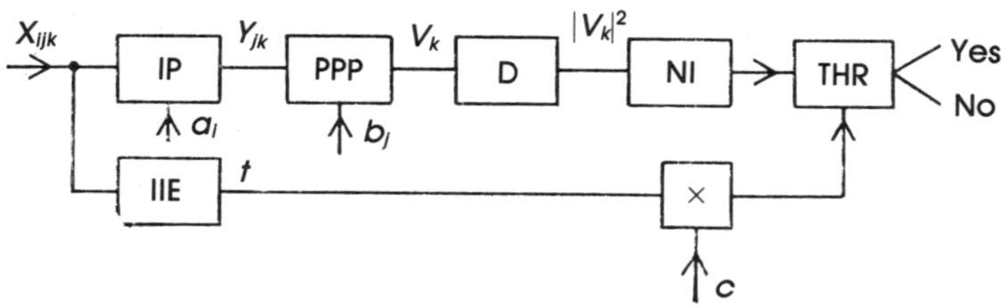

Fig. 7.2 Block diagram of the optimum decision rule for detecting a slowly fluctuating signal in uniform interference with unknown intensity. IPP = intrapulse processing; PPP = pulse-to-pulse processing; D = square law detector; NI = noncoherent integration; IIE = interference intensity estimate; X = multiplication by constant C.

As is known, the optimum detector (7.13) provides the maximum probability of detection for any signal-to-interference ratio $\gamma = \nu/\epsilon$, while limiting the probability of false alarm ($F \leqslant F_0$). At the same time, within this class of Neyman-Pearson rules, this rule minimizes the threshold level of the signal-to-interference ratio γ_0 required to provide any given probability of detection D_0. It should be noted that γ is a ratio of the signal power to interference power, in the statistic V_{11k}, i.e., the signal-to-interference ratio before the detector at the output of the linear portion of the processor where the coherent summation is performed.

Such parameters of the signal as its time delay and Doppler frequency are not usually known *a priori*, and are subject to estimation at the same time as the signal is detected. This situation may be described as that of detection with selection of one of L mutually orthogonal signals, which are associated with different values of the indicated parameters, i.e., as a symmetric L-alternative detection problem. It may be shown, in accordance with Korado [7], that the optimum algorithm for such L-alternative detection indicates the presence of the lth form of the signal, if the following inequalities are all met simultaneously:

$$\sum_{k}^{K} \left| \sum_{j\perp}^{m} b_{lj}^{*} \sum_{i}^{n} a_{li}^{*} X_{ijk} \right|^2 \geqslant C \sum_{i,j,k}^{n,m,K} |X_{ijk}|^2 \sum_{k}^{K} \left| \sum_{j}^{m} b_{lj}^{*} \sum_{i}^{n} a_{li}^{*} X_{ijk} \right|^2$$
$$\geqslant \sum_{k}^{K} \left| \sum_{j}^{m} b_{qj}^{*} \sum_{i}^{n} a_{qi}^{*} X_{ijk} \right|^2, \quad q \neq l \tag{7.14}$$

where C is determined from the condition

$$F = \sum_{l=0}^{L} F_l = F_0$$

Actually, in this case, the minimax rule for L-alternative detection, in a background of interference with unknown intensity, makes a decision on the presence of the lth signal using

$$f_l \geqslant C \quad \text{and} \quad f_l \geqslant f_q, \quad l \neq q$$

where f_l is the likelihood ratio for the lth signal, with the least favorable *a priori* distributions of the unknown parameters γ_l and ϵ. Amongst the L probabilities of detection for the different signals, this rule maximizes the minimum value D_l, while limiting the probability of false alarm for any signal to a given value F_0. The structure of the optimum decision rule as modified for the L-alternative case is shown in Fig. 7.3 (see [7]), with only one of the L matched filters shown. The notations in this and the following figures are the same as those in Fig. 7.2.

Unlike the case of detecting a single signal, the L-alternative detection processor has multiple channels, L of which are the usual signal processing channels for the L different signal forms, and the usual "interference" channel is used to estimate the interference intensity, so as to stabilize the probability of false alarm. Here, the intrapulse processing is realized in the form of an intrapulse processor matched to the l signal forms, and the pulse-to-pulse processor in the form of a discrete Fourier transform.

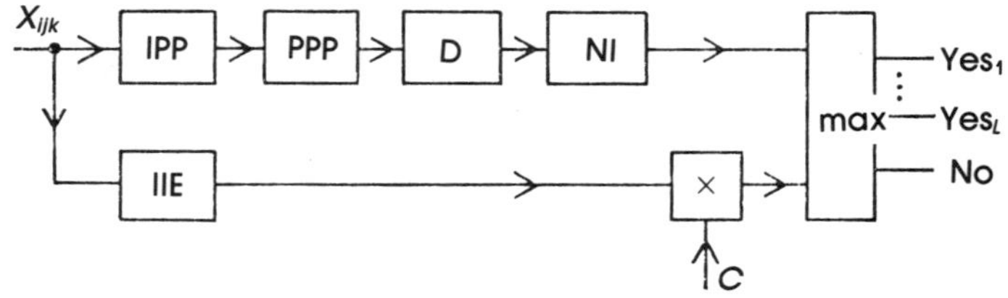

Fig. 7.3 Block diagram of the optimum decision rule for detection/discrimination of slowly fluctuating signals in interference with unknown intensity, with symmetric alternatives (the notation is the same as that in Fig. 7.2). Only one of the L channels is shown.

In this case, with a combination of orthogonal signal forms, a small modification to the decision rules makes it possible to solve the multiple detection problem with multiple alternatives, that is, to decide on the presence of each of the L possible signals individually and simultaneously. In this case, the decision on the presence of the lth signal is made in accordance with (7.13), where a_i and b_j are replaced with a_{li} and b_{lj}, matched to the lth signal. The analogous processing remains effective in the practical case of a quasi-orthogonal combination of signals. The structure of the multiple detection problem is shown in Fig. 7.4. In the terminology of Shirman [21], the detector in Fig. 7.4 performs practically complete detection resolution of a group of L targets.

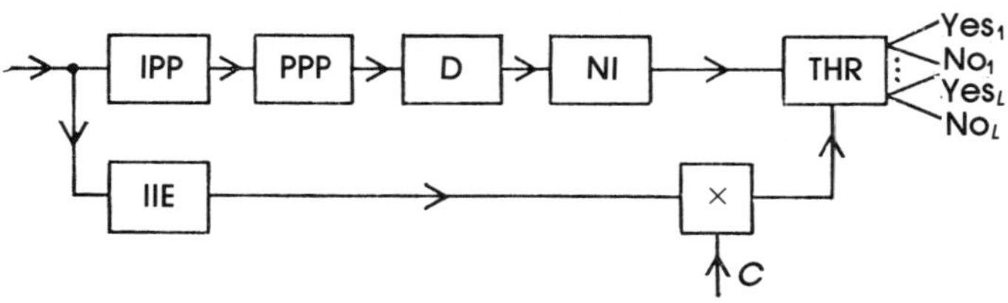

Fig. 7.4 Block diagram of a quasioptimum decision rule for multiple-target detection with a slowly fluctuating signal (D is a linear or square-law detector).

When the number of signal resolution channels in range and velocity is close to the number of input samples in an interval T_k, i.e. $L \approx nm$, it is useful to increase the number of resolution channels to nm. In this case, the structure of the quasioptimal decision rules will take the form shown in Fig. 7.5. When there are resolution elements for which it is known that there will be no useful signals, the interference intensity estimate may be obtained by summing the signals in these channels noncoherently. If it is not certain that any of the channels will be free, however, the interference intensity should be estimated using all of the resolution elements. The structure of the multiple target detection decision rule which has been presented is not optimal. We will show how the rule may be improved. The multiple target problem of detecting L signal forms, or the problem of completely resolving groups of L targets, may be viewed as a multiple alternative detection problem, for various sets of all the possible combinations of the various signals being present or absent [21]. In the case of orthogonal signals, there is a unitary linear transformation of the input sample with the matrix B, such that it converts each of the signal forms

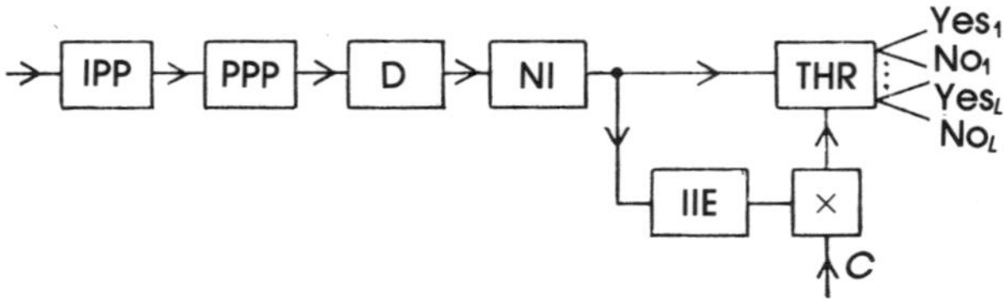

Fig. 7.5 Another version of the multiple-target decision rule.

with unknown intensity into one of the elements of the transformed sample $\mathbf{V}_k = B\mathbf{X}_k$. A transformation with such properties is usually called a reduction to a canonical form. Here, the independent component of the first signal will be present in the first of the L elements of $\mathbf{V}_k$, of the second signal in the following elements, and so on.

It is not difficult to show [7] that the optimum (*minimax*) decision rule for multiple target detection indicates the presence of the sth group of signals when the following inequalities are satisfied simultaneously:

$$f_s(Q_s) + \alpha_s \geqslant 0, \quad f_s(Q_s) + \alpha_s \geqslant f_r(Q_r) + \alpha_r \qquad (7.15)$$
$$f_s(Q_s) = (\ln 1 - \gamma_{0s}(1 + \gamma_{0s})^{-1} Q_s)$$

where

$$Q_s = \frac{\sum_l^{L_s} \sum_k^{K} \left| \sum_j^m b_{li}^* \sum_i^n a_{li} X_{ijk} \right|^2}{\sum_{i,j,k}^{n,m,K} |X_{ijk}|^2}$$

Here γ_{0s} is the threshold signal-to-interference ratio in the L_s sample values, forming signals in the sth group (the thresholds are the same within each group); α_s are constants, determined from the requirement for a given probability of false alarm and the equality $D_s(\gamma_{0s}) = D_r(\gamma_{0r})$, $s, r = 1, \ldots, S$, $S = 2^{L-1}$.

Actually, in this case, the density distribution for $\mathbf{X}$, for the sth group $l \in J_s$ takes the form:

$$p(\mathbf{X}|\gamma_s, \epsilon) = (\pi\epsilon)^{-nmK} (1 + \gamma_s)^{-KL_s}$$
$$\times \exp\left[-\epsilon^{-1}\left(\sum_{i,j,k}^{n,m,K} |X_{ijk}|^2 - \frac{\gamma_s}{1 + \gamma_s}\right.\right.$$
$$\left.\left.\times \sum_{k=1}^{K} \sum_{l \in J_s}^{L_s} \left| \sum_j^m b_{lj}^* \sum_i^n a_{li}^* \times X_{ijk} \right|^2\right)\right]$$

which after integration in (7.8) leads to (7.15). For a given probability of false alarm, this rule maximizes min $[D_s(\gamma_s)]$ in the region $\gamma_s \geqslant \gamma_{0s}$. A block diagram for the decision rule (7.15) is shown in Fig. 7.6. When detecting weak signals, $\gamma_{0s} \to 0$ and $f_s(Q_s) \to \gamma_{0s} Q_s$, which significantly simplifies the decision rule.

The rule (7.15) differs from multiple-alternative single-target detection described in (7.14). A varying number of different signal components, entering each of the alternative groups, are summed noncoherently in each interval T_k, and the invariant statistics Q_s are compared with the threshold and with each other only after their optimum nonlinear transformation, taking into account the different values of γ_{0s}. We note that an analogous structure applies to the optimum rule for multiple-alternative detection of signals from identical targets, distributed simultaneously and uniformly over several radar resolution elements, in both time delay and Doppler shift. The causes of this situation may be the large range extent of some of the targets, the effects of multipath propagation, and also the nonstationarity of the parameters of the reflecting objects and the path. In these situations, after coherent intrapulse and pulse-to-pulse processing and detection, the processor performs post-detection integration of the signal energy in those (usually neighboring) resolution cells in which the signals may be present simultaneously. With identical γ_{0s}, the decision rule (7.15) is simplified and differs from (7.14) only in the additional summation of the statistics

$$\sum_k^K \left| \sum_j^m b_{lj}^* \sum_i^n a_{li}^* X_{ijk} \right|^2, \quad l \in J_s$$

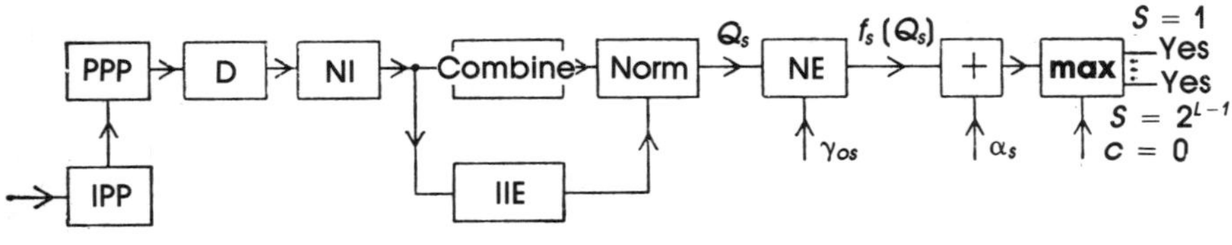

Fig. 7.6 Block diagram of the optimum multiple-target decision rule for slowly fluctuating signals (Norm is normalization, NE is a nonlinear element).

7.3.3 Detection of a Rapidly Fluctuating Signal in a Background of Interference With Unknown Intensity

A rapidly fluctuating signal fluctuates independently in different repetition periods, and is therefore determined statistically only by the form of the intrapulse modulation, taking account of the differences caused by the time delay and Doppler shift. As follows from Korado [10], in this case the optimum algorithm for detection, in terms of the transformed sample (with $B_2 = I$), takes the form:

$$\sum_{j,k}^{m,K} |V_{ijk}|^2 \geq C \sum_{i,j,k}^{n,m,K} |V_{ijk}|^2 \tag{7.16}$$

where $V_{1jk} = \Sigma_i^n a_i^* X_{ijk}$ is the result of the intrapulse transformation with the matrix B_1. Accordingly, in terms of the input vector, in place of (7.13) we have

$$\sum_{j,k}^{m,K} \left| \sum_{i}^{n} a_i^* X_{ijk} \right|^2 \geq C \sum_{i,j,k}^{n,m,K} |X_{ijk}|^2 \tag{7.17}$$

and in place of (7.14):

$$\begin{aligned} \sum_{j,k}^{m,K} \left| \sum_{i}^{n} a_{li}^* X_{ijk} \right|^2 &\geq C \sum_{i,j,k}^{n,m,K} |X_{ijk}|^2 \\ \sum_{j,k}^{m,K} \left| \sum_{i}^{n} a_{li}^* X_{ijk} \right|^2 &\geq \sum_{l,k}^{m,K} \left| \sum_{i}^{n} a_{si}^* X_{ijk} \right|^2 \end{aligned} \tag{7.18}$$

The structure of the quasioptimal decision rule for multiple-target signal detection takes the form shown in Fig. 7.7. In the case of rapidly fluctuating signals, coherent processing and integration are performed only during the intervals T_p, and for pulsed signals, only during the interval corresponding to the pulse duration. The energy of each lth signal, included in the various T_p, as in optimum detection in a background of interference with known intensity, is summed noncoherently with post-detection integration.

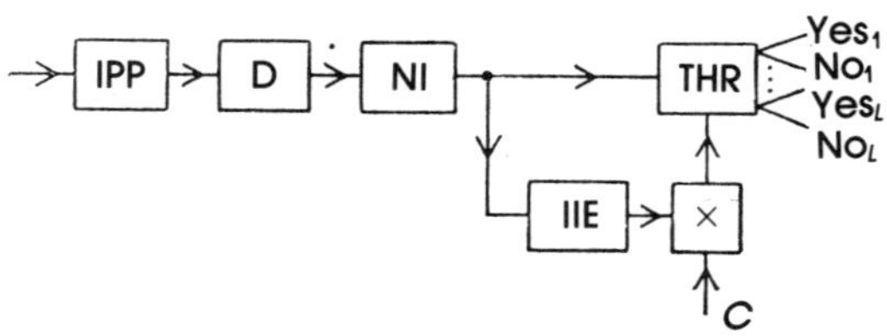

Fig. 7.7 Block diagram of a quasioptimum decision rule for multiple-target detection of rapidly fluctuating signals in interference with unknown intensity.

If the magnitude of the Doppler shift of the received signal $|\Delta f_D| \leqslant 1/T_p$, then the reference signals a_{li} for the intrapulse processing will differ from the transmitted signal only by the time delay. If $|\Delta f_D| > 1/T_p$, they will also differ in the frequency shifts. The discrete arrangement of the processing channels in τ and f is determined, as usual, by the width of the spectrum of the transmitted signal and the pulse duration.

7.4 DETECTING SIGNALS IN A BACKGROUND OF NONSTATIONARY INTERFERENCE FROM RADIO STATIONS

7.4.1 A Model of Radio Frequency Interference (Active Interference)

The main features which are characteristic of radio frequency interference are its temporal nonstationarity over intervals exceeding several seconds, and its practically uniform power distribution within the time intervals T_k or within range-velocity resolution elements (channels). Therefore, in contrast with the case of the models of cosmic and atmospheric interference, we will consider the intensity ϵ_k of this interference to be different in different intervals T_k, and unknown *a priori*. On the other hand, within each of the intervals T_k, the interference strength will be considered to be constant, i.e., $\mathbf{E}[|Z_{ijk}|^2] = \epsilon_k$, for all i and j. In particular, the intervals T_k may correspond to intervals of operation at various frequencies.

7.4.2 Optimal Detection of a Slowly Fluctuating Signal in a Background of Nonstationary Interference

The input data is split into independent groups of samples with different unknown parameters ν_k and ϵ_k. In this case, as opposed to that considered in Sec. 7.2, the detection problem entails the optimal combination of signal information from different sample groups [11]. Here, the transformation $V_k = BX_k$ is performed independently for each T_k, and the density distribution of the transformed kth sample V_k takes the form:

$$p(V_k|\gamma_k, \epsilon_k) = (\pi\epsilon)^{-nm}(1 + \gamma_k)^{-1} \times \exp\left[-\epsilon_k^{-1}\left(\sum_{i=2,j=2}^{n,m} |V_{ijk}|^2 - \frac{\gamma_k}{1 + \gamma_k}|V_{11k}|^2\right)\right]$$

Using this expression in (7.9) and integrating, we arrive at the minimax decision rule in the region $\Omega_1 = \cup_k \Omega_{1k}$ with Ω_{1k}: $\gamma_k \geqslant \gamma_0$:

$$\sum_{k=1}^{K} \ln(1 - \alpha Q_k)^{-1} \geq C \tag{7.19}$$

where

$$Q_k = \frac{|V_{11k}|^2}{\sum_{i,j}^{n,m} |V_{ijk}|^2} = \frac{\left| \sum_{j=1}^{m} b_j^* \sum_{i=1}^{n} a_i^* X_{ijk} \right|^2}{\sum_{i,j}^{n,m} |X_{ijk}|^2} \tag{7.20}$$

and $\alpha = \gamma_0/(1 + \gamma_0)$. Here γ_0 is the threshold value of the signal-to-interference ratios $\gamma_k = \nu_k/\epsilon_k$, limiting from below the subspace Ω_{1k}: $\gamma_k \geq \gamma_0$.

In accordance with (7.19), the structure of the optimum decision rule for detecting a slowly fluctuating signal takes the form shown in Fig. 7.8, and is a particular case of the rule for optimum combinations of the channels for random signal detection from Korado [11], with $\mathbf{M}_{pk} = N^{-1}I_N$, $\mathbf{M}_{sk} = gg^*$, and $g = a \otimes b$, where $\otimes$ denotes the Kronecker product. With $K = 1$, this rule is the complex version of the rule presented in Repin and Tartakovskiy [19].

As may be seen, in contrast with Fig. 7.2, the diagram in Fig. 7.8 contains a normalizing device (Norm) after the square-law detector D, where the result of the coherent signal processing $|V_k|^2 = |V_{11k}|^2$, obtained every interval T_k, is normalized by the estimate of the interference intensity in the same interval, i.e., divided by the statistic $t_k = \sum_{i,j}^{n,m} |X_{ijk}|^2$. The resulting statistics Q_k are invariant relative to scale transformations in the sample groups, and their distributions are independent of the interference intensity ϵ_k, for a fixed γ_k, both when the signal is present and when it is absent. This ensures a fixed probability of false alarm, independent of $k = 1, \ldots, K$. After normalization, the statistics Q_k undergo optimum combination through a nonlinear transformation in a nonlinear element (NE), which forms $f(Q_k) = \ln(1 - \alpha Q_k)^{-1}$, and noncoherent integration with subsequent comparison to the constant threshold. The value of the threshold C is wholly determined by the desired probability of false alarm. In accordance with the optimization criterion, this maximizes the minimum probability of detection of all the decision rules in the region Q_1, while maintaining the given probability of false alarm F_0. Here, the least favorable distribution of the unknown parameters (γ_k, ϵ_k) is equal to $\Pi P(\gamma_k)P(\epsilon_k)$, where $P(\gamma_k)$ is concentrated at the point $\gamma = \gamma_0$, and the least favorable distribution for ϵ_k is $d\epsilon_k/\epsilon_k$ i.e., the minimum probability of detection in Ω_k is at the point $\gamma_k = \gamma_0$, $k = 1, \ldots, K$. At this point, the minimax rule is the most powerful likelihood or invariant decision rule.

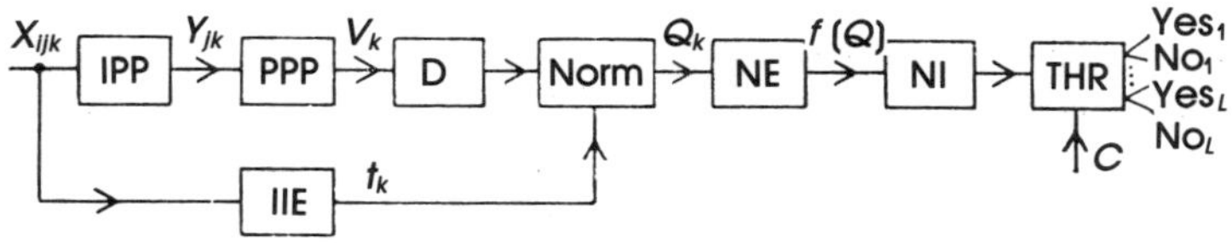

Fig. 7.8 Block diagram of the optimum decision rule for detecting a slowly fluctuating signal in nonstationary interference.

In the case of weak signals, $\gamma_0 \to 0$, $\alpha \to 0$, $f(Q) \to \alpha Q$, and the nonlinear element may be excluded. It is possible to show that such a decision rule is a minimax rule at $\gamma_{0k} = \gamma_k$ (see [17]). For strong signals, $\gamma_0 \to \infty$, $\alpha \to 1$, and $f(Q) \to \ln(1 - Q)^{-1}$. In this case, the rule (7.19) tends to the asymptotic limit minimax decision rule. For given values of D, the cases $\gamma_0 \to 0$ and $\gamma_0 \to \infty$ correspond to large and small numbers of samples mn. In this sense, the locally ($\gamma_0 \to 0$) optimum decision rule is also asymptotically ($mn\sqrt{K} \to \infty$) optimum.

We note that the maximum likelihood decision rule for this case differs from the decision rule of Fig. 7.8 only in the form of the nonlinear transformation. [See: Zakharov, S.I. and Korado, V.A. *Ob'edinenie nezavisimykh kanalov obnaruzheniya signala na fone pomekh s neizvestnymi intensivnostyami po kriteriyu maksimal'novo pravdopodobiya* (*Combining independent signal detection channels in interference with unknown strength, using the maximum likelihood criterion*). Radiotekhnika i Elektronika, (1982), no. 1, pp. 61–64.]

$$f(Q) = \rho - (mn - 1)\ln(1 - Q) - \ln Q \quad \text{for } Q > 1/mn$$
$$f(Q) = 0 \quad \text{for } Q \leq 1/mn,$$

where $\rho = -mn \ln(mn) - (mn + 1)\ln(mn - 1)$.

Setting the function $f(Q)$ in a Taylor series about the point $1/mn$, we obtain a simplified version of the function for the nonlinear element for the maximum likelihood decision rule:

$$f(Q) = \begin{cases} (Q - 1/mn)^2, & Q > 1/mn \\ 0, & Q \leq 1/mn \end{cases}$$

In analogy with the case considered in Sec. 7.2, it is not difficult to find that the optimum decision rule for m-ary detection in a background of nonstationary interference differs from the decision rule shown in Fig. 7.3 only in the individual interference intensity estimates ϵ_k in each T_k, the normalization by the estimates ϵ_k, and the nonlinear transformation prior to the noncoherent integration. The analogous decision rule (see Fig. 7.5) takes the form shown in Fig. 7.9.

In conclusion, it should be noted that this decision rule may be generalized to the case in which the intervals of quasistationary interference with $\epsilon_{ijk} = \epsilon_s$ last for several, say K', intervals T_k, and the overall observation interval T_0 includes several, say S, such larger intervals T_s. In this case, the decision rule is different in that between the nodes D and Norm, preliminary noncoherent integration NI_1 is performed at intervals T_s; the second noncoherent integration NI_2, performed after normalization and nonlinear transformation, combines the results of the signal processing in the intervals T_s. An optimum multiple target decision rule of this type is shown in Fig. 7.10.

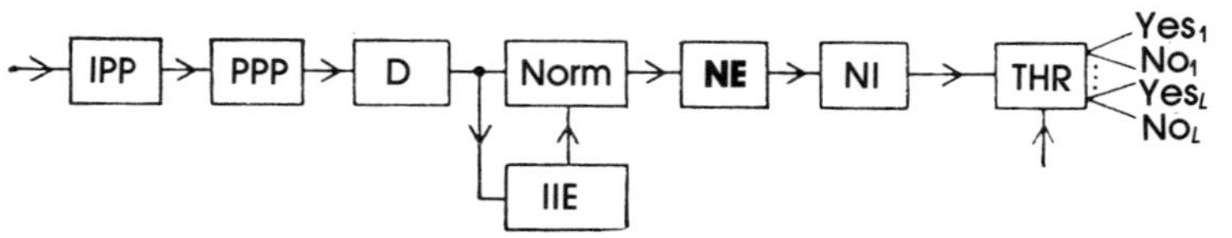

Fig. 7.9 Block diagram of a quasioptimum decision rule for multiple-target detection of slowly fluctuating signals in nonstationary interference.

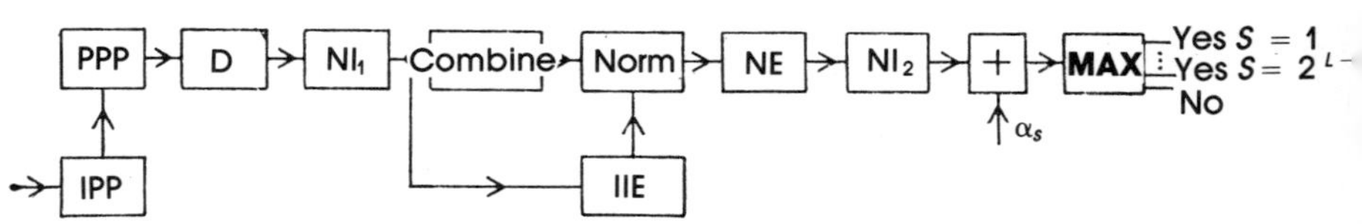

Fig. 7.10 Version of the optimum decision rule for multiple-target detection with noncoherent integration before and after normalization.

7.4.3 Optimum Detection of Rapidly Fluctuating Signals in a Background of Nonstationary Interference

Using the method of the previous section, it is not difficult to show that the optimum decision rule for detecting rapidly fluctuating signals differs from the optimum decision rule (7.17) in that the coherent integration of the signal processing results in the intervals T_k is replaced with noncoherent integration, and the statistics Q_k (see (7.20)), are replaced with the statistics:

$$Q_k = \frac{\sum_{i=1}^{m} \left| \sum_{i=1}^{n} a_i^* X_{ijk} \right|^2}{\sum_{i,j}^{n,m} |X_{ijk}|^2} \tag{7.21}$$

In accordance with (7.19) and (7.21), the structure of the optimum decision rule for detecting rapidly fluctuating signals takes the form shown in Fig. 7.11. For a rapidly fluctuating signal, the nonlinear transformation of the statistics Q_k for the maximum likelihood decision rule may be placed in the form:

$$f(Q_k) = \begin{cases} \rho - m(n-1)\ln(1-Q_k) - m\ln Q & \text{for } Q_k \geqslant 1/n \\ 0 & \text{for } Q_k < 1/n \end{cases} \tag{7.22}$$

where $\rho = m\ln m + m(n-1)\ln[m(n-1)] - mn\ln(mn)$.

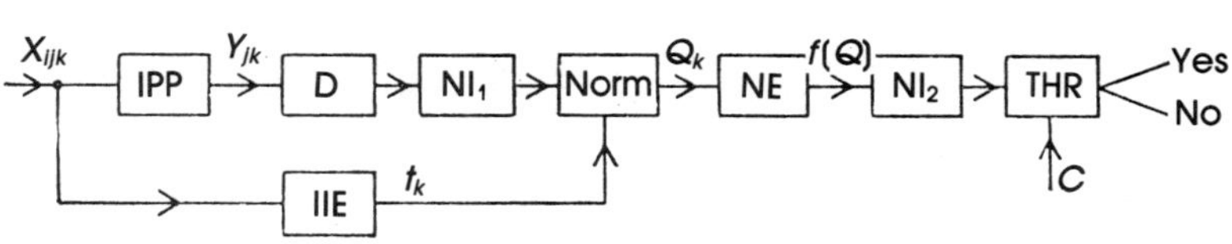

Fig. 7.11 Block diagram of the optimum decision rule for detecting rapidly fluctuating signals in nonstationary active interference.

When detecting rapidly fluctuating signals in a background of highly non-stationary interference, the optimum detection algorithm takes the form

$$\sum_{j,k} f(Q_{jk}) \geqslant C \tag{7.23}$$

where

$$Q_{jk} = \frac{\left| \sum_{i=1}^{n} a_i^* X_{ijk} \right|^2}{\sum_{i=1}^{n} |X_{ijk}|^2}$$

and

$$f(Q_{jk}) = \ln(1 - \alpha Q_k)^{-1} \quad \text{and} \quad \alpha = \gamma_0/(1 + \gamma_0)$$

The optimum decision rules for multiple target detection with rapidly fluctuating signals are designed analogously. In particular, the decision rule shown in Fig. 7.9 replaces the rule shown in Fig. 7.12.

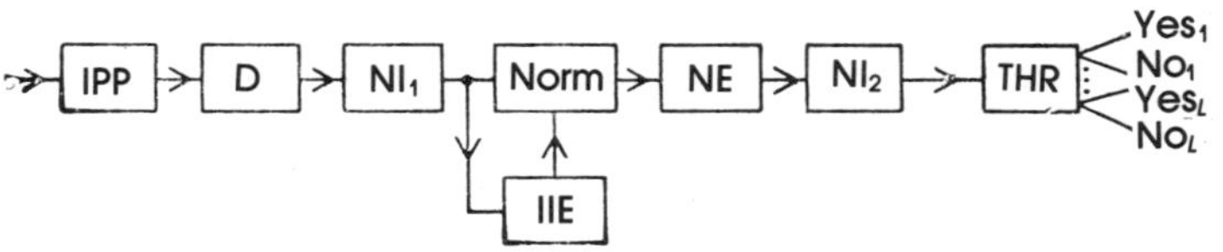

Fig. 7.12 Block diagram of the decision rule for multiple-target detection of rapidly fluctuating signals in nonstationary interference.

7.5 THE DETECTION OF A SIGNAL IN CLUTTER

7.5.1 The Clutter Model

The main difference between clutter (signals reflected from the earth) and wide-band active interference, which needs to be considered when deriving decision rules, is its significant pulse-to-pulse correlation. We will consider two cases: (*a*) the pulse-to-pulse correlation matrix for the clutter over its quasistationarity interval is known *a priori*, with accuracy to the scale factors ϵ_k, and (*b*) this matrix is unknown *a priori*. In both cases, we will assume that the *a priori* pulse-to-pulse correlation matrix for the clutter, at least over some interval within a period, is proportional to the unit matrix. When this matrix is proportional to a known matrix which is not the unit matrix, then the corresponding linear whitening transformation may be applied. With the help of the appropriate (not necessarily unitary) preliminary linear transformation, the first case may be transformed to the problem already examined in Sec. 7.4, and will not be covered in more detail here.

Considering that the clutter fluctuations from period to period represent a stationary random process (over each interval of its stationarity T_k), we will simplify the analysis of the second case (*b*) by assuming that the frequency components resulting from the pulse-to-pulse discrete weighted Fourier transform of the clutter at intervals T_k are mutually uncorrelated, and their powers ϵ_{jk} are unknown *a priori*. The same is assumed for the rapidly fluctuating useful signals. The slowly fluctuating signal is assumed to be nonzero for only one of the Fourier components.

More generally, the power spectrum of the pulse-to-pulse fluctuations may be nonzero in a group of $M' \ll mn$ neighboring Fourier components. In the simplest case, $M' = 1$ (corresponding to a signal which is fully coherent over the interval T_k). We note that the clutter model presented above also satisfactorily describes a mixture of clutter, active interference and fluctuating noise with unknown intensities.

7.5.2 Detection of a Slowly Fluctuating Signal

In accordance with the model adopted above, after a unitary intrapulse linear transformation B_1, and a pulse-to-pulse discrete Fourier transform B_2 on the samples from each T_k, the problem becomes one of detecting a random signal from the transformed discrete sample $\mathbf{V} = \{V_{ijk}\}$, where $\mathbf{E}[|V_{ijk}|^2] = \epsilon_{jk}$ in the absence of the signal, and $\mathbf{E}[|V_{1lk}|^2] = \epsilon_{lk} + \nu_{lk}$, $\nu_{lk} > 0$ when the signal is present, where l is the number of the Fourier component (frequency resolution element) containing the signal; ϵ_{jk} is the intensity of the clutter, unknown *a priori*; and ν_{lk} is the signal intensity, also unknown *a priori*.

The problem consists of checking the complex hypothesis H_0: $\gamma_{lk} = 0$, $l = 1, \ldots, m$, $k = 1, \ldots, K$, against the complex alternative H_1: $\gamma_{lk} \geqslant \gamma_0$ and $\gamma_{qk} = 0$, $q \neq l$, where $\gamma_{lk} = \nu_{lk}/\epsilon_{lk}$. This problem has a large number of unknown clutter parameters, and is invariant relative to the group G, equal to the direct product of the transformation groups G_{jk}, corresponding to independent scale transformations in each of the samples V_{jk}.

Integrating the probability density of the signal and clutter mixture, and just the clutter, over ϵ_{jk} relative to the least favorable invariant $\Pi_{j,k}(d\epsilon_{jk}/\epsilon_{jk})$ (see [11]), it is not difficult to find that the minimax decision rule for the transformed sample V has the form (7.19), where in place of (7.20) we have

$$Q_{kdl} = \frac{\left| \sum_{j=1}^{m} b_{lj}^* \sum_{i=1}^{n} a_{di}^* X_{ijk} \right|^2}{\sum_{l}^{n} \left| \sum_{j=1}^{m} b_{lj}^* X_{ijk} \right|^2} \tag{7.24}$$

Here a_{di} and b_{lj} are elements of the weighted intrapulse processing and pulse-to-pulse processing vectors.

In accordance with (7.19) and (7.24), the structure of the rule for detecting a slowly fluctuating signal is given in Fig. 7.13. In the case

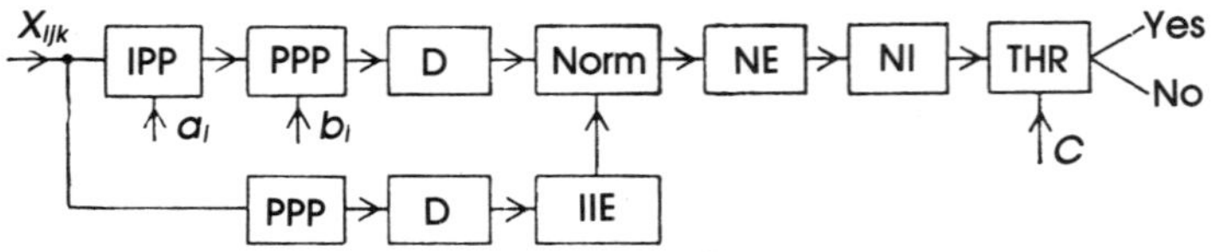

Fig. 7.13 Block diagram of the optimum decision rule for detecting a slowly fluctuating signal in clutter with unknown fluctuating spectrum.

$k = n = 1$, this rule differs from the rule (12.8.15) in Repin and Tartakovskiy [19] only in the form of the nonlinear transformation.

As opposed to Fig. 7.8, in the decision rule shown in Fig. 7.13, the interference (clutter) intensity estimate used for normalization is formed after pulse-to-pulse processing, individually for that frequency resolution element which might contain the useful signal. Due to this normalization, the probability of false alarm for the receiver of Fig. 7.13 does not depend on ϵ_{jk}, i.e., does not depend on the change in the intensity and correlation properties of the clutter over the intervals T_k.

In the case of multiple target detection, the decision rule is analogous to that of Fig. 7.9, and is shown in Fig. 7.14. Here, as in Fig. 7.9, there is a version in which the interference intensity is estimated after the coherent processing. This corresponds to Q_k of the form:

$$Q_{kdl} = \frac{\left| \sum_{j=1}^{m} b_{lj}^* \sum_{i=1}^{n} a_{di}^* X_{ijk} \right|^2}{\sum_{d}^{n'} \left| \sum_{j=1}^{m} b_{lj}^* \sum_{i=1}^{n} a_{di}^* X_{ijk} \right|^2} \tag{7.25}$$

where $\mathbf{b}_1 = \{b_{l,j}\}$ and $\mathbf{a}_d = \{a_{di}\}$ are the weighted vectors for the discrete weighted pulse-to-pulse Fourier transform and intrapulse processing respectively; for example, the dth time resolution element and the lth pulse-to-pulse frequency processing resolution element. Here, unlike the decision rule shown in Fig. 7.9, the interference intensity estimate is calculated independently for each l, from those n' intrapulse resolution elements for which the ϵ_{lk} are identical. The decision rule for optimum multiple target detection with slowly fluctuating signals is designed analogously, as is the version of the optimum decision rule of Fig. 7.10 for the case, when the interval over which the interference is quasistationary exceeds the interval during which the signal is coherent. We note that in light of the basic equality:

$$\sum_{j=1}^{m} b_{ij}^{*} \sum_{i=1}^{n} a_{di} X_{ijk} = \sum_{i=1}^{n} a_{di}^{*} \sum_{j=1}^{m} b_{lj}^{*} X_{ijk}$$

the sequence of intrapulse and pulse-to-pulse processing may be reversed in the decision rules, without changing the results of the processing.

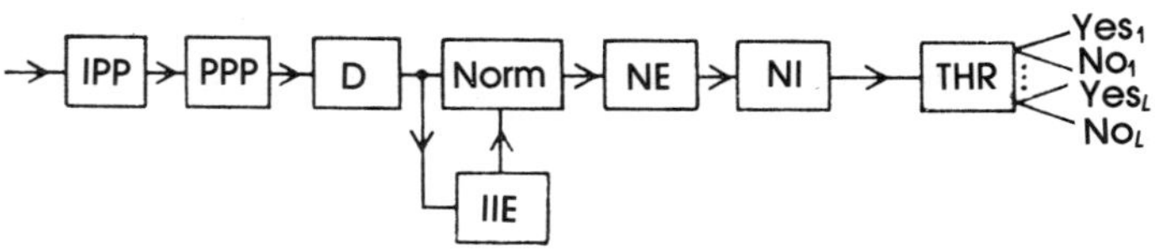

Fig. 7.14 Block diagram of a quasioptimum decision rule for multiple-target detection of a slowly fluctuating signal in clutter.

7.5.3 Detecting a Rapidly Fluctuating Signal

In this case, the amplitude and initial phase of the signal reflected from the target, in different repetition periods, take random and independent values, and when the signal is present in V_{djk}, $\mathbf{E}[|V_{djk}|^2 = \epsilon_{jk} + \nu_{jk}$, where $\nu_{jk} \geq 0$ for all j and k. Thus, after the discrete Fourier transform, the signal and interference may be present simultaneously in each pulse-to-pulse frequency resolution element, i.e., for each value of j. Here, it is sufficient to consider those frequencies within the limits of the repetition frequency of the transmitted signal. As before, ϵ_{jk} and ν_{jk} are unknown *a priori,* and may be different for every j and k, and the problem becomes that of checking the composite hypothesis $H_0{:}\gamma_{jk} = 0$, $j = 1, \ldots, m$, against the composite alternative $H_1{:}\gamma_{jk} \geq \gamma_0$, where $\gamma_{jk} = \nu_{jk}/\epsilon_{jk}$. The problem, as before, is invariant relative to scale transformations in the samples T_{jk}.

After integration of the probability density V over ϵ_{jk} with the least favorable *a priori* distribution $\Pi_{j,k}(d\epsilon_{jk}/\epsilon_{jk})$, the likelihood ratio of the maximum invariant for detection of the signal in the dth resolution element leads to the minimax decision rule of the form:

$$\sum_{l,k}^{m,K} \ln(1 - \alpha Q_{kdl})^{-1} \geq C \tag{7.26}$$

where $\alpha = \gamma_0/(1 + \gamma_0)$ and

$$\frac{\left|\sum_{j=1}^{m} b_{lj}^{*} \sum_{i=1}^{n} a_{di}^{*} X_{ijk}\right|^{2}}{\sum_{i}^{n} \left|\sum_{j=1}^{m} b_{lj}^{*} X_{ijk}\right|^{2}} \tag{7.27}$$

The structure of the corresponding optimum decision rule for the detection of a rapidly fluctuating signal is shown in Fig. 7.15. Unlike the case in Fig. 7.13, after the nonlinear transformation (NE) in Fig. 7.15, in the block labeled "IAF", there is noncoherent addition of the frequency resolution elements, resulting from the pulse-to-pulse processing (on the index l). In this case, Doppler frequency resolution is possible only at the intrapulse processing (IPP) block. Normalizing the signal by the interference power estimate in each frequency resolution element has two effects. It ensures that the probability of false alarm will be independent of the actual form of the pulse-to-pulse clutter fluctuation spectrum. Simultaneously, it maintains effective noncoherent addition of the signal information in the various frequency elements, which contain unknown and widely varying clutter powers. This effects automatic rejection of those frequency channels in which the clutter is most intense, and in particular, rejection of clutter spectrum spikes corresponding to reflections from fixed or slowly moving water or land surfaces. Such clutter peaks may be displaced in frequency due to changes in the height of the reflecting layers of the ionosphere, and also reflections from large natural irregularities moving in the ionosphere (meteor trails, for instance); these will not reduce the rejection of clutter in the processor described above, as they would in a moving target indicator. The use of a nonlinear element, analogous to (7.22), corresponding to the maximum likelihood form, makes it possible to exclude from the processing those frequency channels in which the signal-to-interference ratio γ_{jk} is too small or even equal to zero, which results when the target signal is too weak, or the clutter too strong, in the corresponding frequency components. It should be noted that the same condition is encountered when detecting signals in a background of active interference.

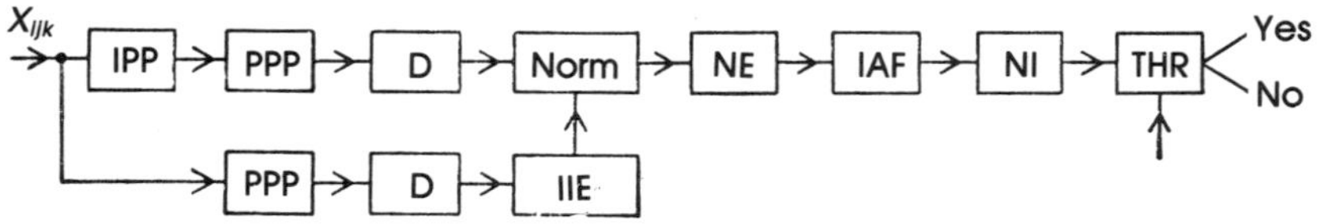

Fig. 7.15 Block diagram of the optimum decision rule for detecting a rapidly fluctuating signal in clutter with unknown fluctuation spectrum.

In the case of multiple target detection, the structure of Fig. 7.14 becomes that shown in Fig. 7.16. In this case:

$$Q_{kdl} = \frac{\left| \sum_{j=1}^{m} b_{lj}^* \sum_{i=1}^{n} a_{dl}^* X_{ijk} \right|^2}{\sum_{d}^{n} \left| \sum_{j=1}^{m} b_{lj}^* \sum_{i=1}^{n} a_{di}^* X_{ijk} \right|^2}$$

where b_{ij} and a_{di} represent the same quantities. The optimum decision rule for multiple target detection with rapidly fluctuating signals is designed analogously, and takes the form shown in Fig. 7.17.

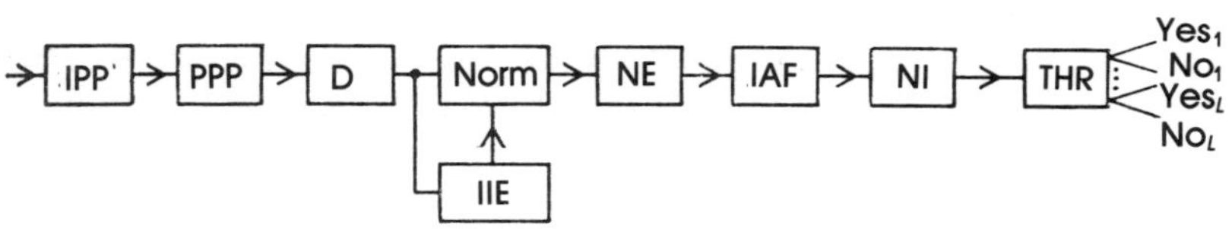

Fig. 7.16 Block diagram of the decision rule for multiple-target detection of a rapidly fluctuating signal in clutter.

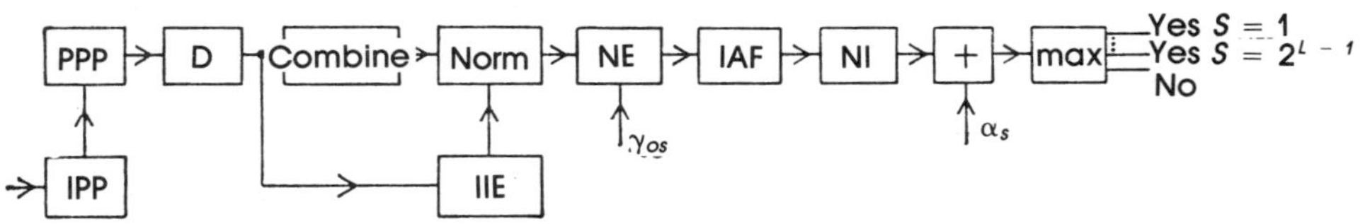

Fig. 7.17 Block diagram of the optimum decision rule for multiple-target detection of a rapidly fluctuating signal in clutter.

7.6 SIGNAL PARAMETER ESTIMATION

7.6.1 Features of The Signal Model

After detecting a target, the next most important problem for any radar is to measure the target angles, a process which may occur simultaneously with detection. From a mathematical point of view, this problem is one of estimating the parameters of a signal in a background of interference, the latter being associated with some family of probability distributions. As with detection, it is convenient to divide the angle measurement process into two stages: preliminary estimation of the target angles (prior to trajectory estimation), and secondary estimation of the parameters of the target trajectory. We will be considering only the first case, keeping in mind that the final estimation will be performed in the second stage, and consists of fitting the primary estimates along some trajectory, and determining the target trajectory parameters.

We will first examine the features of the model of a point target signal and interference, from the point of view of parameter estimation. The following signal parameters are estimated in an over-the-horizon radar: the time delay of the received signal relative to the transmission time, which depends on the target range; the Doppler shift within the band limited by the pulse repetition frequency f_p (when the width of the fluctuation spectrum is much less than f_p), or in a band much greater than f_p (for the case of a wideband rapidly fluctuating signal).

Using a multichannel version of the signal and interference model, we place the input vector of complex sample amplitudes over the observation interval T_0 corresponding to the duration of preliminary (pretrack) processing (see Fig. 7.1), in the form $\mathbf{X} = \{X_{ijk}\} = \mathbf{S} + \mathbf{V}$, where $\mathbf{S} = \{S_{ijk}\}$ is the input vector of the useful signal samples at the output of the receiver's IF amplifier with elements of the reference vectors $\mathbf{S}_{rijk}$; $\mathbf{Z} = \{Z_{ijk}\}$ is the analogous sample of additive Gaussian interference; i is the index of the sample within the repetition period T_{jk} (see Fig. 7.1); j is the index of the repetition period within the larger interval; k is the index of the interval T_k within the observation period T_0; and r is the number of the receiver channel. The signal may then be described over the observation interval T_0 as

$$S_{rijk} = U_{jk}a_i(\tau, \omega)b_f(\omega')\alpha_r(\beta)$$

where U_{jk} are complex Gaussian random quantities with $\mathbf{E}[U_{jk}] = 0$ and $\mathbf{E}[U_{jk}|^2] = \nu_{jk}$, determining the signal strength over the intervals T_{jk}; $\mathbf{a}(\tau, \omega) = \{a_i(\tau, \omega)\}$ is the column vector determining the intrapulse signal modulation over the intervals T_{jk} and characterizing the time position or delay of the signal (in the case of a rapidly fluctuating signal, these coefficients may also characterize the Doppler shift, when this shift is large); $\mathbf{b}(\omega') = \{b_j(\omega')\}$ is the column vector, determining the pulse-to-pulse signal modulation, and characterizing the Doppler shift for slowly fluctuating signals; $\overline{\alpha}(\beta) = \{\alpha_r(\beta)\}$ is the column vector, determining the spatial amplitude phase modulation and characterizing the angle of arrival of the signal in azimuth; r is the receiver channel number. Some additional properties of the signals and interference will be examined below, in the design of specific algorithms.

7.6.2 Criteria for the Optimum Estimation of Signal Parameters in Interference With Incompletely Known Statistical Characteristics

Let the p-channel input sample vector $\mathbf{X} = \{X_{ijk}\}$ correspond to the observation interval T_0. Each sample may be viewed as a point in a $2pNK$-dimensional Euclidean space E_{2pNK}, i.e., $\mathbf{X} \in E_{2pNK}$. Here, in each input channel r, the scalar sample $\mathbf{X}_r = \{X_{rijk}\}$ belongs to a $2NK$-dimensional real Euclidean space E_{2NK}. Also, let $p(\mathbf{X}|\theta)$ be the distribution for $\mathbf{X}$, known to within some determined finite set of unknown parameters $\theta \in \Omega$. The set of parameters θ may include, as before, various signal parameters including its energy, and interference parameters which are not known *a priori*.

From the point of view of estimating the received signal parameters, the unknown parameters are divided into the useful θ_u, and the interfering θ_i, i.e. $\theta = (\theta_u, \theta_i)$. The interference parameters belong to the subset of interfering parameters. The statistical problem of signal parameter estimation consists of finding some statistic $\hat{\theta}_u = f(\mathbf{X})$ (which is a vector, in general), the value of which is close to the desired parameter. The quality of the estimate is naturally characterized by the closeness of the value $\hat{\theta}_u$ to the actual parameter value $\hat{\theta}_u$, that is, by the accuracy of the estimate. To determine the quality of the estimate, various error measures may be used, or the loss functions K, of which the most commonly used is the square error function of the form:

$$K = K(\hat{\theta}_u, \theta_u) = (\hat{\theta}_u - \theta_u)^* B(\hat{\theta}_u - \theta_u) \tag{7.28}$$

where B is some positive definite matrix. The loss function, with a minus sign, is sometimes called the utility function. To obtain the characteristics of the quality of some estimate $f(\mathbf{X})$, it is necessary, setting $\hat{\theta}_u = f(\mathbf{X})$, to average the loss K relative to the probability density for each of the unknown parameters $\theta \in \Omega$. These average losses are called risk functions, $R[\theta, f(\mathbf{X})] = E_\theta K(\theta_i, f(\mathbf{X})]$.

One of the possible approaches to parameter estimation when there is *a priori* uncertainty, as in the detection problem, is the minimax approach. The minimax criterion for estimation consists of selecting that estimate f_0 from the set of possible estimates $f \in \mathcal{F}$, such that

$$\min_{f \in \mathcal{F}} \max_{\theta \in \Omega} R(\theta, f) = \min_{\theta \in \Omega} R(\theta, f_0) \tag{7.29}$$

Thus, the minimax approach to the estimation problem amounts to choosing that estimate which minimizes the greatest risk (or largest mean square error) as the parameter $\theta \in \Omega$ varies over its full range.

Another approach, which is particularly widely used when there is *a priori* uncertainty, is the maximum likelihood principle, by which the estimate of the required parameter θ_u is taken to be those values which maximize the likelihood function $L(\theta, \mathbf{X})$, which is a nonnegative real function of θ and $\mathbf{X}$, proportional to the density distribution $p(\mathbf{X}|\theta)$, i.e., $L(\theta, \mathbf{X}) = kp(\mathbf{X}|\theta)$. In other words, the maximum likelihood estimate is that value of θ which makes the observed value $\mathbf{X}$ most likely. The estimate of the useful signal parameter $\hat{\theta}_u = f(\mathbf{X})$ corresponds to the value of θ which maximizes $L(\theta, \mathbf{X})$, i.e., $\max L(\theta, \mathbf{X}) = L(\hat{\theta},\mathbf{X})$, $\hat{\theta} = (\hat{\theta}_u, \hat{\theta}_i)$, where $\hat{\theta}_i$ is the estimate of the interfering parameter. It is known that the maximum likelihood estimate is asymptotically (with unlimited growth of the size of the homogeneous independent sample) efficient and unbiased.

With a finite discrete sample, maximum likelihood estimates usually turn out to be quite good, and in many cases these estimates coincide with minimax estimates [19]. The Cramer-Rao lower bound, which sets a lower limit on the error variance for any estimate, is usually used when analyzing the efficiency of maximum likelihood estimates [12].

7.6.3 Measuring Signal Parameters With Space-Time Processing Systems

The maximum likelihood criteria will be used in the ensuing discussion of the design of signal parameter estimation algorithms. If the estimated parameters take continuous values in some interval, then when forming the function proportional to $L(\theta, \mathbf{X})$ and finding its maximum over θ, corresponding to the estimate of the parameter θ_u, a large number (theoretically, infinitely large) number of calculations of $L(\theta, \mathbf{X})$ is required to cover the set of various values of θ. The technical realization of such a device is therefore quite complex. To ease the requirements on the receiver without incurring an unacceptable loss in accuracy, various methods are used, based on the use of multichannel receivers with linear channels, designed only for some "reference" values of the parameters [15]. With the appropriate choice of these reference values, the output statistics of these channels will in practice be sufficient statistics for calculating the likelihood function, and estimating the unknown parameter θ. It is possible to use these channels directly to obtain a coarse estimate of θ, on the basis of the channel with the greatest output level, or to obtain a more accurate estimate using various methods, one of which is the maximum likelihood method.

We will consider a set of receiver channels matched to the reception of signals, based on a particular selection of values for some parameter θ. Each channel is characterized by the so-called channel discrimination function $\lambda(\theta)$, i.e., the dependence of the amplitude of the useful signal at the channel output on the deviation of the parameter from the corresponding value. As is well known, in order to refine the estimate of any fixed value of a parameter, at least two processing channels are necessary, tuned to reference values close to the value of the parameter. It is possible to use so-called "sum" and "difference" channels as the two channels, for which the signal in the sum channel is proportional to $\lambda(\theta)$, and that in the difference channel is proportional to $-\partial\lambda/\partial\theta$. To obtain a refined estimate of parameters over a wide range, it is possible to use several pairs of such channels, tuned to separate values of the parameters.

7.6.4 Estimating the Parameters of Signals in Stationary Interference

We will examine the estimation of parameters with the example of estimating the angle of arrival β of signals in isotropic interference. To reduce the number p of channels to two, we perform a linear transformation on the $p \times N_K$ matrix $\mathbf{X}$ with the $2 \times p$ matrix A_0, the rows of which are orthonormal and proportional to $\alpha_0 = \{\alpha_r^*(\beta_0)\}$ and $\dot{\alpha}_0 = \{(\partial\alpha_r^*/\partial\beta)|_{\beta 0}\}$ respectively (β_0 is the reference direction). Then, with small deviations from the reference direction, i.e., $|\beta - \beta_0| \ll 1$, we have the following linear model of the dependence of the input data on β. It is assumed that after the transformation A_0, the interference in the channels will be independent:

$$\mathbf{x} = \mathbf{s} + \mathbf{z}, \quad \mathbf{y} = k_0(\beta - \beta_0)\mathbf{s} + \mathbf{w}, \quad k_0 = (\dot{\alpha}_0^* \dot{\alpha}_0)^{1/2}$$

where $\mathbf{x}$, $\mathbf{y}$ are the samples in the two channels, obtained as a result of the transformation $(\mathbf{x}/\mathbf{y}) = A_0\mathbf{X}$; $\mathbf{s} = \{s_{ijk}\}$, $s_{ijk} = U_{jk}a_ib_j$; U_{jk} are independent Gaussian complex random quantities; $E[U_{jk}] = 0$; $E[|U_{jk}|^2] = \nu$; k_0 is a proportionality coefficient, characterizing the actual antenna system; $\mathbf{a} = \{a_i\}$ is the form of the intrapulse signal modulation; $\mathbf{b} = \{b_f\}$ is the form of the pulse-to-pulse signal modulation; ν is the signal power over the interval T_0, unknown *a priori*. For slowly fluctuating signals, $U_{jk} = U_k$ and $s_{ijk} = U_ka_ib_j$, and for rapidly fluctuating signals, $b_f = 1$ and $s_{ijk} = U_{jk}a_i$. The interference samples $\mathbf{z} = \{z_{ijk}\}$ and $\mathbf{w} = \{w_{ijk}\}$ are independent Gaussian complex random quantities, with $E[z_{ijk}] = E[w_{ijk}] = 0$, and $E[|z_{ijk}|^2] = E[|w_{ijk}|^2] = \epsilon$, where ϵ is the interference intensity, unknown *a priori*. The sufficiency of the statistics $\mathbf{x}$, $\mathbf{y}$ for obtaining a maximum likelihood estimate of the angle β will be shown below.

7.6.5 Estimating the Parameters of a Slowly Fluctuating Signal

As in the detection problem, we introduce the preliminary orthogonal transformation of the input samples in each channel $B = B_2B_1$, where B_1 is the matrix of simultaneous intrapulse unitary transformations, such that the first row of B_1 is proportional to $\mathbf{a}^*$; B_2 is the matrix of simultaneous pulse-to-pulse unitary transformations, such that the first row is proportional to $\mathbf{b}^*$; * denotes transposition and complex conjugation. Then B_1 takes the signal to the first sample values of each T_{jk}, and B_2 takes the signal from these values to the first value of each T_k.

As a result of this transformation, the signal correlation matrix in each channel will have a diagonal structure with K nonzero elements on the main diagonal, equal to $\nu\mathbf{a}^*\mathbf{a}\mathbf{b}^*\mathbf{b} = \nu$ in the $\mathbf{x}$ channel, and $k_0^2(\beta - \beta_0)^2\nu$ in the $\mathbf{y}$ channel. The signals in these two channels remain fully correlated. If the input samples $\mathbf{V} = B\mathbf{x}^\tau = \{V_{ijk}\}$ and $\phi = B\mathbf{y}^\tau = \{\phi_{ijk}\}$ are jointly Gaussian, and the complex signal amplitude has a Gaussian distribution $p(U_k|\nu) = (\pi\nu)^{-1}\exp(-|U_k|^2/\nu)$, the likelihood function for the two-channel estimation problem will take the form:

$$p(\mathbf{V}, \Phi|\beta, \nu, \epsilon) = h(\mathbf{V}, \Phi, \epsilon)\{\pi^2[\epsilon + \nu(1 + \delta^2)]\}^k \times \exp\Bigg\{-\epsilon^{-1}[\epsilon + \nu(1 + \delta^2)]^{-1} \times \Bigg[(\epsilon + \nu\delta^2)\sum_{k=1}^{K} V_{11k}^*V_{11k} + 2\nu\delta\,\mathrm{Re}\left(\sum_{k=1}^{K} V_{11k}^*\Phi_{11k}\right) + (\epsilon + \nu)\sum_{k=1}^{K}\Phi_{11k}^*\Phi_{11k}\Bigg]\Bigg\}$$

where $\delta = k_0(\beta - \beta_0)$, and $h(\mathbf{V}, \Phi, \epsilon)$ does not depend on ν or β.

The equation for the maximum likelihood estimate of β is obtained by equating the partial derivatives of the likelihood function with respect to ν and β to zero, for an arbitrary fixed ϵ, and reducing the system from two equations to a quadratic equation, which may be written in the form analogous to that in Moska [22]:

$$\delta^2\,\mathrm{Re}\left(\sum_{k=1}^{K} V_{11k}^*\Phi_{11k}\right) - \delta\sum_{k=1}^{K}\Phi_{11k}^*\Phi_{11k} + \delta\sum_{k=1}^{K} V_{11k}^*V_{11k} - \mathrm{Re}\left(\sum_{k=1}^{K} V_{11k}^*\Phi_{11k}\right) = 0$$

We then obtain the estimate from the transformed sample:

$$\hat{\beta} - \beta_0 = \frac{\sum_{k=1}^{K} (\Phi_{11k}^* \Phi_{11k} - V_{11k}^* V_{11k})}{k_0 2 \operatorname{Re}\left(\sum_{k=1}^{K} V_{11k}^* \Phi_{11k}\right)} + \frac{\left\{\left[\sum_{k=1}^{K} (\Phi_{11k}^* \Phi_{11k} - V_{11k}^* V_{11k})\right]^2 4 \operatorname{Re}^2\left(\sum_{k=1}^{K} V_{11k}^* \Phi_{11k}\right)\right\}^{1/2}}{k_0\, 2 \operatorname{Re}\left(\sum_{k=1}^{K} V_{11k}^* \Phi_{11k}\right)} \tag{7.30}$$

In this form, however, the estimate is both inconvenient in terms of practical applicability, and rather complex for analysis. Therefore, we take the asymptotic limit $\gamma \gg 1$ (where $\gamma = \nu/\epsilon$ is the signal-to-interference ratio). From (7.30) we obtain

$$\hat{\beta} - \beta_0 = \frac{\operatorname{Re}\left(\sum_{k=1}^{K} V_{11k}^* \Phi_{11k}\right)}{k_0 \sum_{k=1}^{K} V_{11k}^* V_{11k}} \tag{7.31}$$

With $K = 1$, (7.31) corresponds to the estimate (15) from Moska [22]. Here, for $\delta \ll 1$ and as γ tends to infinity asymptotically, we use

$$\Phi_{11k} \simeq \delta V_{11k}$$

As may be seen, the estimate $\hat{\beta}$ is invariant relative to the group of scale transformations on the input sample, and for fixed γ does not depend on ϵ. Consequently, $\hat{\beta}$ is the maximum likelihood estimate for the p-channel problem. We note that when the interference in **x** and **y** are correlated and the correlation matrix of **x** and **y** is known, the analysis may be reduced to that examined above.

If we go from the transformed sample to **x** and **y**, then the algorithm for estimating β for a slowly fluctuating signal may be expressed as follows:

$$\hat{\beta} - \beta_0 = \frac{\operatorname{Re}\left[\sum_{k=1}^{K}\left(\sum_{j=1}^{m} b_j^* \sum_{i=1}^{n} a_i^* x_{ijk}\right)^* \left(\sum_{j=1}^{m} b_j^* \sum_{i=1}^{n} a_i^* y_{ijk}\right)\right]}{k_0 \sum_{k=1}^{K} \left|\sum_{j=1}^{m} b_j^* \sum_{i=1}^{n} a_i^* x_{ijk}\right|^2}$$

Such an estimate corresponds to one signal with known $\{a_i\}$ and $\{b_j\}$. When there are L signals with known $\{a_{li}\}$ and $\{b_{lj}\}$, it is possible to estimate β for each lth channel in accordance with the formula:

$$(\hat{\beta} - \beta_0)_l = \frac{\operatorname{Re}\left(\sum_{k=1}^{K} V_{1lk}^* V_{2lk}\right)}{k_0 \sum_{k=1}^{K} V_{1lk}^* V_{1lk}}$$

where $V_{1lk} \equiv V_{l11k}$ and $V_{2lk} \equiv \phi_{l11k}$. When $\mathbf{a}_l$ and $\mathbf{b}_l$ are orthogonal, this estimate remains the asymptotic $(\gamma \to \infty)$ maximum likelihood estimate.

We note that the accuracy of this estimate may be increased if the linear model is replaced with a nonlinear model with $\delta = (\beta - \beta_0)$, where f is a known monotonic function. The estimate will then be $\beta - \beta_0 = f^{-1}(\delta)$.

With one signal and unknown $\{a_i\}$ and $\{b_j\}$, it is possible to apply processing which is analogous to that used for multiple alternative detection, forming L_1 channels in $\mathbf{b}$ and L_2 channels for $\mathbf{a}$. Here, β is estimated for b_l and a_d, corresponding to the channels (l_0, d_0) in which the maximum output signal is found. The set of channels in τ and ω', formed for intrapulse and pulse-to-pulse processing, may also be used for preliminary range and velocity estimation. The number of these channels is chosen on the basis of the necessary accuracy and resolution. [See Tsar'kov, N.M. *Mnogokanal'nie radiolokatsionnie izmeriteli (Multichannel radar)*. Moscow: Sovietskoe Radio, 1980. 192 pp.] The block diagram of the processing described above, derived directly from (7.31), is shown in Fig. 7.18, in which we use the notation:

$$X = V_{1k} = \sum_{j=1}^{m} b_j^* \sum_{i=1}^{n} a_i^* x_{ijk}, \quad Y = V_{2k} = \sum_{j=1}^{m} b_j^* \sum_{i=1}^{n} a_i^* y_{ijk}$$

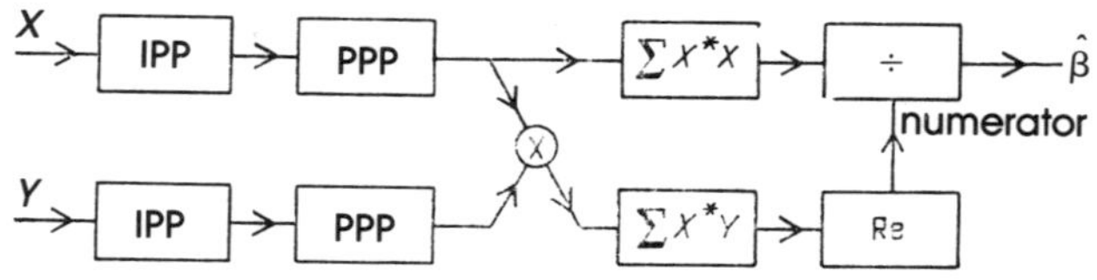

Fig. 7.18 Block diagram of an algorithm for calculating the maximum likelihood estimate of the angle of arrival β of a slowly fluctuating signal in stationary interference with unknown strength.

When it is necessary to refine the estimates of τ and ω', it is possible to use algorithms analogous to that for estimating the angle of arrival β. Expressions for the estimates of τ and ω, analogous to (7.31) for the case of L signals, are written as follows:

$$(\hat{\tau} - \tau_0)_l = \frac{\mathrm{Re}\left[\sum_{k=1}^{K}\left(\sum_{j=1}^{m} b_{jl}^{*} \sum_{i=1}^{n} a_{il}^{*} x_{ijk}\right)^{*}\left(\sum_{j=1}^{m} b_{jl}^{*} \sum_{i=1}^{n} \dot{a}_{il}^{*} x_{ijk}\right)\right]}{k_{\tau 0} \sum_{k=1}^{K}\left|\sum_{j=1}^{m} b_{jl}^{*} \sum_{i=1}^{n} a_{il}^{*} x_{ijk}\right|^{2}}$$

$$(\hat{\omega} - \omega_0)_l = \frac{\mathrm{Re}\left[\sum_{k=1}^{K}\left(\sum_{j=1}^{m} b_{jl}^{*} \sum_{i=1}^{n} a_{il}^{*} x_{ijk}\right)^{*}\left(\sum_{j=1}^{m} \dot{b}_{jl}^{*} \sum_{i=1}^{n} a^{*}\; x_{ijk}\right)\right]}{k_{\omega 0} \sum_{k=1}^{K}\left|\sum_{j=1}^{m} b_{jl} \sum_{i=1}^{n} a_{il}^{*} x_{ijk}\right|^{2}}$$

where $\dot{a}_i = [\partial a_i(\tau)/\partial\tau]|\tau_0$; τ_0 is the reference delay; $b_j = [\partial b_j(\omega')/\partial\omega']|\omega_0'$; ω_0' is the reference frequency.

7.6.6 Estimating the Parameters of a Rapidly Fluctuating Signal

In this case, just as in the detection problem, the difference between rapidly and slowly fluctuating signals is represented by replacing V_{11k} with

$$\sum_{j=1}^{m} V_{1jk}$$

and the estimate of β takes the corresponding form:

$$(\hat{\beta} - \beta_0) = \frac{\mathrm{Re}\left[\sum_{k,j}^{K,m}\left(\sum_{i=1}^{n} a_i^{*} x_{ijk}\right)^{*}\left(\sum_{i=1}^{n} a_i y_{ijk}\right)\right]}{k_0 \sum_{k,j}^{K,m}\left|\sum_{i=1}^{n} a_i^{*} x_{ijk}\right|^{2}} \qquad (7.32)$$

The corresponding block diagram is shown in Fig. 7.19, where

$$X = V_{1jk} = \sum_{i=1}^{n} a_i^{*} x_{ijk}, \quad Y = V_{2jk} = \sum_{i=1}^{n} a_i^{*} y_{ijk}$$

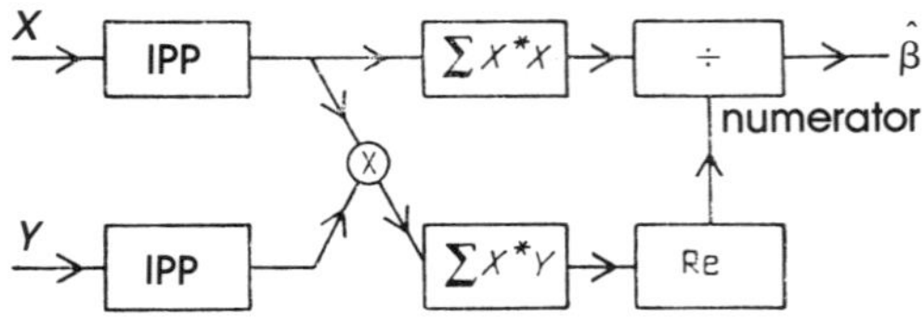

Fig. 7.19 Block diagram of an algorithm for estimating the angle of arrival β for a rapidly fluctuating signal.

7.6.7 Estimating the Parameters of a Signal in a Background of Nonstationary Active Interference

As before, we will assume that the interference intensities ϵ_k in different intervals T_k are different and unknown *a priori*, i.e., $E[|Z_{ijk}|^2] = E[|w_{ijk}|^2] = \epsilon$ for all i and j.

We will split the input sample into a set of independent subsamples with different unknown parameters ν_k and ϵ_k (for a slowly fluctuating signal). Considering the interference within each such sample to be stationary, it is possible to obtain K estimates, analogous to (7.31). There then arises the problem of combining these estimates optimally, such that the estimate of the parameter in the complete sample is the most accurate. This estimate may be more accurate than an individual estimate, due to the use of all the information in the sample. The combination of partial estimates is optimal according to some criterion if the estimate of the unknown parameter is optimal by that criterion.

The optimal combination of the estimates $\hat{\beta}_k$ of the form (7.31), using the maximum likelihood criterion, presents difficulties associated with the complexity of the probability density of the estimates $\hat{\beta}_k$, and especially of their joint distribution. Therefore, we will use a quasioptimal combination method, obtaining an estimate which minimizes the mean square error in the class of linear estimates [15], for the set of unbiased partial estimates $\hat{\beta}_{ku}$. In this case:

$$\hat{\beta} = \sum_{k=1}^{K} \alpha_k \hat{\beta}_{ku}, \quad \sum_{k=1}^{K} \alpha_k = 1$$

where $\hat{\beta}_{ku}$ is the angle estimate for an interval of quasistationary interference.

When the estimates are uncorrelated,

$$\alpha_k = D_k^{-1}\left(\sum_{k=1}^{K} D_k^{-1}\right)^{-1}$$

where D_k is the variance of the kth estimate. For untransformed samples of a slowly fluctuating signal with unknown γ_k (see [22]):

$$D_k = 0.5(1 + \gamma_k)^{-1}, \quad E(\hat{\beta}_k - \beta_0) = (\beta - \beta_0)\gamma_k(1 + \gamma_k)^{-1}$$

$$(\hat{\beta} - \beta_0)_{d_0,l_0} = \frac{1}{k_0 \sum_{k=1}^{K} (1 + \gamma_k)} \sum_{k=1}^{K} \frac{(\gamma_k + 1)^2}{\gamma_k} \frac{\mathrm{Re}(V_{1k0}^* V_{2k0})}{V_{1k0}^* V_{1k0}} \tag{7.33}$$

where we use the locally unbiased estimates $\beta_k - \beta_0 = (1 + \gamma_k)/\gamma_k$. Here γ_k is the signal-to-interference ratio, and V_{1k0} and V_{2k0} correspond to V_{1k} and V_{2k} for the channels d_0, l_0. In place of γ_k it is possible to use its maximum likelihood estimate for $\beta = \beta_0$, which in this case takes the form:

$$\hat{\gamma}_k|_{\beta-\beta_0} = \frac{(2mn - 1)V_{1k0}^* V_{1k0}}{\sum_{d,l}^{n,m} (V_{1kdl}^* V_{1kdl} + V_{2kdl}^* V_{2kdl}) - V_{1k0}^* V_{1k0}} - 1$$

where d_0 and l_0 correspond to the range-velocity coordinates at the point where there is a signal. Other estimates are possible, fixing the γ_k at the level γ_0, with guaranteed accuracy characteristics. The block diagram of the processing in (7.33) is shown in Fig. 7.20, for the case of a slowly fluctuating signal.

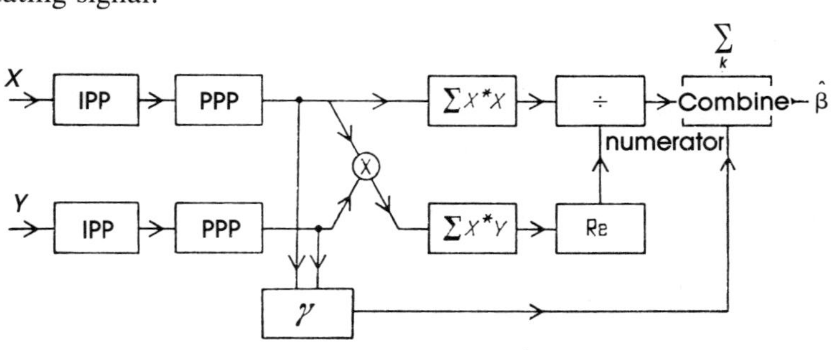

Fig. 7.20 Block diagram of an algorithm for calculating the maximum likelihood estimate of the angle of arrival β of a slowly fluctuating signal in nonstationary interference with unknown strength.

In the case of a rapidly fluctuating signal, the interference will be stationary over the intervals T_k. In these intervals we may use the estimate (7.30), and then produce a combined estimate of the form (7.33). Finally, we obtain

$$\hat{\beta} - \beta_0 = \frac{1}{k_0 \sum_{k=1}^{N} (1 + \gamma_k)} \sum_{k=1}^{K} \frac{(\gamma_k + 1)^2}{\gamma_k} \frac{\mathrm{Re}\left(\sum_{j=1}^{m} V_{1k0}^* V_{2jk0} \right)}{\sum_{j=1}^{m} V_{1jk0}^* V_{1jk0}}$$

In this case, we use the estimate:

$$\hat{\gamma}_k|_{\beta=\beta_0} = \frac{(2n - 1) \sum_{j=0}^{m} V_{1jk0}^* V_{1jk0}}{\sum_{d=1}^{n} \sum_{j=1}^{m} (V_{1jkd}^* V_{1jkd} + V_{2jkd}^* V_{2jkd}) - \sum_{j=1}^{m} V_{1jk0}^* V_{1jk0}} - 1$$

This is different from the preceding case in the summation over j and in that V_{1k0} and V_{2k0} are replaced with

$$V_{1jk0} = \sum_{i=1}^{n} a^* i x_{ijk}|_{d0} \quad \text{and} \quad V_{2jk0} = \sum_{i=1}^{n} a^* i y_{ijk}|_{d0}$$

7.6.8 Estimating Parameters in Clutter

The important difference of clutter, as was noted in Sec. 7.3, is the significant pulse-to-pulse correlation. As in the detection problem, we assume that the components of the weighted discrete Fourier transform (the transformation B_2) in the intervals T_k of the clutter are uncorrelated, and unknown *a priori,* for both slowly and rapidly fluctuating signals.

When estimating the parameters of a slowly fluctuating signal, after the transformations B_1 and B_2, the problem is reduced to that of estimating the parameter β in interference with $E[|Z_{ijk}|^2] = \epsilon_{jk}$. This problem differs from that considered earlier only in the large number of interference parameters, which is reflected mainly in the estimate of γ_k, which must be performed individually for each velocity cell. Thus,

$$(\hat{\beta} - \beta_0)_{d_0,l_0} = \frac{1}{k_0 \sum_{k=1}^{N} (1 + \gamma_k)} \sum_{k=1}^{K} \frac{(\gamma_k + 1)^2}{\gamma_k} \frac{\mathrm{Re}(V_{1k0}^* V_{2k0})}{V_{1k0}^* V_{2k0}}$$

$$\hat{\gamma}_k|_{\beta=\beta_0} = \frac{V_{1k0}^* V_{1k0}(2n - 1)}{\sum_{d=1}^{n} (V_{1kdl_0}^* V_{1kdl_0} + V_{2kdl_0}^* V_{2kdl_0}) - V_{1k0}^* V_{1k0}} - 1$$

The estimate for a rapidly fluctuating signal has an analogous form. The difference from the foregoing case consists in the combination of information obtained in l_1 velocity channels, corresponding to the width of the signal fluctuation spectrum, using the usual combination formula. The block diagrams of estimation of the parameter β for slowly and rapidly fluctuating signals are analogous to the corresponding diagrams for the cases of parameter estimation in nonstationary interference.

7.7 THE CHARACTERISTICS OF SIGNAL DETECTION AND PARAMETER ESTIMATION IN INTERFERENCE WITH UNKNOWN INTENSITY

7.7.1 Signal Detection Characteristics in Stationary Interference With Unknown Intensity

Detection characteristics for slowly and rapidly fluctuating signals using the decision rules (7.12) and (7.15) may be obtained with the use of statistics which are invariant relative to scale transformations in the input sample space. These statistics have the form:

$$f_1(V) = \frac{\sum_{k}^{K} |V_{11k}|^2}{\sum_{i,j,k}^{n,m,K} |V_{ijk}|^2}$$

and

$$f_2(V) = \frac{\sum_{j,k}^{m,K} |V_{1jk}|^2}{\sum_{i,j,k}^{n,m,K} |V_{ijk}|^2}$$

When there is no signal, these statistics have beta distributions, with K, $K(mn - 1)$ and mK, $mK(n - 1)$ degrees of freedom, respectively. The decision rules (7.12) and (7.15) may also be expressed through the statistics:

$$f_3(V) = \frac{\sum_{k}^{K} |V_{11k}|^2}{\sum_{i,j,k}^{n,m,K} |V_{ijk}|^2 - \sum_{k}^{K} |V_{11k}|^2}, \quad i \neq 1, \quad j \neq 1$$

$$f_4(V) = \frac{\sum_{j,k}^{m,K} |V_{1jk}|^2}{\sum_{i,j,k}^{n,m,K} |V_{ijk}|^2 - \sum_{j,k}^{m,K} |V_{1jk}|^2}$$

The probabilities of false alarm F and detection D for the decision rules $f_3(V) \geqslant C$ and $f_4(V) \geqslant C$ are

$$F = 1 - G\left(C\frac{M}{R}, 2R, 2M\right) = I_{1/(1+C)}\,(M, R) \tag{7.34}$$

$$D = 1 - G\left(C_1\frac{M}{R}, 2R, 2M\right) = I_{1/(1+C_1)}\,(M, R) \tag{7.35}$$

where R and M are the numbers of "signal" and "interference" samples, respectively; $G(\mathbf{X}, 2R, 2M)$ is the integral Fisher-Snedekor distribution, tabulated in Bol'shev and Smirnov [1]; $I_y[M, R]$ is the incomplete beta function, tabulated in Pearson [23]; C is a threshold constant; $C_1 = C/(1 + \gamma)$; and γ is the signal-to-interference ratio.

For the decision rule $f_3(V) \geqslant C$, the number of samples are $R = K$ and $M = (mn - 1)K$, and for the decision rule $f_4(V) \geqslant C$, $R = mK$, and $M = (n - 1)mK$. For the case $R = 1$, often encountered in practice, (7.34) and (7.35) are simplified: $F = (1 + C)^{-M}$ and $D = C_1^{-M}(1 + C_1^{-1})^{-M}$. With sufficiently large numbers of "interference" samples M, the detection characteristics are the same as those for a rapidly fluctuating signal in stationary interference with known intensity [2]. The threshold signal-to-interference ratio for the given decision rule is determined from the formula:

$$\gamma_0 = p_F/p_D - 1 \tag{7.36}$$

where p_F and p_D are the $100F$- and $100D$-percent points of the χ^2-distribution with $2R$ degrees of freedom.

The threshold signal-to-interference ratio as a function of the integration parameter R is plotted in Figs. 7.21 and 7.22, calculated for the optimum decision rule (7.12) using (7.36) and the tables in Bol'shev and Smirnov [1] and Pearson [23].

With small parameter values, the estimate of the interference intensity has a large variance, which leads to additional energy losses in the threshold signal, so that the signal-to-interference ratio must be increased to maintain a given value of D, as compared with the case when the interference strength is known *a priori* and ordinary receivers are used. As the shown by the analysis in Dmitrenko and Korado [5], the energy losses depend on the ratio M/R (see Fig. 7.23) and, for $R \geqslant 5$ and $M/R \geqslant 10$, do not exceed 1 dB.

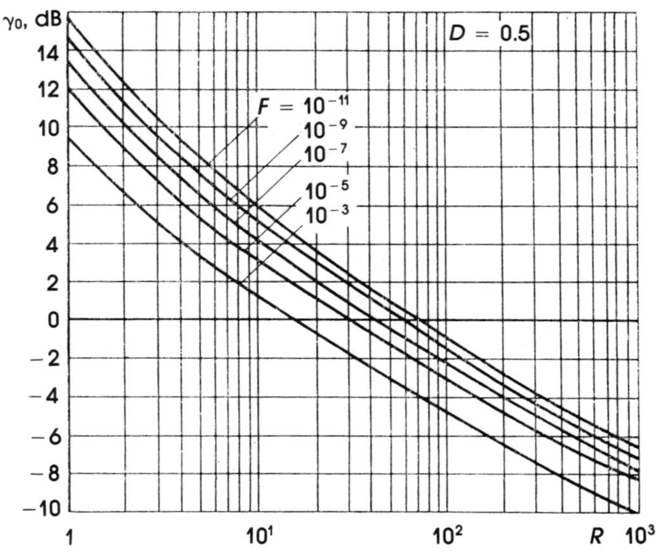

Fig. 7.21 Detection characteristics for the optimum decision rule (7.37) with $D = 0.5$.

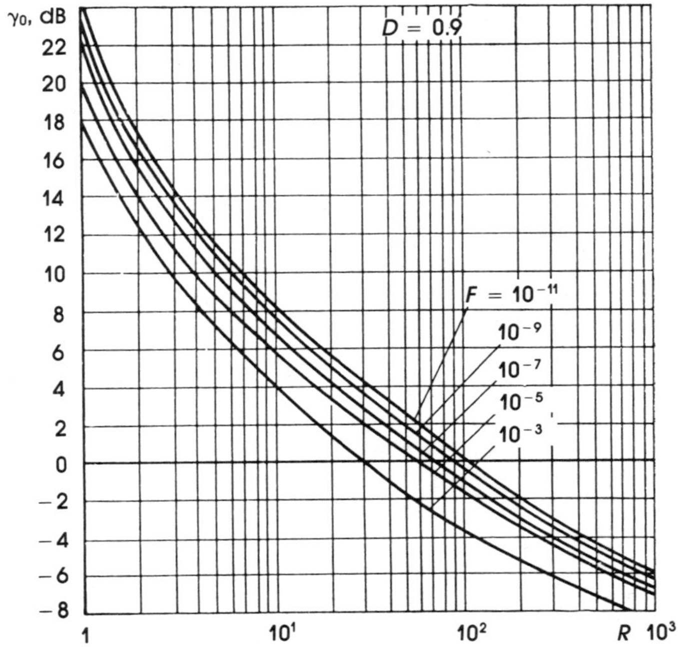

Fig. 7.22 Detection characteristics for the optimum decision rule (7.37) with $D = 0.9$.

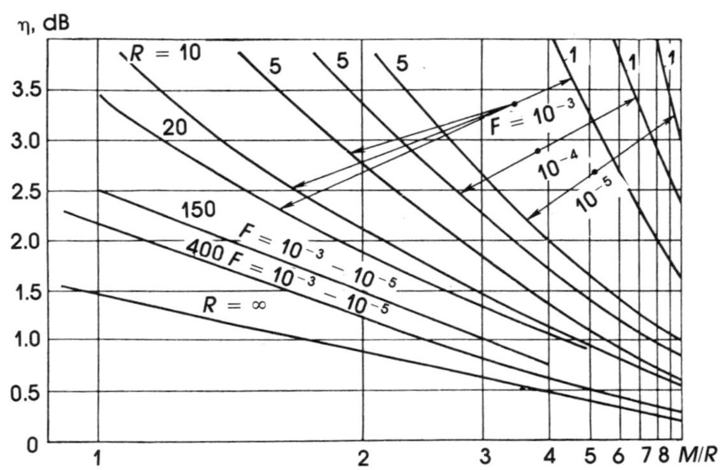

Fig. 7.23 The loss in the threshold signal as a function of the quantity M/R for $D = 0.95$ and various values of R and F.

7.7.2 Signal Detection Characteristics in Nonstationary Interference and Clutter

As was shown in Sec. 7.5, the decision rules for signal detection in nonstationary interference and clutter differ in the form of the function $f(Q_k)$ given the nonlinear transformation of the statistic:

$$Q_k = \frac{\sum_j^R |Y_{jk}|^2}{\sum_j^R |Y_{jk}|^2 + \sum_j^M |Z_{jk}|^2}$$

where Y_{jk} and Z_{jk} are elements of the corresponding signal and interference transformed samples, and R and M are integration parameters. The forms of the function $f(Q_k)$ for various decision rules are shown in Table 7.1.

Table 7.1

Decision Rule	Form of the function $f(Q_k)$
Minimax	$\ln(1 - \alpha Q_k)^{-1}, \ \alpha = \gamma_0/(1 + \gamma_0)$
Locally minimax	Q_k
Maximum Likelihood	$\begin{cases} \ln \dfrac{B_0}{(1 - Q_k)^M Q_k^R}, & Q_k > \dfrac{R}{M + R} \\ 0, & Q_k \leqslant \dfrac{R}{M + R} \end{cases}$ $B_0 = M^M R^R (M + R)^{-(M+R)}$
Quasi-optimal	$\begin{cases} \left(Q_k - \dfrac{R}{M + R}\right)^2, & Q_k > \dfrac{R}{M + R} \\ 0, & Q_k \leqslant \dfrac{R}{M + R} \end{cases}$

When detecting slowly and rapidly fluctuating signals in interference, the statistic Q_k is determined with (7.20) and (7.21), where in the first case, $R = 1$ and $M = (n - 1)m$, and in the second, $R = m$ and $M = (n - 1)m$. When detecting slowly and rapidly fluctuating signals in clutter, the statistic Q_k is determined from (7.24) and (7.27), where $R = 1$ and $M = n - 1$.

It is extremely difficult to obtain accurate analytic expressions for the detection characteristics using optimum incoherent integration. As a limiting case, for detection in nonstationary interference, we may consider the characteristics of the following decision rule with $K \to \infty$:

$$\sum_{j,k}^{R,K} \frac{\gamma_k}{(1 + \gamma_k)\epsilon_k} |Y_{jk}|^2 \geqslant C \tag{7.37}$$

which is optimum for the case of unknown interference intensities ϵ_k and signal-to-interference ratios γ_k, $k = 1, \ldots, K$, in different time intervals T_k or at the output of different discrete Fourier transform filters.

The threshold signal-to-interference ratio γ_0 for the rule (7.37), with identical values of γ_k, is determined with (7.34), and the dependence of γ_0 on the integration parameter R is shown in Figs. 7.21 and 7.22. As analysis and modeling show, the detection characteristics for signals in nonstationary interference, with γ_k, may be estimated with sufficient accuracy using the characteristics of the algorithm based on choosing K' of K results of the comparison of the statistics Q_k, or the identical statistics

$$t_k = \frac{\sum_j^R |Y_{jk}|^2}{\sum_j^M |Z_{jk}|^2}$$

with the threshold C_0. An estimate of the loss in the threshold signal-to-interference ratio for optimum noncoherent integration, in comparison with (7.37), may be produced using the loss η of the algorithm of preliminary threshold processing (see Fig. 7.23) relative to the algorithm $\Sigma_i^R |Y_i|^2 \geqslant C$ for noncoherent integration, using the formula:

$$\eta = 10 \ln \frac{Q_{F_1}/Q_{D_1} - 1}{p_{F_1}/p_{D_1} - 1} \tag{7.38}$$

where Q_D and p_D are the $100D_1$-percent points of the corresponding F-distribution with $2R$ and $2M$ degrees of freedom and the χ^2-distribution with $2R$ degrees of freedom; Q_F and p_F are the $100F_1$-percent points of the distributions $F_{2R,2M}(X)$ and $\chi^2_{2R}(X)$.

The values D and F for the K' of K algorithm, with identical γ_k, are determined from the equations:

$$D = I_{D_1}(K, K - K' + 1) = \sum_{n=K'}^{K} C_K^n D_1^n (1 - D_1)^{K-n}$$

$$F = I_{F_1}(K, K - K' + 1) = \sum_{n=K'}^{K} C_K^n F_1^n (1 + F_1)^{K-n}$$

where $I_{D,F}(K, K - K' + 1)$ is the incomplete beta function; D_1 and F_1 are the probabilities of exceeding the threshold level C_0, with the signal present and absent, respectively, with random values t_k, $k = 1, \ldots, K$. The values F_1 and D_1 are determined with (7.34) and (7.35), and may also be partly determined from the graphs in Figs. 7.21 and 7.22, taking account of the losses shown in Fig. 7.23 (see [5]). As a result, the signal-to-interference ratio in decibels, for the case of detection in nonstationary interference, is found to be $\gamma_{thr} = \gamma_0 + 2$, where γ_0 and η are the values calculated with (7.36) and (7.38).

With a large number of parameters K ($K \gg 1$), it is convenient to use a Gaussian approximation for the statistic (7.37). Then the base characteristics of the decision rule (7.37) will be determined by the equation:

$$D = 1 - \Phi \frac{\sqrt{\sum_k^K \gamma_k^2/(1 + \gamma_k)^2}\, \Phi^{-1}(1 - F) - \sqrt{R} \sum_k^K \gamma_k^2/(1 + \gamma_k)}{\sqrt{\sum_k^K \gamma_k^2}} \tag{7.39}$$

where $\Phi(x) = (1/\sqrt{2\pi}) \int \exp(-t^2/2)\, dt$ is the cumulative probability integral and F is the given probability of false alarm. The characteristics of the locally optimum decision rule $\sum_k^K Q_k \geq C_\alpha$, which is the simplest of the decision rules for $K \gg 1$, $\gamma \ll 1$, $k = 1, \ldots, K$, when using a Gaussian approximation to the statistic $\sum_k^N Q_k$, is determined from the equation:

$$D = 1 - \Phi \left[\Phi^{-1}(1 - F) - \sqrt{\frac{MR}{K(M + R + 1)}} \sum_k^K \frac{\gamma_k}{1 + \gamma_k} \right] \tag{7.40}$$

When the signal-to-interference ratios ($\gamma_k = \gamma \ll 1$, $k = 1, \ldots, K$) are identical, (7.39) and (7.40) are simplified and take the respective forms

$$D = 1 - \Phi[\Phi^{-1}(1 - F) - \sqrt{RK}\,\gamma]$$
$$D = 1 - \Phi\left[\Phi^{-1}(1 - F) - \sqrt{\frac{RMK}{R + M + 1}}\,\gamma\right] \qquad (7.41)$$

Equation (7.41) makes it possible to determine the asymptotic ($K \to \infty$, $\gamma \to 0$) losses in the threshold signal-to-interference ratio:

$$\eta = 10 \ln \sqrt{\frac{R + M + 1}{M}} \qquad (7.42)$$

The values of the loss η obtained with (7.42) are shown in Table 7.2.

Table 7.2

	Losses η, dB, for the ratio M/R				
# Samples	1	5	10	20	100
1	2.39	0.73	0.40	0.21	0.04
5	1.7	0.47	0.25	0.13	0.03
10	1.61	0.43	0.23	0.12	0.02
20	1.56	0.41	0.22	0.11	0.02
100	1.52	0.40	0.21	0.11	0.02

With $M \to \infty$ (interference samples) and a finite number of signal samples R, $\eta \to 0$, with $R = M \to \infty$, $\eta \to \sqrt{2}$ (1.5 dB), and with $R = M = 1$, $\eta = \sqrt{3}$ (2.39 dB). This loss in the signal-to-interference ratio is the cost of not knowing the actual interference strength and signal-to-interference ratios γ_k. As the number of interference samples M increases, the accuracy of the estimate of the interference strength increases, and the losses decrease. We note that the decision rule of (7.37), with any fixed values of γ_k and ϵ_k, is not usable if the interference strengths are unknown *a priori*, because under these conditions this decision rule will not stabilize the probability of false alarm.

To estimate the detection characteristics in nonstationary interference with $R \gg 1$, $M \gg 1$ and small values of K, the corresponding asymptotic characteristics of the maximum likelihood decision rules may be used [17].

When there is no signal, the distribution of the statistic for this decision rule tends to a central χ^2-distribution, with K degrees of freedom. The threshold constant C_F is determined from the equation $\int_{C_F}^{\infty} p(y)\, dy = F$, where $p(y)$ is the density of the central χ^2-distribution with K degrees of freedom and F is the given probability of false alarm. The probability of detection is then

$$D(\lambda) = \int_{C_F}^{\infty} p(y|\lambda)\, dy \tag{7.43}$$

where $p(y|\lambda)$ is the density of the noncentral χ^2-distribution with K degrees of freedom and noncentrality parameter:

$$\lambda = R \sum_{k=1}^{K} \frac{\gamma_k^2}{(1 + \gamma_k)^2}$$

The losses in the threshold signal-to-interference ratio for this case are approximately 1 dB, as compared with the baseline decision rule (7.36), with, for example, $K = 2$, $R = 50$, $F = 10^{-3}$, and identical γ_k. With small values of the parameters M, K, and R, when the asymptotic expressions in (7.39) and (7.43) cannot be used, the detection characteristics may be obtained using the Monte Carlo method. The probability of detection $D(\gamma)$ for a slowly fluctuating signal for various decision rules is shown in Fig. 7.24, obtained with the Monte Carlo method. This case corresponds to the detection of signals with a uniform energy spectrum in clutter with a uniform energy spectrum, after a preliminary Fourier transform and the combination of information from the outputs of $K = 10$ filters, with one signal and $M = 10$ interference samples in each filter. As may be seen, the minimax, asymptotic minimax, and locally-minimax decision rules exceed the maximum likelihood decision rule in $D(\gamma)$. The losses in the threshold signal-to-interference ratio for the indicated decision rules, as compared with the decision rules (7.37), are 1–2 dB, with $D = 0.9$, for example.

The case of nonstationary signal and interference, or signal and clutter with nonuniform energy spectra, was modeled by assigning different values of the signal-to-interference ratio for different channels. Figure 7.25 presents the detection characteristics for the following example: a slowly fluctuating signal in clutter with a nonuniform distribution in the velocity channels (filters), with identical signal-to-interference ratios $\gamma > 0$ for two channels and $\gamma = 0$ for the other eight of ten channels, integration parameters $R = 1$, $M = 10$, and $F = 10^{-3}$. As may be seen from Figs. 7.24

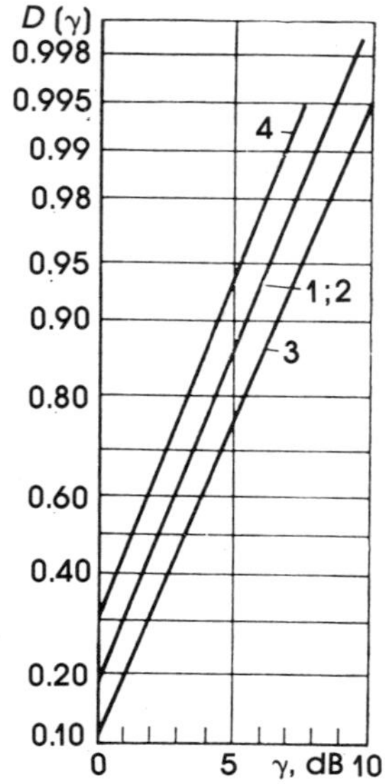

Fig. 7.24 Detection characteristics with $\gamma_k = \gamma$, $R = 1$, $k = 10$, $M = 10$, and $F = 10^{-3}$;
1, 2– minimax and locally minimax rules
3 – maximum likelihood decision rule
4 – the decision rule (7.37)

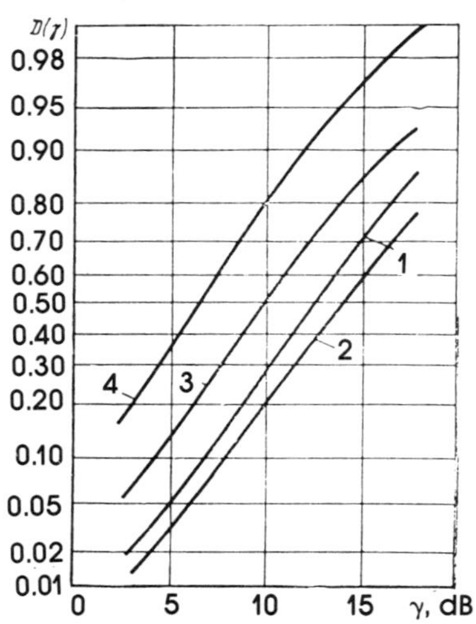

Fig. 7.25 Detection characteristics for decision rules with $\gamma_k = \gamma \neq 0$ in two channels, $k = 10$, $R = 1$, $M = 10$, $F = 10^{-3}$ (the curves are numbered as in Fig. 7.24).

and 7.25, the absence of the signal in a large fraction of the combined detections channels leads to a substantial loss in comparison with the decision rules (7.37). For the maximum likelihood decision rule, this loss increases by 2–4 dB. The maximum likelihood decision rule is in this case somewhat more effective than the minimax and locally-minimax decision rules. This is explained by the fact that with the maximum likelihood decision rule, the signal-to-interference ratio is estimated for each of the combined channels, and those channels in which there is no signal are practically excluded (rejected) from further processing. In the minimax decision rule, the controlled alternative hypothesis region is limited by the condition $\gamma_k \geqslant \gamma_0$, i.e., the signal is expected in each channel, which does not occur in this example. However, as the number of channels in which the signal is received increases relative to the total number of channels which are combined, the detection characteristics of the minimax decision rule rapidly approach those of the maximum likelihood decision rule, and the two are virtually identical with the signal being in even just four of $K = 10$ channels. The detection characteristics of the much simpler quasi-optimal decision rule (see Table 7.1) are in practice no worse than those of the maximum likelihood decision rule.

7.7.3 The Accuracy of Signal Parameter Estimation

The characteristics of estimates of signal parameters in stationary interference, of the sort in (7.31), are examined in Moska [22]. The mean value of the estimate for $|\beta - \beta_0| \ll 1$ is equal to $E[\hat{\beta} - \beta_0] = \delta\gamma/(1 + \gamma)$, where $\delta = \beta - \beta_0$; γ is the signal-to-interference ratio $\gamma = \nu/\epsilon$ with unit discrimination slope ($k_0 = 1$); β is the exact value of the angle of arrival.

As may be seen in the graph in Fig. 7.26 (see [22]), this estimate is asymptotically unbiased, although for small and mid level values of γ there is a bias, independent of the number of samples K. There is no second moment for this estimate, but the spread of the estimate may be characterized by the width S of the probability distribution at the level 0.7 and the corresponding variance for the Gaussian distribution. For this estimate, $S_\gamma^{(k)} = S_{\gamma=0}^{(k)}(1 + \gamma)^{-1/2}$.

Figure 7.27 shows $S_\gamma^{(k)}$ and the Cramer-Rao bound $\sigma_{CR}^2 = 1/2K(1 + \gamma)$ for the variance of $\hat{\beta}$, as a function of the number of transformed samples K. It is clear from the figure that as $K \rightarrow \infty$, the estimate is asymptotically efficient. When combining the estimates of signal parameters in nonstationary interference, the potential accuracy (variance) of the estimate with a large number of samples and large signal-to-interference ratio is

$$D_\Sigma = \left[\sum_{k=1}^{K} D_k^{-1}\right]^{-1} = \left[2K_1 \sum_{k=1}^{K} (1 + \gamma_k)\right]^{-1}$$

here D_k and γ_k have the same meaning as in Sec. 7.6, and in this case $D_k = \sigma_{CR}^2 = 1/2K_1(1 + \gamma)$, where K_1 is 1 or m, for slowly and rapidly fluctuating signals. These characteristics were obtained for the case when the coefficient $k_0 = 1$ in the signal model in two channels (see Sec. 7.6.4).

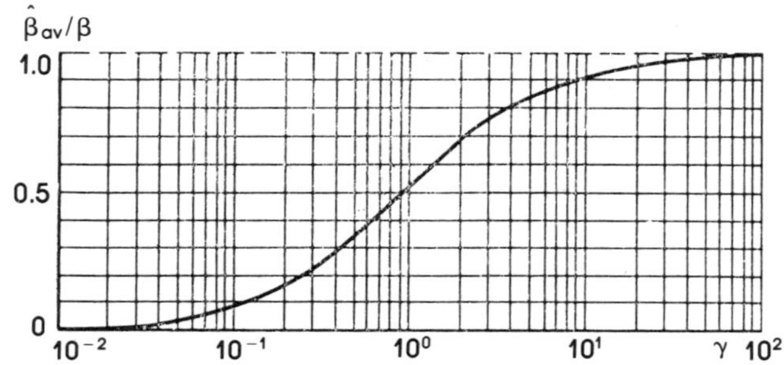

Fig. 7.26 $\hat{\beta}_{av}/\beta$ as a function of the signal-to-interference ratio $\overline{\gamma}$ (see [22]).

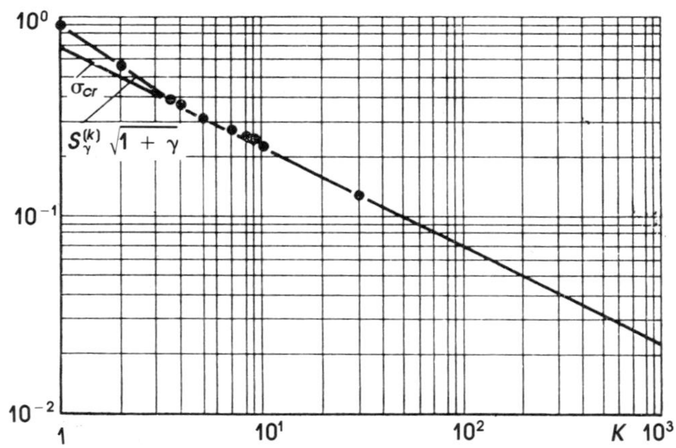

Fig. 7.27 Relative estimates of the quantities $S_\gamma^{(k)}$ and σ_{CR} as functions of the number of samples K.

7.8 TRACK PROCESSING

In general, track processing consists of processing the received signal for sufficiently large time intervals, during the course of which the variations in the parameters of the useful signal (subject to detection) are many times larger than the radar resolution in these parameters.

The main goal of track processing for any surveillance radar is the detection and discrimination of signal "trajectories," estimating the current values of the signal parameters, and also of parameters of track models. The mathematical model of a signal during a track processing interval differs from those considered earlier in that the parameters of one or a group of several signals are not considered to be constant, but continuously varying in time, according to some regular or random pattern, which itself may possess some parameters which are unknown *a priori*.

The solution to this problem is optimized using the well-known methods of detection theory, signal parameter discrimination, and estimation in interference with some *a priori* uncertainty. The problem is similar to that of the preliminary processing discussed earlier, except that now the detection problem is viewed as a whole over the entire observation interval. In particular, the problem of track detection-discrimination is viewed as one of verifying a "track hypothesis," that is, a hypothesis about the existence of several concrete trajectories from some finite set of signal trajectories in interference with unknown parameters.

The approach to solving the track problem outlined above is comparatively complicated in a digital implementation, and the problem is therefore usually divided into two parts [13], solved independently: primary (before track) processing, and secondary (track) processing. In primary processing, the radar information obtained from coherent and preliminary incoherent processing includes the detection of useful (target) signals in the radar resolution cells, the formation of radar "pips" in the radar resolution channels at the signal parameter coordinates, and also estimating the parameters of these detections [13]. Primary processing is performed over an interval during which the signal parameters vary so slightly that they may be considered to belong to a single resolution element.

This solves the problems of initial signal detection, and the resolution and estimation of signal parameters in various interference conditions, and also the stabilization of the probability of false alarm. This part of the problem has been examined in Ch. 7, except for the separation of the marks. The track processing problem in its simplest version is analogous to traditional methods of secondary processing in radar, and is examined in detail in a number of works [6,13], and others. We will limit ourselves here to enumerating some methods of track processing, referring the reader to the corresponding literature.

The block diagram of a typical secondary (track) processor contains blocks [13], performing the following operations: automatic acquisition (detection) of the trajectories, tracking the trajectories, smoothing the coordinates, selecting trajectories on the basis of additional indications that they are useful trajectories, and verifying the criteria for preliminary track detection.

Track processing algorithms may include determination of the correlation matrices of the errors in extrapolating the coordinates, estimating the beginning and end of a track maneuver (i.e., sharp changes in the track parameters), estimating the reliability of track detection and the accuracy of track parameter estimation [13].

A rigorous theory for estimating the characteristics of track processing, taking into account interfering parameters, is rather complex, and the corresponding characteristics are usually estimated through modeling or full-scale experiment (see, for example, [18]). However, if the system parameters are optimally selected, then its effectiveness will not be much different from that of an optimum trajectory processor for the case of a trajectory known *a priori*. Therefore, the baseline characteristics may be calculated approximately with the equations $F = F_{ft}N_t$, $D = D_{ft}$, where N_t is the number of different allowable independent trajectories, starting at a given time of the processing; F_{ft} and D_{ft} relate to a given fixed trajectory, in which F is the probability of false alarm at one time in the processing. The value D_{ft} is determined in essence by the characteristics of optimum detection of a random Gaussian signal in random Gaussian interference for n independent samples, corresponding to the full length of the track processing. Therefore, to estimate the baseline threshold ratio γ of pretrack processing for given values of D,F, and n, the curves of Figs. 7.21 and 7.22 may be used (with $n = R$).

The number N_t of different acceptable trajectories (i.e., those trajectories which might be followed by real targets) at one point in the processing may be expressed as $N_t = N_{re}N'_{\tau}$, where N_{re} is the number of signal parameter resolution elements, and N'_t is the number of different possible trajectories, beginning at a given time in the processing in a given resolution element. The actual threshold signal-to-interference ratio γ_Σ will exceed the initial value γ_0 by relatively small amounts due to the additional energy losses associated with the *a priori* uncertainty in the signal and interference parameters.

In more complex interference conditions, including those considered earlier, the appropriate selection of signal waveforms, pretrack and track processing tends to keep the detection characteristics at the level obtained for detection in uniform interference and clutter, in a radar frequency channel relatively free of interference.

The accuracy with which the real-time trajectory values may be determined, using the maximum likelihood method, is shown in Fig. 7.28. The figure shows the relative mean square error ϵ in the current estimate of the coordinate y and its derivative V_n as a function of the number of processing periods, for linear and quadratic track hypotheses. For comparison, the inversely proportional relation typical of pretrack processing is plotted; this relation is also typical for track processing when the trajectories are known *a priori* except for the initial value of a single estimated coordinate. As may be seen, in the usual case where several unknown independent parameters are estimated jointly, the accuracy is somewhat degraded due to the large amount of *a priori* uncertainty in the estimation problem.

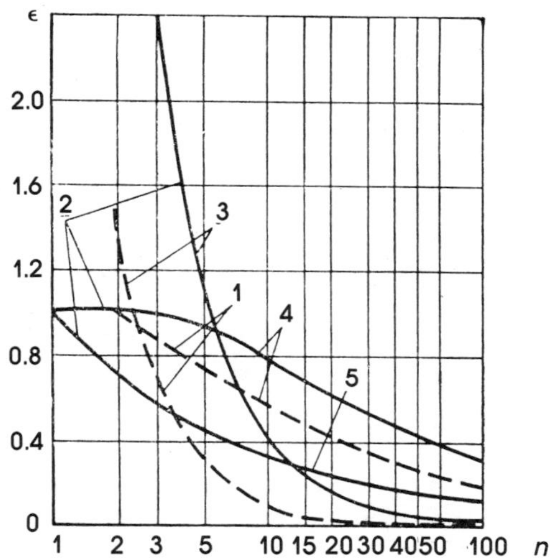

Fig. 7.28 The relative mean-square error ϵ in the coordinate and its derivative as a function of the number of track processing points.

1 — linear hypothesis
2 — quadratic hypothesis
3 — $\sigma_V T_0/\sigma_y$
4 — σ_{yn}/σ_y
5 — $n^{-1/2}$

The data obtained for the "signal" trajectory is then transformed into a target trajectory in geographic or some other appropriate coordinates. This part of the secondary processing problem, with the need to integrate information obtained from several displaced radars, is examined in detail in Kuz'min [13] and Van Koyk and Pandikov [3]. For these calculations, the radar must assess the significant influence of the radio wave propagation conditions along the path on the signal parameters, and, in particular, estimate the azimuthal deviation and the factor by which the range is lengthened, that is, the ratio of the range calculated from the time delay on the basis of straight-line propagation in a vacuum, to the actual range to the object along the arc of a large circle. [See Mishchenko, Yu.A. *Zagorizontnaya radiolokatsiya* (Over-the-horizon radar). Moscow: Voenizdat, 1972. 96pp.]

REFERENCES

1. Bol'shev, L. and Smirnov, N.V. *Tablitsa matematicheskoy statistiki (Table of mathematical statistics)*. Moscow: Nauka, 1965. 464 pp.
2. Vaynshteyn, L.A. and Zubakov, V.D. *Vydelenie signalov na fone sluchaynykh pomekh (The extraction of signals from random interference)*. Moscow: Sovietskoe Radio, 1960. 447 pp.
3. Van Koyk and Pandikov. *Statisticheskoe opisanie strategiy ob'edineniya trass pri nablyudenii neskol'kimi radiolokatorami (A statistical description of a strategy for combining paths with observations by several radars)*. Zarubezhnaya Radioelektronika, (1971), no. 11, pp. 69–84.
4. Gutkin, L.S. *Teoriya optimal 'nykh metodov radiopriema pri fluktuatsionnykh pomekhakh (The theory of optimal methods for radio reception in fluctuating interference)*. Moscow: Sovietskoe Radio, 1972. 448 pp.
5. Dmitrienko, A.N. and Korado, V.A. *Kharaketristiki obnaruzheniya paketa nezavisimo fluktuiruyushchukh impul'sov na fone gaussovskoy pomekhi s neizvestnoy intensivnost 'yu (the characteristics of detecting a packet of independently fluctuating pulses in gaussian interference with unknown intensity)*. Radiotekhnika i Elektronika, vol. 12 (1967), no. 7, pp. 1272–1274.
6. Zhdanyuk, B.F. *Osnovy statisticheskoy obrabotki traektornykh izmereniy (Fundamentals of statistical trajectory measurement processing*. Moscow: Sovietskoe Radio, 1978. 284 pp.

7. Korado, V.A. *Minimaksnoe mnogoal 'ternativnoe obnaruzhenie signalov na fone pomekh s neizvestnymi parametrami (Minimax multiple alternative detection of signals in interference with unknown parameters).* Moscow: 1971 (Dep. rukop. NIIEIR, D-2950. Moscow, 1971).
8. ———*Ob optimal'nom obnaruzhenii signalov pri vozdeystvii pomekh s neizvestnymi parametrami (On the optimum detection of signals in interference with unknown parameters).* Radiotekhnika i Elektronika, vol. 14 (1969), no. 2, pp. 239–248.
9. ———*Ob optimal'nom obnaruzhenii signalov na fone pomekh s neizvestnymi parametrami pri ogranichennoy veroyatnosti lozhnoy trevogi (On the optimum detection of signals in interference with unknown parameters with a limited probability of false alarm).* Radiotekhnika i Elektronika, vol. 15 (1970), no. 7, pp. 1419–1427.
10. ———*Optimal 'noe obnaruzhenie sluchaynykh signalov na fone sluchaynikh pomekh neizvestnoy intensivnosti pri uclovii postoyanstva veroyatnosti lozhnoy trevogi (The optimum detection of random signals in random interference with unknown strength with a constant probability of false alarm).* Radiotekhnika i Elektronika, vol. 13 (1968), no. 5, pp. 832–841.
11. ———*Optimal 'noe ob'edinenie nezavisimykh kanalov obnaruzheniya signalov na fone gaussovykh pomekh s neizvestnymi intensivnostyami (The optimum combination of independent signal detection channels in gaussian interference with unknown intensity).* Radiotekhnika i Elektronika, vol. 17 (1972), no. 3, pp. 618–623.
12. Kramer, G. *Matekaticheskie metody statistiki (Mathematical methods of statistics).* Moscow: Mir, 1975. 648 pp.
13. Kuz'min, S.Z. *Osnovy teorii tsifrovoy obrabotki radiolokatsionnoy informatsii (The fundamentals of the theory of digital processing of radar information).* Moscow: Sovietskoe Radio, 1974. 432 pp.
14. Kuz'min, S.A. *Tsifrovaya obrabotka radiolokatsionnoy informatsii (Digital processing of radar information).* Moscow: Sovietskoe Radio, 1967. 400 pp.
15. Kulikov, E.I. and Trifonov, A.P. *Otsenka parametrov signalov na fone pomekh (Estimating the parameters of signals in interference).* Moscow: Sovietskoe Radio, 1978. 296 pp.
16. Levin, B.R. *Teoreticheskie osnovy statisticheskoy radiotekhniki (Theoretical fundamentals of statistical radio engineering).* Moscow: Sovietskoe Radio, vol. 3, 1976. 288 pp.
17. Lehman, E. *Testing statistical hypotheses.* New York: John Wiley, 1959.

18. Leonov, A.I., Vasenev, V.N., Baydukov, Yu.I., *et al. Modelirovanie v radiolokatsii* (A.I. Leonov, Ed.). Moscow: Sovietskoe Radio, 1979. 264 pp.
19. Repin, V.G. and Tartakovskiy, G.P. *Statistichekiy sintez pri apriornoy neopredelennosti i adaptatsiya informatsionnykh sistem (Statistical synthese with a priori uncertainty and adaptivity in the information systems)*. Moscow: Sovietskoe Radio, 1977. 432 pp.
20. Slutskiy, E.E. *Tablitsy dlya vychisleniya nepolnoy b-function and χ^2 probability function (Tables for calculating the incomplete b-function and χ^2 probability function,* A.N. Kolmogorov, Ed.). Moscow and Leningrad: Izd-vo AN SSSR, 1950. 71 pp.
21. Shirman, Ya.D. *Rasreshenie i szhatie signalov*. Moscow: Sovietskoe Radio, 1974. 360 pp.
22. Moska, E. *Angle estimation in amplitude comparison monopulse system*. IEEE Trans., vol. AES-5 (1969), no. 2, pp. 205–212.
23. Pearson, K. *Tables of the incomplete beta functions*. Cambridge, England, 1934. 553 pp.

Chapter 8

Signal Detection and Parameter Estimation in Interference With Unknown Angular Distribution

8.1 INTRODUCTION

It was assumed in Ch. 7 that the detection of a signal in various forms of interference was optimized for a single receiver channel. The receiver antenna and its pattern were treated as fixed. As is well known, however, the angular distribution of interference and clutter in the HF band is not constant in time, and depends on the operating frequency, and therefore is not completely known beforehand. This situation results in the requirement that the antenna-feed system be able to adapt to the changing spatial characteristics of the interference [16, 18].

If an adaptive beam-forming system is included when optimizing the signal processing, it is possible to calculate, and to some extent cancel, instability in the parameters of the individual antenna elements, and also refine the estimates of these parameters. This significantly improves the performance of an adaptive beamformer, not only for constant, but also complex interference conditions which are unknown *a priori*.

The modeled interference in the different channels has a correlation matrix which is unknown *a priori*, due to the *a priori* uncertainty in the angular distribution of interference power, the possible correlation of the individual angular components (due to multipath), and the lack of complete data on the characteristics of the receiving elements and their connections in the operating band [2].

The problem of optimizing the adaptive beam-forming system is approached by seeking the best extraction of the signal from the noise, one measure of which is the signal-to-interference ratio at the output of the antenna system. In signal detection systems, however, other factors are also important, such as limiting the probability of false alarm, independently of the actual interference conditions, and maximizing the probability of detection for a given probability of false alarm. This requires not only the extraction of the signal, but also the formation of some statistic to be compared with a threshold. The distribution of this statistic and its variation with various interference parameters, unknown *a priori*, are quite significant in this processing.

There are a large number of publications, both domestic and foreign, treating the theory and practice of adaptive space-time signal processing, including some specifically addressing over-the-horizon radar [1, 7, 8, 12, 13, 15–18]. The majority of these works deal with iterative algorithms for adaptive space-time processing with feedback. In this chapter, we concentrate on questions concerning optimal space-time processing using direct estimates of the interchannel interference correlation matrix. This method provides the best performance with a limited adaptation time [7, 17, 2].

8.2 THE SIGNAL AND INTERFERENCE MODEL

Let there be p receiving channels as considered in Ch. 7, connected to different antennas or antenna elements. We will assume that the amplitude-phase distribution of the useful signal in the receiver channels is known for each given direction of reception. After discrete time sampling, the complex input sample vector at the output of the receiver channels is $\mathbf{X} = \mathbf{S} + \mathbf{Z}$, where $\mathbf{X} = \{X_{rijk}\}$ is the total input sample; $\mathbf{S} = \{S_{rijk}\}$ is the useful target signal; $\mathbf{Z} = \{Z_{rijk}\}$ is additive interference, such that $X_{rijk} = S_{rijk} + Z_{rijk}$, where $r = 1, \ldots, p$ is the receiver channel number.

The input sample vector $\mathbf{X}_k$ and its components $\mathbf{S}_k$ and $\mathbf{Z}_k$ will be viewed as $p \times N$ matrices, where $N = mn$ is the number of samples during the interval T_k. It is assumed that for a slowly fluctuating signal $S_{rijk} = U_k \alpha_r a_i b_j$; $U_k = u_k \exp(i\phi k)$, and for a rapidly fluctuating signal, $S_{rijk} = U_{jk} \alpha_r a_i$; $U_{jk} = u_{jk} \exp(i\phi k)$, where $\vec{\alpha} = \{\alpha_r\}$ is a known complex p-vector, normalized such that $\vec{\alpha}^* \vec{\alpha} = 1$, describing the amplitude-phase relations of the signals in the different channels. The meaning of the vectors $\mathbf{a} = \{a_i\}$, $\mathbf{b} = \{b_j\}$, and the parameters U_{jk} and $\nu_{jk} = E[|U_{jk}|^2]$ remain as before (see Sec. 7.2), where it is assumed that $\mathbf{a}$ and $\mathbf{b}$ are normalized so that $\mathbf{a}^*\mathbf{a} = \Sigma_i |a_i|^2 = 1$ and $\mathbf{b}^*\mathbf{b} = \Sigma_j |b_j|^2 = 1$. The model of the interference $\mathbf{Z}$ is defined for each form of interference in the corresponding paragraphs.

8.3 DETECTING A SIGNAL IN TEMPORALLY UNCORRELATED STATIONARY GAUSSIAN INTERFERENCE WITH UNKNOWN INTERCHANNEL CORRELATION MATRIX

In this case, Z_{rijk} are independent complex random quantities with expected value $E[Z_{ijk}] = 0$ and $E[Z_{rijk}^* Z_{tijk}] = m_{rt}$, where $\mathbf{M} = \{m_{rt}\}$ is the $p \times p$ interchannel correlation matrix, which is unknown *a priori*. We will first consider detection of a slowly fluctuating signal for the simple case $K = 1$, i.e., when the signal is coherent over the entire observation interval $T_0 = T_k$ (see Fig. 7.1), and is known *a priori* up to the amplitude and initial phase (see Ch. 7).

We will transform the input sample to a form which is more convenient for further analysis. We apply to the input sample a preliminary linear unitary transformation $\mathbf{W} = \mathbf{X}\mathbf{T}^T$ in the time domain, using that matrix $\mathbf{T}$ which transforms the rows of $\mathbf{X}$ so that S_{rij} is transformed to the single value $N^{1/2}U\alpha_r$ for all i and j. The interchannel correlation matrix $\mathbf{M}$ remains constant under the transformation $\mathbf{W}$.

The problem consists of checking the composite hypothesis $H_0{:}\gamma = 0$ against the composite alternative $H_1{:}\gamma \geqslant \gamma_0$, where $\gamma = \nu\vec{\alpha}^*\mathbf{M}^{-1}\vec{\alpha}$ is a generalized signal-to-interference ratio, taking spatial selectivity into account, and $\nu = E[|U|^2]$. The parameter γ is numerically equal to the signal-to-interference energy ratio at the output of an optimum space-time processing filter, matched to the signal and the spatial distribution (correlation properties) of the interference.

As follows from the theory of optimum detection of a quasideterministic signal in interference with an unknown correlation matrix [2], the optimum decision rule takes the form:

$$f(Q_p, Q_{p-1}) \geqslant C \tag{8.1}$$

i.e., it depends on two invariantly sufficient statistics Q_p and Q_{p-1}, related to the sample $\mathbf{W}$ by the relations:

$$\begin{aligned} Q_p &= N\overline{\mathbf{W}}^*(\mathbf{W}\mathbf{W}^*)^{-1}\overline{\mathbf{W}} \\ \Delta Q &= Q_p - Q_{p-1} = N\frac{|\vec{\alpha}^*(\mathbf{W}\mathbf{W}^*)^{-1}\overline{\mathbf{W}}|^2}{\vec{\alpha}^*(\mathbf{W}\mathbf{W}^*)^{-1}\vec{\alpha}} \end{aligned} \tag{8.2}$$

where $\overline{\mathbf{W}}$ is a p-vector with components $\overline{\mathbf{W}}_r = N^{-1}\Sigma_{i,j}W_{rij}$.

The form of the function f may be different for signals with deterministic or random amplitude U and for different signal detection methods or optimization criteria. In particular, the minimax decision rule (invariant, with the highest probability of detection) for detecting a signal with a random complex Gaussian amplitude U, has the form:

$$[1 + \gamma_0(1 - Q_{p-1})]^{N-p}[1 + \gamma_0(1 - Q_p)]^{-(N-p+1)} \geqslant C \tag{8.3}$$

Here γ_0 is the threshold level of the generalized signal-to-interference ratio $\gamma = \nu\vec{\alpha}^*\mathbf{M}^{-1}\vec{\alpha}$, where $\nu = E[|U|^2]$. The local minimax decision rule is simplified to

$$(N - p)(Q_p - Q_{p-1}) + Q_p \geqslant C \tag{8.4}$$

Among all rules which are invariant relative to the group G_r of all linear transformations with lower triangular matrices, the maximum likelihood decision rule decides on the presence of the signal with (see [3])

$$(Q_p - Q_{p-1})(1 - Q_{p-1})^{-1} \geqslant C \tag{8.5}$$

This rule is equivalent to a test of the likelihood ratio to verify a hypothesis on the average value of a random vector with unknown correlation matrix [14]. The asymptotically optimum decision rule ($\gamma_0 \rightarrow \infty$), which is close to the maximum likelihood rule, is also of interest:

$$(1 - Q_{p-1})^{N-p}(1 - Q_p)^{-(N-p+1)} \geqslant C \tag{8.6}$$

along with various simplified quasi-optimal decision rules, for example:

$$Q_p - Q_{p-1} \geqslant C \tag{8.7}$$

It is not difficult to verify that the statistics Q_p and Q_{p-1} which appear in these decision rules, and, consequently, the left-hand sides of these rules, are G_T–invariant, and when the signal is not present, their distribution does not depend on the actual interference correlation matrix. Due to this characteristic, the probability of false alarm for all such detectors is independent of the actual angular distribution of interference power, and also of the cross-correlation of the interference being received from different directions. At the same time, effective signal detection is maintained in the direction determined by the vector $\vec{\alpha}$, whenever the signal is present.

We will examine a detailed block diagram of the decision rule (8.1). First, we note that $\mathbf{WW}^* = \mathbf{XT}^T\mathbf{T}^T\mathbf{X}^* = \mathbf{XX}^*$, where $\hat{\mathbf{M}} = \mathbf{XX}^*$ is a matrix, proportional to the sample interchannel correlation matrix $\mathbf{N}^{-1}\mathbf{XX}^*$, which, if there is no signal, is the estimate of the interchannel interference correlation matrix $\mathbf{M}$.

Let $\mathbf{g}^T$ be normalized such that $\mathbf{g}^*\mathbf{g} = 1$, and $\mathbf{N}$ be the signal row-vector, contained in the rows of the matrix $\mathbf{X}$. Then, by definition, $\mathbf{Tg} = N^{-1/2}(1, \ldots, 1)^T = N^{-1/2}\mathbf{e}$ and

$$N^{1/2}\overline{\mathbf{W}} = N^{-1/2}\,\mathbf{We} = N^{-1/2}\mathbf{XT}^T\mathbf{e} = \mathbf{Xg}^{*T} = \sum_j b_j^* \sum_i a_i^* \mathbf{X}_{ij} \tag{8.8}$$

where $\mathbf{X}_{ij}$ is the p-element column vector of the ith sample during the jth interval T_j; a_i and b_j are normalized intrapulse and pulse-to-pulse signal vectors: ^{*T} denotes complex conjugation and transposition. Using the values of $\overline{\mathbf{W}}$ and $\mathbf{WW}^*$ in (8.2), we obtain the block diagram of the decision rule for optimal p-channel detection of a slowly fluctuating quasideterministic signal in Gaussian interference with unknown interchannel correlation matrix, which may be placed in the form shown in Fig. 8.1, where IPP is intrapulse (intraperiod, for CW operation) processing $\mathbf{v}_j = \sum_i a_i^* \mathbf{X}_{ij}$, PPP is pulse-to-pulse processing, $\overline{\mathbf{W}} = N^{-1/2} \sum_j b_j^* \mathbf{v}_j$; CME is estimation of the interchannel correlation matrix; ABF is adaptive beamforming $\mathbf{Y} = \mu^* \overline{\mathbf{W}}$; ABFPC is calculation of the adaptive beamformer parameters $\mu = \mathbf{M}^{-1}\vec{\alpha}/(\vec{\alpha}^*\mathbf{M}^{-1}\vec{\alpha})^{1/2}$; Q_p is calculation of the statistic Q_p; D is square-law detection; $f(Q) = f(Q_{p-1}, Q_p)$ is the decision function of the invariantly sufficient statistics; *THR* is threshold comparison; C is a constant threshold, determined by the given probability of false alarm; and $\vec{\alpha}$ is the signal p-vector.

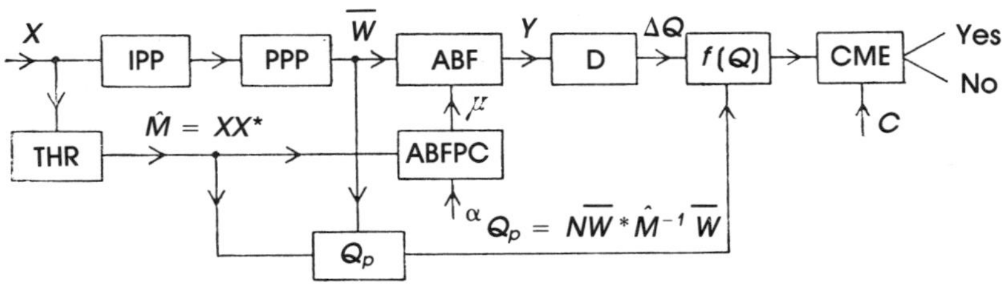

Fig. 8.1 Block diagram for the optimum decision rule when detecting a slowly fluctuating signal. IPP = intrapulse processing; PPP = pulse-to-pulse (period-to-period) processing; ABF = adaptive beam-former; D = detector; THR = threshold comparison; CME = correlation matrix estimation; ABFPC = adaptive beam-former parameter calculation.

As may be seen from the drawing, the adaptive beamforming is controlled by the vector μ, which is calculated from the estimate of the interchannel correlation matrix of the same finite discrete input sample $\mathbf{X}$, which might contain the signal. In general, optimum processing requires complex intrapulse and pulse-to-pulse processing in each receiver channel, to form the statistic Q_p, related to the complex Hotelling statistic T_c^2 by the relation $Q_p = T_c^2/(1 + T_c^2)$.

In a quasioptimal version of the decision rule (8.7), Q_p does not need to be generated, and the spatial processing in the adaptive beamforming system may be moved to the front end of the processor. Actually, from (8.8) we find that

$$Y = \frac{\vec{\alpha}^*(\mathbf{WW}^*)^{-1}\,\overline{\mathbf{W}}}{\sqrt{\vec{\alpha}^*(\mathbf{WW}^*)^{-1}\,\vec{\alpha}}} = N^{-1/2}\sum_j b_j^* \sum_i a_i^* \mu^* \mathbf{X}_{ij} \tag{8.9}$$

where $\mu = \hat{\mathbf{M}}^{-1}\vec{\alpha}/(\vec{\alpha}^*\hat{\mathbf{M}}^{-1}\vec{\alpha})^{1/2}$, $\hat{\mathbf{M}} = \mathbf{WW}^* = \mathbf{XX}^*$.

The block diagram of the decision rule (8.7), corresponding to (8.9), is shown in Fig. 8.2. If apart from the primary sample $\mathbf{X}$ there may be obtained an additional interference sample ξ with the same interchannel correlation matrix as in $\mathbf{X}$, then, combining both parts of the sample in one signal vector $\mathbf{g}$ with the corresponding zeroes, we find that the optimum processing will be different only in that the estimate $\hat{\mathbf{M}}$ should be calculated from the entire sample. In cases when the matrix $\mathbf{M}$ is relatively constant, then it is sufficient to estimate $\hat{\mathbf{M}}$ from one previously obtained interference sample ξ, which justifies the essence of the block diagram in Fig. 8.2.

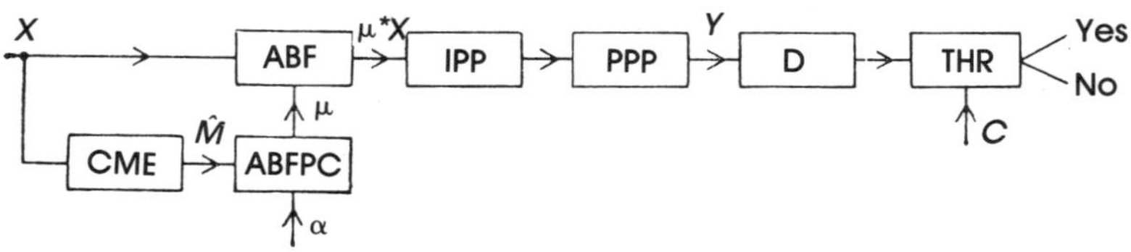

Fig. 8.2 Block diagram for a quasioptimal decision rule when detecting a slowly fluctuating signal.

As is evident, the block diagram in Fig. 8.2 has only a single channel for intrapulse and pulse-to-pulse processing. In order to operate, however, this system generally needs a delay of the input information by the interval T, which is necessary to calculate the weight vector μ^*. Both optimal and quasioptimal decision rules for detecting quasideterministic signals may be generalized to the case of multiple-alternative detection, i.e., detection and discrimination of signals differing, for example, in time delay or Doppler shift [3].

For a detection problem which is symmetric relative to the alternatives, the generalized algorithm (8.1) will take the form

$$\max_l f(Q_p^l, Q_{p-1}^l) \geqslant C \tag{8.10}$$

The corresponding block diagram differs from that shown in Fig. 8.1 in that the processing before the block $f(Q)$ is performed in parallel for each possible lth signal form, determined by the vectors $\mathbf{a}^l$ and $\mathbf{b}^l$, and then the largest statistic is used to compare $f(Q^1)$ with the threshold.

For the case of multiple-target detection with an unknown number of contributing signals and their parameters (see Sec. 7.2), the front end of the decision rule is changed accordingly, to include combination of signals with different parameters, which may be present simultaneously. Quasioptimal algorithms for multiple-target detection may be designed according to the principles studied in Sec. 7.2. The block diagram of such detectors will be the same as before, except for the additional noncoherent processing.

Up to this time we have considered the detection of a slowly fluctuating signal with $K = 1$. Using methods for combining independent detection channels as in Ch. 7 and Korado [2.4], it is not difficult to show that when detecting a rapidly fluctuating signal, the local optimum decision rule differs from that in Fig. 8.1, in that the statistics Q_P and Q_{p-1} are calculated in each intrapulse processing interval, and the coherent pulse-to-pulse processing becomes noncoherent post-detection combination (integration) of the corresponding statistics. The block diagram of such a detector is shown in Fig. 8.3, where NI is noncoherent integration of the statistics Q_P and $Q_P - Q_{p-1}$, obtained in sequential repetition periods. The interchannel correlation matrix is calculated at the same intervals as in Fig. 8.1. The structure of the simplified quasi-optimal detector of (8.7) is altered similarly, and is shown in Fig. 8.4 for the detection of a rapidly fluctuating signal. As may be seen, the receiver in this design differs from the corresponding receiver with no adaptive processing, in that it provides for the estimation of the interchannel correlation matrix and calculation of the adaptive beamformer weighting parameters.

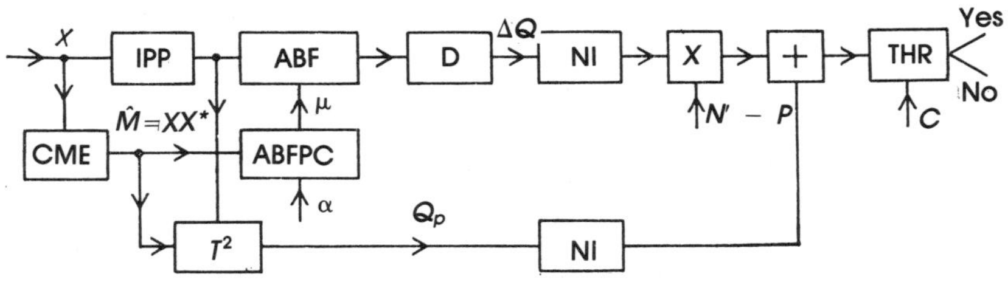

Fig. 8.3 Block diagram for a locally optimum decision rule, when detecting a rapidly fluctuating signal. NI = noncoherent integration.

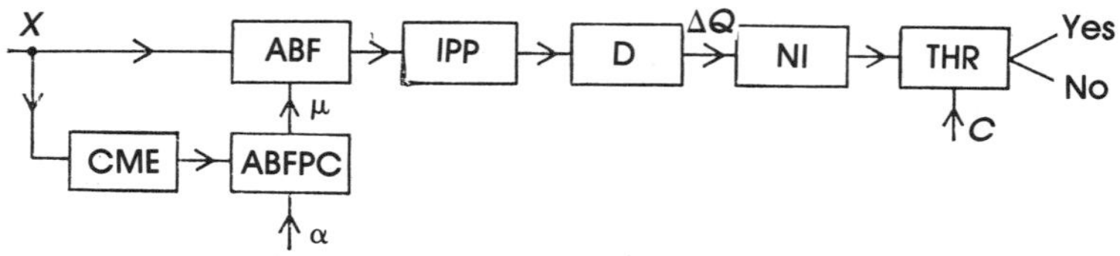

Fig. 8.4 Block diagram for a quasioptimal decision rule, when detecting a rapidly fluctuating signal.

The basic structure of the algorithms presented above (see Figs. 7.3 and 8.4) does not change for $K > 1$. In this case, the noncoherent integration (NI) should encompass the entire interval of pretrack processing. For the case which has been discussed, when the interference is stationary, the interchannel correlation matrix should be estimated using information from the entire observation interval.

An analogous approach to the $K > 1$ case is justified for the detection of a slowly fluctuating signal using a locally optimum approach, considered earlier. In this case, post-detection integration and refinement of the correlation matrix estimate are added to Fig. 8.1. For the quasioptimum receiver of Fig. 8.2., the estimate need not be refined, and μ may be calculated using any sufficient number of samples of the signal-plus-interference mixture, or of just the interference, using the same correlation matrix $\mathbf{M}$ as in the primary sample.

The normalized factor $(\vec{\alpha}^*\mathbf{M}^{-1}\vec{\alpha})^{-1/2}$ in the vector μ may be calculated from the data at the output of the adaptive beam-forming system with the weighted vector $\mathbf{h} = \hat{\mathbf{M}}^{-1}\vec{\alpha}$. Actually, if we denote the row vector for the unnormalized output of the adaptive beamformer with $\mathbf{V}_\alpha^* = \mathbf{h}^*\mathbf{X} = \vec{\alpha}^*\mathbf{M}^{-1}\mathbf{X}$, then

$$\mathbf{V}_\alpha^*\mathbf{g}^{*T}\ \mathbf{h}^*\mathbf{X}\mathbf{g}^{*T} = N^{1/2}\mathbf{h}\overline{\mathbf{X}} = N^{1/2}\alpha^*\hat{\mathbf{M}}^{-1}\overline{\mathbf{W}}$$

and

$$\mathbf{V}_\alpha^*\mathbf{V}_\alpha = \vec{\alpha}^*\hat{\mathbf{M}}^{-1}\vec{\alpha}$$

In this case, the block diagram of Fig. 8.2 is replaced by that in Fig. 8.5, for $K > 1$. When detecting-discriminating single signals or groups of signals

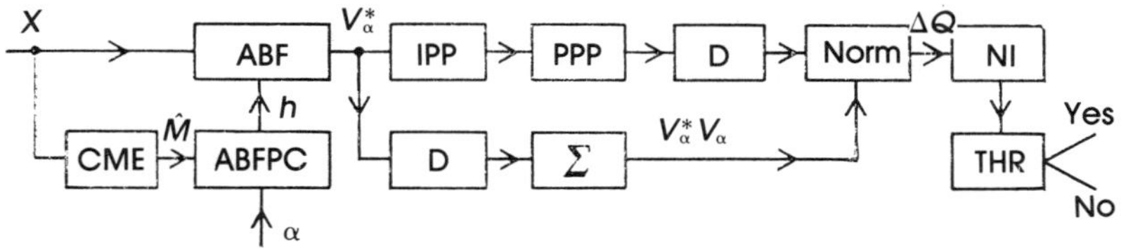

Fig. 8.5 Block diagram for a quasi-optimal decision rule, when detecting a slowly fluctuating signal for $k > 1$. NORM = normalization.

with optimally normalized vectors $\mathbf{a}_l$ and $\mathbf{b}_l$, the intrapulse and pulse-to-pulse processors perform unitary linear transformations of the form $U\mathbf{V}_\alpha = \mathbf{V}_U$, for which $Tt_\alpha = \mathbf{V}_\alpha^*\mathbf{V}_\alpha = \mathbf{V}_U^*\mathbf{V}_U = \vec{\alpha}^*\hat{\mathbf{M}}^{-1}\vec{\alpha}$.

This last equation may be used to calculate the normalizing factor from an estimate of the interference strength $\mathbf{V}_U^*\mathbf{V}_U$ after intrapulse and pulse-to-pulse processing and square-law detection, that is, by post-detection integration in the range-velocity resolution cells, as was done in Sec. 7.2. The corresponding version of the block diagram in Fig. 8.5 is shown in Fig. 8.6. In practice, the system of vectors $\mathbf{a}_l$ and $\mathbf{b}_l$, which determine the spacing of the resolution elements, may be incomplete and not orthogonal; in this case, however, they are generally approximately orthogonal, and the normalizing factor may be estimated as in Fig. 8.6. There is the additional possibility that the normalization may be performed with respect to separate groups of elements, for which the interchannel correlation matrix values are almost identical. The interference power estimate and normalization may also be performed after noncoherent integration NI, wherein the normalization is replaced by automatic control of the threshold, proportional to the estimate of the interference power.

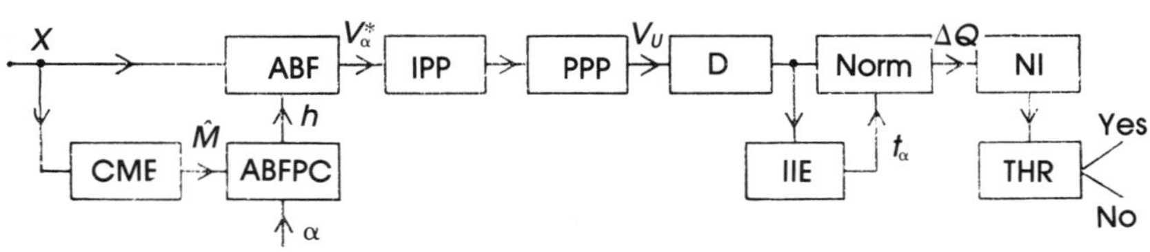

Fig. 8.6 A version of the quasi-optimal decision rule for detecting a slowly fluctuating signal for $k > 1$, in which the normalization factor is calculated after detection. IIE = interference intensity estimation.

8.4 DETECTING SIGNALS IN NONSTATIONARY ACTIVE INTERFERENCE

For nonstationary interference, Z_{rijk} are independent complex random quantities with $E[Z_{rijk}] = 0$ and $E[Z^*_{rijk}Z_{tijk}] = m_{rtk}$, where $\mathbf{M}_k = \{m_{rtk}\}$ are different $p \times p$ interchannel correlation matrices, unknown *a priori*. For the case of very nonstationary interference, $E[Z^*_{rijk}Z_{tijk}] = m_{rtjk}$, for which $\mathbf{M}_{jk} = \{m_{rtjk}\}$ are different correlation matrices, unknown *a priori*.

When detecting a slowly fluctuating signal, assuming, as in Sec. 7.3, that $\nu_{jk} = \nu_k$ are different and unknown *a priori*, one of the methods for optimum combination of the radar information when there are repeated samples [4] is the optimum decision rule for detection of a quasideterministic signal in nonstationary interference with unknown interchannel correlation matrix, which has the form:

$$\sum_k^K \ln f(Q_{p,k}, Q_{p-1,k}) \geqslant C \tag{8.11}$$

Here $Q_{p,k}$ and $Q_{p-1,k}$ are statistics, calculated separately with (8.2) for each of the k samples, i.e.:

$$\begin{aligned} Q_{p,k} &= N\overline{\mathbf{W}}_k^*\hat{\mathbf{M}}_k^{-1}\overline{\mathbf{W}}_k \\ Q_{p,k} - Q_{p-1,k} &= N|\vec{\alpha}^*\hat{\mathbf{M}}_k^{-1}\overline{\mathbf{W}}_k|^2/\vec{\alpha}^*\hat{\mathbf{M}}_k^{-1}\vec{\alpha} \end{aligned} \tag{8.12}$$

where

$$\begin{aligned} \hat{\mathbf{M}}_k &= \mathbf{W}_k\mathbf{W}_k^* = \mathbf{X}_k\mathbf{X}_k^*, \quad N^{1/2}\overline{\mathbf{W}}_k \\ &= \sum_j b_j^* \sum_i a_i^*\mathbf{X}_{ijk}, \quad N = mn \end{aligned} \tag{8.13}$$

The function f in (8.11) is analogous to $f(Q_p, Q_{p-1})$ in (8.3).

The block diagram for the optimal detection of a slowly fluctuating signal in nonstationary active interference differs from that in Fig. 8.1 in the optimal noncoherent integration of the results of processing in each of the k samples, which is performed directly before the threshold comparison. This block diagram is shown in Fig. 8.7, where $f(Q_k)$ is an invariant statistic, having the following forms for the different decision rules which are analogous to (8.3)–(8.7):

- for the minimax decision rule:

$$f(Q_k) = (N - p)\ln[1 + \gamma_0(1 - Q_{p-1,k})] - (N - p + 1)\ln[1 + \gamma_0(1 - Q_{p,k})] \quad (8.14)$$

- for the local minimax decision rule:

$$f(Q_k) = (N - p)(Q_{p,k} - Q_{p-1,k}) + Q_{p,k} \quad (8.15)$$

- for the asymptotically optimum decision rule ($\gamma \to \infty$):

$$f(Q_k) = (N - p)\ln(1 - Q_{p-1,k}) - (N - p + 1)\ln(1 - Q_{p,k}) \quad (8.16)$$

- for the maximum likelihood decision rule, among those which are G_T-invariant:

$$f(Q_k) = f_1(I_k) = \begin{cases} \rho - (N - p)\ln(1 - I_k) - \ln I_k & \text{for } I_k \geqslant 1/(N - p + 1) \\ 0 & \text{for } I_k < 1/(N - p + 1) \end{cases} \quad (8.17)$$

where $N = mn$, $I_k = (Q_{p,k} - Q_{p-1,k})/(1 - Q_{p-1,k})$; $\rho = (N - p)\log(N - p) - (N - p + 1)\ln(N - p + 1)$;

- for the simplified quasioptimal decision rule:

$$F(Q_k) = Q_{p,k} - Q_{p-1,k} \quad (8.18)$$

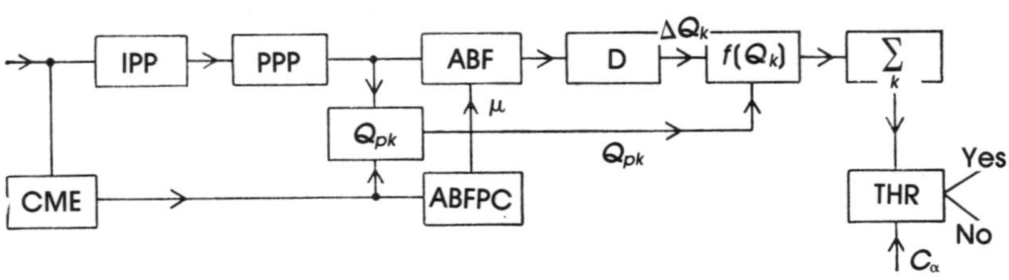

Fig. 8.7 Block diagram of the optimum decision rule for detecting a slowly fluctuating signal in nonstationary interference.

As may be seen, the decision rules (8.15) and (8.18) differ from (8.4) and (8.7) only in the summation of the k statistics over k; (8.14) and (8.16) differ in the logarithmic nonlinear transformation and summation over k; and (8.17) in the specific nonlinear transformation and summation over k. The nonlinear transformation for the maximum likelihood decision rule is plotted in Fig. 8.8, using natural logarithms. [See Zakharov, S.I. and V.A. Korado, Combining independent detection channels in interference with unknown strength suing the maximum likelihood criterion. *Radiotekhnika i elektronika,* 1982, Vol. 64 No. 1, p. 61.] This curve is well approximated by a simple quadratic transformation with a shift, of the form:

$$f_2(I_k) = A[I_k - 1/(N - p + 1)]^2 \quad \text{for } I_k \geqslant 1/(N - p + 1),$$

where A = const, and $f_2(I_k) = 0$ for $I_k < 1/(N - p + 1)$.

As may be seen in Fig. 8.8, just as in Sec. 7.3, the nonlinear transformation excludes the contribution of I_k not exceeding the threshold value $I_0 = 1/(N - p + 1)$, and strongly accentuates those statistics with high I_k. This facilitates more effective noncoherent integration of the signals when the generalized signal-to-interference ratios γ_k are widely varying over k, which may be caused by changes in the strength of both the interference and the useful target signal. An analogous generalization for the case of nonstationary interference holds for locally-optimum detection of rapidly fluctuating signals, and multiple-alternative and multiple-target signal detection.

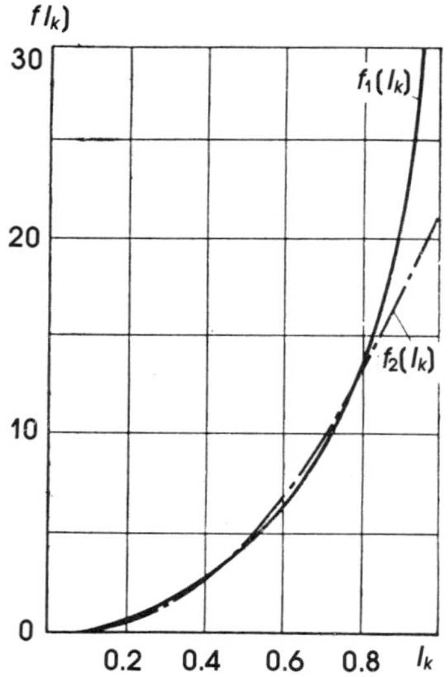

Fig. 8.8 Response of the nonlinear element for the maximum likelihood decision rule with $p = 10$, $N = 20$.

As an example, Fig 8.9 shows the block diagram of a quasi-optimal multiple-target receiver for rapidly fluctuating signals in very nonstationary interference. The notations in the diagram are: $Q^l_{j,k} = (Q^l_{p,j,k}\, Q^l_{p-1,j,k})$ are invariantly sufficient statistics for the (j, k)th interval T_{jk} of intrapulse processing for the lth signal form; $\Delta Q^l_{j,k} = Q^l_{p,j,k} - Q^l_{p-1,j,k}$ is noncoherent pulse-to-pulse signal integration; Σ_k is a noncoherent combination of the information from the intervals T_k. The form of the function $f(Q^l_{jk})$ is analogous to the functions $f(Q_k)$ in (8.14)–(8.18). Here, l is the number of one of L signal forms, different combinations of which correspond to different groups of targets.

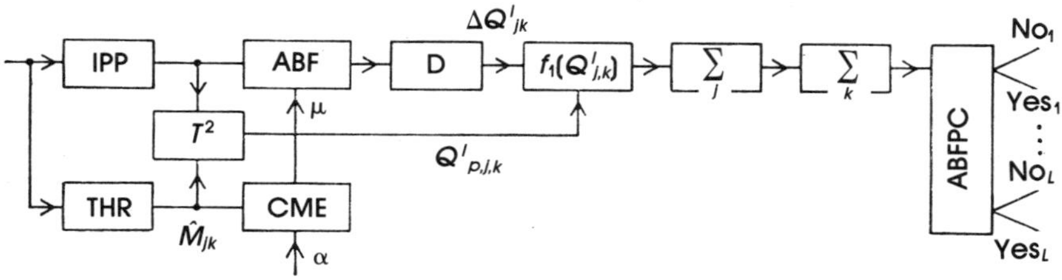

Fig. 8.9 Block diagram for a quasioptimal decision rule for multiple-target detection of rapidly fluctuating signals.

In cases when it is necessary to process signals arriving from different directions, the processing described above should be performed for each of the vectors $\vec{\alpha}_\beta$, $\beta = \beta_1, \ldots, \beta_q$, where q is the number of possible directions from which the signal may be arriving, equal to the number of partial adaptive beams. The possible number of signal forms S increases accordingly. It is possible to use a single estimate of the interchannel correlation matrix of the form $\hat{\mathbf{M}}_k = \mathbf{X}_k\mathbf{X}^*_k$ to form the adaptive beams in all q interesting directions β. With a large number of angle (azimuth, for example) processing channels, a realization of the optimum algorithm (see Fig. 8.9), similar to those in Figs. 8.1 and 8.7, may make sense, since in this case the number of spatial processing channels is usually not reduced after adaptive beamforming, and the correlation matrices $\mathbf{M}_k$ may be estimated simultaneously during intrapulse and pulse-to-pulse processing. With a complete orthonormal set of vectors ($\mathbf{a}_l$ and $\mathbf{b}_l$), it is possible to estimate $\mathbf{M}_k$ even after the intrapulse and pulse-to-pulse processing. In fact, let

$$N^{1/2}\overline{\mathbf{W}}_l = \mathbf{X}\mathbf{g}^{*T}_l = \sum_j b^*_{jl} \sum_i a^*_{il}\mathbf{X}_{il} \tag{8.19}$$

Then, considering the completeness and orthogonality of the system of vectors $\mathbf{a}_l$ and $\mathbf{b}_l$, we have

$$\hat{\mathbf{M}} = \mathbf{X}\mathbf{X}^* = \sum_l \overline{\mathbf{W}}_l \overline{\mathbf{W}}_l^* \tag{8.20}$$

In particular, the structure of the simplified quasioptimal algorithm (see Fig. 8.2) for detecting slowly fluctuating signals for q directions in a multiple-target version, and with $K > 1$, takes on the form shown in Fig. 8.10.

In practice, the system of vectors $\mathbf{a}_l$ and $\mathbf{b}_l$, determining the separation of the range (time) and velocity (frequency) resolution cells, are not always complete and strictly orthonormal. In these cases, however, the processing of Fig. 8.10 remains feasible, and with $N \geqslant 10p$, is sufficiently effective.

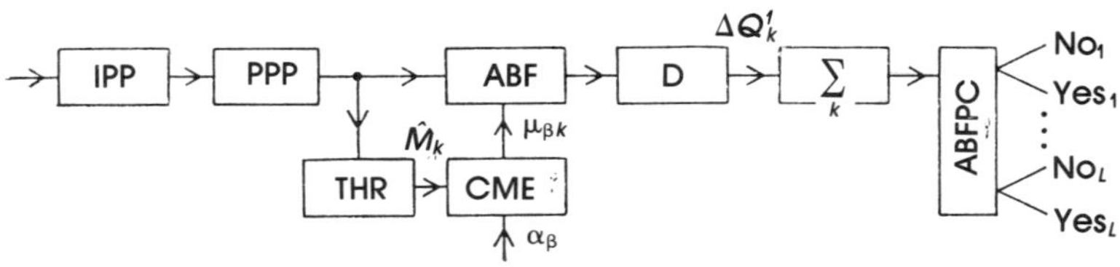

Fig. 8.10 Block diagram for a quasioptimal decision rule for multiple-target detection of slowly fluctuating signals.

8.5 DETECTING A SIGNAL IN CLUTTER WITH UNKNOWN FLUCTUATION SPECTRUM

In this case, the quantities Z_{rijk} are correlated in j, with the corresponding correlation matrix unknown *a priori*. Using the clutter model of Sec. 7.4, when detecting a slowly fluctuating signal, after quasiunitary linear intrapulse transformation $\mathbf{B}_1$ and pulse-to-pulse discrete Fourier transformation $\mathbf{B}_2$ with preliminary weighting, which serves to improve the frequency selectivity of the corresponding filters [9], all of which occurs during each interval T_k, the problem is reduced to that of multichannel detection of a random signal, with the transformed discrete sample $\mathbf{V} = \mathbf{X}\mathbf{B}^T = \{\mathbf{V}_{ijk}\}$, where $E[V_{ijk}] = 0$; $E[\mathbf{V}_{ijk}\mathbf{V}_{ijk}^*] = \mathbf{M}_{jk}$ when there is no signal, $\mathbf{V}_{ijk} = U_{jk}\alpha + \xi_{ijk}$ when the signal is present in the lth frequency resolution element, where $E[|U_{jk}|^2] = \nu_{jk} = 0$ for $j \neq l$ and $E[\xi_{ijk}] = 0$; $E[\xi_{ijk}\xi_{ilk}^*] = \mathbf{M}_{jk}$.

The signal intensities $\nu_{lk} > 0$ are different and unknown over k, and the interchannel correlation matrices of the interference $\mathbf{M}_{jk}$ are assumed to be different and unknown *a priori* over j and k. The problem consists of checking the composite hypothesis H_0: $\gamma_{lk} = 0, l = 1, \ldots, m; k = 1, \ldots, K$ against the composite alternative H_0: $\gamma_{lk} \geqslant \gamma_0$ and $\gamma_{qk} = 0, q \neq l$, where $\gamma_{lk} = \nu_{lk}\alpha^*\hat{\mathbf{M}}_{lk}^{-1}\alpha$ with l known *a priori*. For such a problem, the optimum decision rule for detecting a signal for a number of criteria, including the minimax, locally minimax and invariant maximum likelihood criteria, has the form:

$$\sum_k f(Q_{p,l,k}, Q_{p-1,l,k}) \geqslant C \tag{8.21}$$

where

$$Q_{p,l,k} = N\mathbf{V}_{dolk}^*\hat{\mathbf{M}}_{lk}^{-1}\mathbf{V}_{dolk}$$

$$Q_{p,l.k} - Q_{p-1,l,k} = N\,|\vec{\alpha}^*\hat{\mathbf{M}}_{lk}^{-1}\mathbf{V}_{dolk}|^2/\vec{\alpha}^*\hat{\mathbf{M}}_{lk}^{-1}\vec{\alpha} \tag{8.22}$$

$$\hat{\mathbf{M}}_{lk} = \sum_d \mathbf{V}_{dlk}\mathbf{V}_{dlk}^*, \quad \mathbf{V}_{dlk} = \sum_j b_{lj}^* \sum_i a_{di}^*\mathbf{X}_{ijk} \tag{8.23}$$

and $f_1(Q_{l,k})$ are functions, analogous to (8.14)–(8.18). It is assumed that the weighted pulse-to-pulse processing vectors $\mathbf{b}_l$ provide practically non-overlapping gains in the different frequency channels of the pulse-to-pulse processing, and the estimate $\hat{\mathbf{M}}_{lk}$ is calculated by averaging over the range resolution cells, over the index d. Equation (8.21) differs from the decision rule (8.11) in that the spatial processing is performed for each lth Doppler resolution element, such that, in contrast with (8.12), only $\mathbf{V}_{lk} = \{V_{rdlk}\}$ enters the estimate of $\mathbf{M}_{lk}$.

An intermediate version is also possible, in which the spatial processing is performed independently not for separate frequency resolution elements, but for groups of these elements, within which the interchannel correlation matrix values may be considered identical. Taking account of the indicated differences, the block diagrams for optimal and quasi-optimal decision rules are analogous to those for the detection of a slowly fluctuating signal in nonstationary interference with unknown interchannel correlation matrix (see Figs. 8.7 and 8.10).

When detecting a rapidly fluctuating signal with unknown fluctuation spectrum, if the signal is present the values ν_{jk} are nonzero, different, and unknown *a priori* for all j and k. The detection problem consists of checking the composite hypothesis H_0: $\gamma_{lk} = 0, l = 1, \ldots, m; k = 1, \ldots, K$, against the composite alternative:

$$H_1: \gamma_{lk} \geqslant \gamma_0, \quad \text{where } \gamma_{lk} = \nu_{kl}\vec{\alpha}^* \mathbf{M}_{lk}^{-1}\vec{\alpha}$$

Thus, the useful signal is contained in all frequency resolution elements and optimal processing include combining the information over the indices l and k. The corresponding decision rule takes the form:

$$\sum_{k,l} f(Q_{p,l,k}), Q_{p-1,l,k}) \geqslant C \tag{8.24}$$

where $Q_{p,l,k}$ and $Q_{p-1,l,k}$ have the same meaning as in (8.22), and the functions $f(Q_{l,k})$ are analogous to (8.14)–(8.18). The block diagrams for optimal and quasi-optimal detection in this case differ from those for the case of detection in nonstationary interference, and, in particular, from the diagrams of Figs. 8.7 and 8.10, in that $\mathbf{M}_{lk}$ is estimated for each l and k only by the data $\mathbf{V}_{lk}$.

8.6 ESTIMATING THE PARAMETERS OF SIGNALS IN INTERFERENCE WITH UNKNOWN ANGULAR DISTRIBUTION

As before, we will consider the example of estimating the angle of arrival of the signal. The signal model corresponds to that described earlier (see Sec. 7.7), except for the fact that we will consider a signal with unknown deterministic amplitude (to simplify the estimation problem). It may be shown that with a large number of input samples, the estimate for a random signal will be close to the estimate for the case considered here. We will assume additive Gaussian interference with unknown interchannel correlation matrix $\mathbf{M}$. At first, we will take $K = 1$. With normalized $\alpha(\beta)$ (see Sec. 7.1), the column vectors $\alpha(\beta_0)$ and $\alpha_0 = [\partial\alpha(\beta)/\partial\beta]|_{\beta=\beta_0}$ are orthogonal, we introduce the transformation of the input sample $\mathbf{X}$, $\mathbf{V}_0 = \mathbf{AXT}^T = \mathbf{AW}$, where $\mathbf{A}$ is a unitary $p \times p$ matrix, the last two rows of which are proportional to $\alpha^*(\beta_0)$ and $\alpha_0^*(\beta_0)$, respectively, and $\mathbf{T}$ is a preliminary unitary transformation of the signal in the time domain (see Sec. 8.3). After this transformation, the correlation matrix will be equal to the product $\mathbf{AMA}^*$, and only the two last components of the average transformed input vector will be nonzero. We go to new coordinates, and place the useful signal in $\mathbf{W} = \mathbf{XT}^T$ in the form:

$$\mathbf{S}_W = U_1\vec{\alpha}_0 + U_2 \frac{\dot{\vec{\alpha}}_0}{(\dot{\vec{\alpha}}_0^* \dot{\vec{\alpha}}_0)^{1/2}} \tag{8.25}$$

where $\vec{\alpha}_0 = \vec{\alpha}(\beta_0)$ and $\dot{\vec{\alpha}}_0 = [\partial\alpha(\beta)/\partial\beta]|_{\beta_0}$; β_0 is the reference direction.

The signal vector in $\overline{\mathbf{V}}_0$ will then have the components $0, \ldots, 0, U_1, U_2$. Now the problem is reduced to that of estimating the two nonzero components of the average value of the Gaussian random vector $\overline{\mathbf{V}}_0$ with unknown correlation matrix.

This problem, in essence, is solved in Giri [14], which considers maximum likelihood tests for verifying the hypothesis that the p-vector of the average:

$$\mathbf{a} = \begin{pmatrix} \mathbf{a}_1 \\ \mathbf{a}_2 \end{pmatrix}$$

with an unknown complex correlation matrix. A maximum likelihood estimate is made of the nonzero subvector $\mathbf{a}_2$: $\hat{\mathbf{a}}_2 = \xi_2 - \mathbf{S}_{21}\mathbf{S}_{11}^{-1}\xi_1$, where

$$\vec{\xi} = \begin{pmatrix} \xi_1 \\ \xi_2 \end{pmatrix}$$

is the sample average, and

$$\mathbf{S} = \begin{pmatrix} \mathbf{S}_{11} & \mathbf{S}_{12} \\ \mathbf{S}_{21} & \mathbf{S}_{22} \end{pmatrix}$$

is the sample correlation matrix. In our case:

$$\vec{\xi} = \overline{\mathbf{V}}_0, \quad \mathbf{S} = \mathbf{V}_0\mathbf{V}_0^*, \quad \mathbf{a}_2 = \begin{pmatrix} \mathbf{U}_1 \\ \mathbf{U}_2 \end{pmatrix}$$

Transforming to the sample $\mathbf{W} = \mathbf{A}^*\mathbf{V}_0$, we obtain $\mathbf{S} = \mathbf{V}_0\mathbf{V}_0^* = \mathbf{AWW}^*\mathbf{A}^* = \mathbf{AMA}^*$, where $\mathbf{M} = \mathbf{XX}^* = \mathbf{WW}^*$ is the matrix proportional to the sample spatial correlation matrix. Placing the transformation matrix $\mathbf{A}$ in the form

$$\mathbf{A} = \begin{pmatrix} \mathbf{A}_1 \\ \mathbf{A}_2 \end{pmatrix}$$

where $\mathbf{A}_2$ is a $2 \times p$ submatrix of the matrix $\mathbf{A}$, having the form

$$\begin{pmatrix} \alpha_1^* & \ldots & \alpha_p^* \\ k_0^{-1}\dot{\alpha}_1^* & \ldots & k_0^{-1}\dot{\alpha}_p \end{pmatrix}, \quad k_0 = (\vec{\alpha}_0^*\vec{\alpha}_0)^{1/2}$$

we write

$$\mathbf{S} = \begin{pmatrix} \mathbf{S}_{11} & \mathbf{S}_{12} \\ \mathbf{S}_{21} & \mathbf{S}_{22} \end{pmatrix} = \begin{pmatrix} \mathbf{A}_1\hat{\mathbf{M}}\mathbf{A}_1^* & \mathbf{A}_1\hat{\mathbf{M}}\mathbf{A}_2^* \\ \mathbf{A}_2\hat{\mathbf{M}}\mathbf{A}_1^* & \mathbf{A}_2\hat{\mathbf{M}}\mathbf{A}_2^* \end{pmatrix}$$

Placing this in the expression for the estimate $\hat{\mathbf{a}}_2$, we obtain

$$\begin{pmatrix} \hat{U}_1 \\ \hat{U}_2 \end{pmatrix} = \mathbf{A}_2\overline{\mathbf{W}} - \mathbf{A}_2\hat{\mathbf{M}}\mathbf{A}_1^*(\mathbf{A}_1\hat{\mathbf{M}}\mathbf{A}_1^*)^{-1}\mathbf{A}_1\overline{\mathbf{W}},$$

whence, because the matrix $\mathbf{A}$ is unitary,

$$(\mathbf{A}_1\overline{\mathbf{M}}\mathbf{A}_1^*)^{-1} = \mathbf{A}_1\hat{\mathbf{M}}^{-1}\mathbf{A}_1^* \quad \text{and} \quad \mathbf{A}_1^*\mathbf{A}_1 + \mathbf{A}_2^*\mathbf{A}_2 = \mathbf{I}$$

where $\mathbf{I}$ is the unity matrix, we find that

$$\begin{pmatrix} \hat{U}_1 \\ \hat{U}_2 \end{pmatrix} = (\mathbf{A}_2\mathbf{M}^{-1}\mathbf{A}_2^*)^{-1}\mathbf{A}_2\hat{\mathbf{M}}^{-1}\overline{\mathbf{W}} = C^{-1}\begin{pmatrix} \Sigma \\ \Delta \end{pmatrix}$$

where

$$\mathbf{C} = \begin{pmatrix} c_{11} & c_{12} \\ c_{21} & c_{22} \end{pmatrix} = \begin{pmatrix} \vec{\alpha}_0^*\hat{\mathbf{M}}^{-1}\vec{\alpha}_0 & \vec{\alpha}_0^*\hat{\mathbf{M}}^{-1}\vec{\alpha}_0 \\ \vec{\alpha}_0^*\hat{\mathbf{M}}^{-1}\vec{\alpha}_0 & \vec{\alpha}_0^*\hat{\mathbf{M}}^{-1}\vec{\alpha} \end{pmatrix}$$

is the sample interchannel correlation matrix at the output of the generalized sum-and-difference channels $\Sigma = \vec{\alpha}_0^*\hat{\mathbf{M}}^{-1}\overline{\mathbf{W}}$ and $\Delta = \vec{\alpha}_0^*\hat{\mathbf{M}}^{-1}\overline{\mathbf{W}}$ (where the overscore denotes the average). After elementary transformations, we write

$$\hat{U}_1 = (\det \mathbf{C})^{-1}(c_{22}\Sigma - c_{12}\Delta)$$

$$\hat{U}_2 = k_0(\det \mathbf{C})^{-1}(c_{11}\Delta - c_{21}\Sigma)$$

Finally, from the estimates for U_1 and U_2, and the equality $\beta - \beta_0 = k_0^{-1}\,\mathrm{Re}(U_1/U_2)$, we obtain the estimate $\hat{\beta}$ in the form:

$$\hat{\beta} - \beta_0 = \mathrm{Re}\left(\frac{c_{11}\Delta - c_{21}\Sigma}{c_{22}\Sigma - c_{12}\Delta}\right) \tag{8.26}$$

Here, as before, for a slowly fluctuating signal:

$$\overline{\mathbf{W}} = N^{-1/2} \sum_{j=1}^{m} b_j^* \sum_{i=1}^{n} a_i^* \mathbf{X}_{ij}$$

In the case of a rapidly fluctuating signal, the estimate $\hat{\beta}$ over the part of the sample **X** corresponding to a coherent interval T_{jk} is also described by (8.26), where

$$\overline{\mathbf{W}} = N^{-1/2} \sum_{i=1}^{n} a_i^* \mathbf{X}_{ij}$$

The block diagram of the processing for a slowly fluctuating signal is shown in Fig. 8.11.

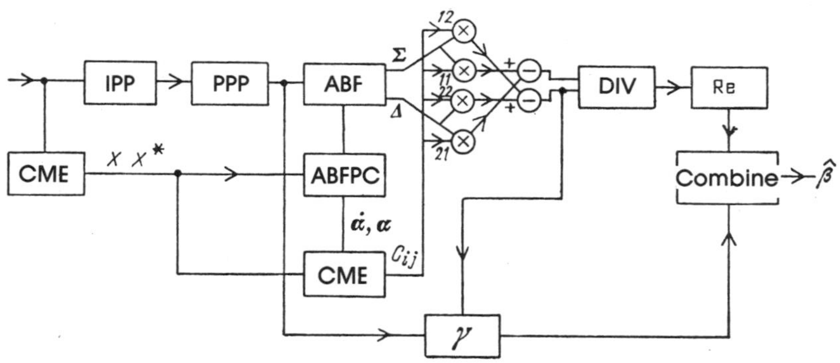

Fig. 8.11 Block diagram of the algorithm for estimating the angle of arrival β of a slowly fluctuating signal in stationary interference with unknown angular distribution. DIV = division.

Due to nonstationary active interference, the interchannel correlation matrices may be different in each interval T_k, and unknown *a priori*. In this case, as above, it is necessary to combine estimates obtained from the different intervals T_k. As before (see Ch. 7), this may be accomplished using the structure of that linear estimate which minimizes the mean square error, i.e., the estimate $\hat{\beta}$ for the range-velocity element $d_0 l_0$ is equal to

$$\hat{\beta} - \beta_0 = \frac{1}{\sum_{k=1}^{K} (1 + \gamma_k)} \sum_{k=1}^{K} \frac{(\gamma_k + 1)^2}{\gamma_k} (\hat{\beta}_k - \beta_0) \tag{8.27}$$

where

$$\hat{\beta}_k - \beta_0 = \mathrm{Re}\left(\frac{c_{11k}\Delta_k - c_{21k}\Sigma_k}{c_{22k}\Sigma_k - c_{12k}\Delta_k}\right)_{d_0l_0}$$

and the index k corresponds to the substitutions in the foregoing equations of $\hat{\mathbf{M}}_k = \mathbf{W}_k\mathbf{W}_k^*$ for $\hat{\mathbf{M}}$, and $\mathbf{W}_k$ for $\mathbf{W}$:

$$\hat{\gamma}_k = \frac{|c_{22k}\Sigma_k - c_{12k}\Delta_k|^2_{d_0,l_0}(mn - 1)}{\left[\sum_{d,l \neq d_0,l_0}^{n,m} |c_{22k}\Sigma_k - c_{12k}\Delta_k|^2_{d,l}\right]} - 1$$

Here the value $\mathbf{W}_k$ for a slowly fluctuating signal is analogous to the previous case (see Fig. 8.11). For a rapidly fluctuating signal, in addition to combining values of k, it is necessary to do so over j. Summation over j is then added to Eq. (8.27), and the values $\mathbf{W}$ correspond to the values $\mathbf{W}$ for the rapidly fluctuating signal in the previous case. The estimate of the angle of arrival in clutter is analogous to that considered earlier (see Sec. 7.6), namely,

$$\begin{aligned} \hat{\beta} - \beta_0 &= \frac{1}{\sum_{k=1}^{K}(1 + \gamma_k)} \sum_{k=1}^{K} \frac{(\gamma_k + 1)^2}{\gamma_k}(\hat{\beta}_k - \beta_0) \\ \hat{\gamma}_k &= \frac{|c_{22k}\Sigma_k - c_{12k}\Delta_k|^2_{d_0,l_0}(n - 1)}{\left[\sum_{d \neq d_0}^{n} |c_{22k}\Sigma_k - c_{12k}\Delta_k|^2_{d,l_0}\right]} - 1 \end{aligned} \tag{8.28}$$

8.7 THE CHARACTERISTICS OF SIGNAL DETECTION AND ESTIMATION IN INTERFERENCE WITH UNKNOWN INTER-CHANNEL CORRELATION MATRIX

We will first consider the characteristics of detecting slowly fluctuating signals in interference with unknown angular distribution for the case $K = 1$, i.e., for a single sampling at intervals T_k. First of all, we note that for all the G_T-invariant decision rules we have analyzed, the detection characteristics depend on the single parameter $\gamma = \nu\vec{\alpha}^*\mathbf{M}^{-1}\vec{\alpha}$. The value of γ depends not only on the average signal energy and the total interference power in the receiver channels, given by the trace of the matrix $\mathbf{M}$, but also on the spatial distribution of the interference power and on the angle of arrival of the signal, determined by the normalized correlation matrix **M**/(tr **M**) and the normalized vector $\vec{\alpha}$. The value of γ may significantly exceed the ordinary signal-to-interference energy ratio $\gamma' = \nu/(\text{tr}\ \mathbf{M})$, i.e., $\gamma \gg \gamma'$. In general $\gamma \geqslant \gamma'$.

The relative effectiveness of each of the receivers may be estimated using the energy loss η in comparison with the optimum receiver for the signal in the same interference, when the correlation matrix **M** is known *a priori*. As is well known, the decision rule with the most power (highest probability of detection) in this idealized case is the decision rule of the form (see [7, 10]):

$$|\vec{\alpha}^* \mathbf{M}^{-1} \overline{\mathbf{W}}| \geqslant C \tag{8.29}$$

for which the dependence of the probability of detection on γ is given by the equation:

$$D(\gamma) = F^{1/(1+r)} \tag{8.30}$$

The function $D(\gamma)$ is plotted for various values of F in Fig. 8.12. The loss η mentioned above depends on D and F, and, expressed logarithmically, has the form $2 = 10 \ln(\gamma/\gamma_\infty)$, where γ and $\gamma_\infty = (\ln F/\ln D) - 1$ are the threshold signal-to-interference ratios for the particular decision rule and the optimum rule (8.29), respectively. Here, γ_∞ is equal to the analogous ratio for the particular decision rule with unknown **M** and infinitely large number $N \to \infty$ of primary or additional interference samples. The value of the energy loss η, in dB, is shown for the decision rules (8.3), (8.4), (8.5), and (8.7) in Tables 8.1–8.4 for $F = 10^{-5}$ and $D = 0.5$ (see [5]). For the rule (8.3), the loss corresponds to the case $\gamma = \gamma_0$. As is clear, the loss η depends on the number of samples N, on the basis of which the interchannel correlation matrix **M** is estimated, and also on the number of receiver channels p.

Being the invariant rule with the greatest power for $\gamma = \gamma_0$, at the points $\gamma = \gamma_0$ the minimax decision rule has the greatest effectiveness amongst the invariant decision rules. In practice, for values of γ_0 which are not too small, corresponding to $D = 0.5$, for example, the performance of the minimax decision rule is equal to the maximum value in a wide range of values of γ, including γ_0, and corresponding to a range of values for D, including $D = 0.5$. The same is true for the maximum likelihood decision rule (8.5). Other, more simple decision rules—the local minimax and the simplified quasioptimal rules—do not perform quite as well as the minimax and the maximum likelihood decision rules. For large values of the ratio N/p, the difference between the various decision rules disappears.

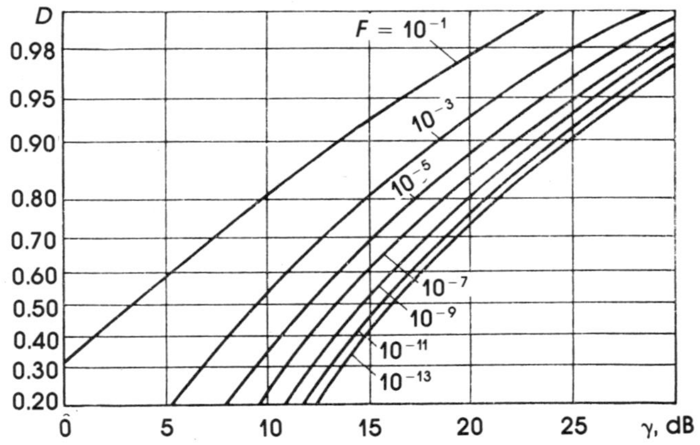

Fig. 8.12 The probability of detection as a function of the signal-to-interference ratio, when the interference correlation matrix is known.

Table 8.1

	Rule (8.3)		
$(N-1)/p$	$p = 2$	$p = 4$	$p = 10$
1	23.1	22.1	22
1.5	18.2	13.0	8.5
2	11.5	7.7	5.0
3	6.5	4.2	2.7
5	3.4	2.2	1.5
10	1.5	1.1	0.7

Table 8.2

	Rule (8.4)		
$(N-1)/p$	$p = 2$	$p = 4$	$p = 10$
1	26	25	24.3
1.5	18.5	15	12.5
2	14	13.8	7.7
3	12.5	6.5	3.2
5	4.2	2.6	1.6
10	1.6	1.1	0.7

Table 8.3

Rule (8.5)			
$(N - 1)/p$	$p = 2$	$p = 4$	$p = 10$
1	23.5	22.4	21.5
1.5	18.2	13	8.5
2	11.5	7.7	5.0
3	6.5	4.2	2.7
5	3.4	2.2	1.5
10	1.5	1.1	0.7

Table 8.4

Rule (8.6)			
$(N - 1)/p$	$p = 2$	$p = 4$	$p = 10$
1	26	25.5	24
1.5	19.5	19.4	16.4
2	16.7	14.4	8.5
3	14.5	7.1	3.3
5	4.5	2.7	1.6
10	1.7	1.2	0.7

The loss η is not great, rapidly decreases with increases in p and N/p, and as is evident from the last rows of the tables, with $p \geqslant 4$ and $N \geqslant 10p$ does not exceed 1.2 dB for all of the decision rules, including (8.6).

Considering that the loss η exhibits a weak dependence on the probability of detection D, the threshold signal-to-interference ratio γ, in dB, may be estimated for all of the decision rules with the equation:

$$\gamma = \gamma_\infty + \eta = 10 \ln(\ln F/\ln D - 1) + \eta \tag{8.31}$$

where η is the value of the loss from Tables 8.1–8.4. The detection characteristics for $\eta = 0$, calculated from (8.30), are shown in Fig. 8.12. We note that for normal operation of the spatial processing algorithms, it is necessary to maintain the condition $N > p$, i.e., with an increase in the number of receiving channels p and the corresponding number of spatially concentrated interference sources which may be canceled, an increase in the number of samples used to estimate the interchannel correlation matrix **M** is also needed. In order that the additional losses η, connected with the *a priori* uncertainty concerning the interference conditions, be kept to a minimum, it is wise to adhere to a more strict requirement, i.e., $N > 10p$. As follows from Secs. 8.3–8.5, the cases of detecting rapidly fluctuating signals, and also the case $K > 1$, differ from that considered above only in the fact that the signals are integrated incoherently. The effectiveness of optimum incoherent integration with $M_k = M$ and $\gamma_k = \gamma_0$ may be estimated satisfactorily as the effectiveness of the method of combining K' of K results of the preliminary threshold comparison, using the best $K' = K'(K)$. Analysis of these characteristics show that the energy losses with optimum incoherent signal integration are close to those already considered above for the case $K = 1$, and values of D_1 and F_1 found from (7.2) and (7.3), and not exceeding them. Therefore, to estimate the detection characteristics for any K, it is possible to use as a baseline the characteristics of the optimum multisample detection algorithm with known signal and interference parameters, i.e., the characteristics of the algorithm of the form:

$$\sum_{k=1}^{K} |\vec{\alpha}^* \mathbf{M}^{-1} \overline{\mathbf{W}}_k|^2 \geqslant C \tag{8.32}$$

subsequently correcting the threshold signal-to-interference ratio by the value of the energy loss η. This same approach is valid for the minimax and particularly for the maximum likelihood decision rules, even in the case when the signal intensities ν_k or the interference intensities ϵ_k significantly differ from one another in the combined samples. For the baseline characteristics we may then use the Neyman-Pearson decision rule with known parameters γ_k and ϵ_k:

$$\sum_{k=1}^{K} \frac{\gamma_k}{1 + \gamma_k} \frac{|\vec{\alpha}^* \mathbf{M}_k^{-1} \overline{\mathbf{W}}_k|^2}{\vec{\alpha}^* \mathbf{M}_k^{-1} \vec{\alpha}} \geqslant C, \quad \mathbf{M}_k = \epsilon_k \mathbf{M} \tag{8.33}$$

The baseline detection characteristics for the decision rules (8.32) and (8.33) are the same as those presented earlier for the decision rules (7.37) in Figs. 7.21 and 7.22. Thus, the characteristics presented in Figs. 8.12, 7.21, 7.22, and Tables 8.1–8.4 may be used to estimate the threshold signal-to-interference ratio for given values of the probability of detection D and probability of false alarm F, in pretrack processing.

The ability to perform spatial selection by the angle of arrival of the signal, in interference with known parameters and a nonuniform angular power distribution, has been demonstrated in a number of works. Thus, Shirman and Manzhos (see references to Ch. 6 [22]), discuss the detection of a signal with a planar uniform antenna array consisting of p elements, nondirectional in azimuth (vertical dipoles, for example). The interference is assumed to be a mixture of isotropic Gaussian interference with intensity ϵ, and random temporally uncorrelated Gaussian interference, received from some fixed azimuth direction ϕ_1 and having intensity ϵ_1.

It was shown that in a wide sector of angles of arrival ϕ_1, the signal-to-interference ratio at the output of the interference canceler was negligibly worse (by tenths of a decibel) than in the case of detecting a signal in isotropic noise, indicating that the locally concentrated interference is practically canceled. Analogous results are obtained for several localized interference sources, the number of which is rather small in comparison with the number of antenna array elements.

Obtaining the required value for γ for a given ratio ν/ϵ is tightly connected with the antenna characteristics. The choice of the number of elements and their geometric arrangement directly determines their influence on the resulting value of γ, and also the economic expenses. As the number of receiving elements p is reduced, the signal-to-interference ratio γ decreases, and extremely rapidly for $p < n$, where n is the number of concentrated interference sources. However, γ grows, if an adaptive beamformer is used with a cancelation beam steered to the interference source.

If the signal is concentrated in a narrow angular sector, then γ will increase without limit if an increase in p is accompanied by an increase in the antenna aperture in the direction of the signal. In practice, however, the finite angular extent of a signal source, $\Delta\beta$, places a limit on the growth of γ. In this case, there is some limit to the width of the antenna aperture, determined by the radius of spatial signal correlation $L_s = \lambda/\Delta\beta$, where λ is the wavelength at which further increase in the antenna aperture is ineffective for a given detection algorithm. The detection capability is reduced due to the loss of signal coherency between distant antenna elements, which is equivalent to a reduction in the antenna gain in the direction to the target. The performance may be improved by using the method of "separated" signal reception, well known in radio communications, in which a large antenna array is divided into spatially separated subarrays, each maintaining coherent processing across its own aperture. The results of this coherent processing in each subarray are combined noncoherently, which is analogous to the method of integrating signals in different resolution elements noncoherently, or noncoherent integration in the time domain after normalization. An analogous effect is provided by the method of detecting a multicomponent signal [6] with optimum combination of the components received from various directions at a large antenna array, or varying the operating frequency or multifrequency operation.

The accuracy of the estimate of the angle of arrival of a signal in uniformly distributed interference will be close to that considered in

Ch. 7. When there are concentrated interference sources, the estimation characteristics will be close to those for uniform interference, if the number of localized interference sources is much less than p, and none of them falls in the main beam. The largest losses will arise do to losses in the adaptive system, as in the case of detection.

In conclusion to these two chapters, it should be noted that the design of an optimum decision rule for signal detection and parameter estimation in interference is almost always the result of compromises between the robustness of the signal and interference models and the actual interference conditions, the accordance of the optimization criteria with the objectives of the designer, the ability to realize the analytical results in hardware, and, finally, the desire to design the system with certain standard approaches. As a result of these compromises, the resulting decision rules will be less than optimal, to some known extent. The decision rules presented in Chs. 7 and 8 are designed to be implemented on specialized computers. The realization of relatively complex decision rules is possible due to the relatively narrow bandwidth of HF receiving equipment, usually on the order of tens of kilohertz.

REFERENCES

1. Kirillov, N.E. *Prostranstbenno-vremennaya obrabotka signalov s podavleniem sosredotochennykh pomekh (Space-time signal processing with suppression of narrow-band interference).* Voprosy Radioelektroniki. Ser. TRS, (1974), no. 3, pp. 21–31.
2. Korado, V.A. *Minimaksnoe obnaruzhenie kvasideterminirovannovo signala na fone gaussovskoy pomekhi s neizvestnoy korrelyatsionnoy matritsey (Minimax detection of a quasideterministic signal in Gaussian interference with unknown correlation matrix).* Radiotekhnika i Elektronika, vol. 23 (1978), no. 2, pp. 326–333.
3. ———. *Optimal'noe mnogokanal 'noe obnaruzhenie-raslichenie kvasideterminirovannykh signalov na fone gaussovskikh pomekh s neizvestnoy mezhkanal 'noy korrelyatsionnoy matritsey (Optimum multichannel detection-resolution of quasideterministic signals in Gaussian interference with unknown interchannel correlation matrix).* Radiotekhnika i Elektronika, vol. 25 (1980), no. 1, pp. 85–92.
4. ———. *Optimal 'noe ob'edinenie nezavisimykh kanalov obnaruzheniya signalov na fone gaussovskikh pomekh s neizvestnymi intensivnostyami (Optimum combination of independent channels in detection of signals in Gaussian interference with unknown intensities).* Radiotekhnika i Elektronika, vol. 17 (1972), no. 3, pp. 618–620.

5. ———. *Kharakteristiki obnaruzheniya kvasideterminirovannovo signala na fone pomekh s neizvestnoy korrelyatsionnoy matritsey (The characteristics of detecting a quasideterministic signal in interference with unknown correlation matrix).* Radiotekhnika i Elektronika, vol. 25 (1980), no. 2, pp. 296–303.
6. Korado, V.A. and Zakharov, S.I. *Obnaruzhenie kvasideterminirovannovo mnogokomponentnovo signala na fone gaussovskoy pomekhi s neizvestnoy korrelyatsionnoy matritsey (The detection of a quasideterministic multiple-component signal in Gaussian interference with unknown correlation matrix).* Radiotekhnika i Elektronika, vol. 25 (1980), no. 3, pp. 647–649.
7. Lavut, A.P. *Obnaruzhenie druzhno fluktuiruyushchevo paketa impul' sov na fone nekorrelirovannoy pomekhi s neizvestnymi parametrami (The detection of a fluctuating packet of signals in uncorrelated interference with unknown parameters).* Radiotekhnika i Elektronika, vol. 10 (1965), no. 8, pp. 1426–1434.
8. Lazutkin, B.A. *Sintez samonastraivayushcheysya po pomekham sistemy obnaruzheniya (The design of detection systems which tune automatically according to the interference).* Radiotekhnika i Elektronika, (1975), no. 11, pp. 2303–2309.
9. Rabiner, L. and Gold, B. *Theory and Applications of Digital Signal Processing.* Prentice-Hall, 1975.
10 Sragovich, V.G. *Ob optimal 'nom obnaruzhenii signala na fone korrelirovannoy gaussovoy pomekhi (Optimum detection of a signal in correlated Gaussian interference).* Radiotekhnika i Elektronika, vol. 4 (1959), no. 5, pp. 745–754.
11. Helstrom, C. *Statistical Theory of Signal Detection.* New York: Pergamon Press, 1968.
12. Shirman, Ya.D. *Rasreshenie i szhatie signalov (Resolution and compression of signals).* Moscow: Sovietskoe Radio, 1974. 360 pp.
13. Davis, R.C., Brennan, L.E., and Reed, L.S. *Angle estimation with adaptive arrays in external noise fields.* IEEE Trans., vol. AES-12 (1976), no. 2, pp. 179–187.
14. Giri, N. *On testing problems concerning mean of multivariate complex gaussian distribution.* Ann. Inst. Statist. Math., vol. 24 (1972), no. 2, pp. 245–250.
15. Griffiths, L.J. *A comparison of quadrature and singlechannel receiver processing in adaptive beamforming.* IEEE Trans., vol. AP-25 (1977), no. 2, pp. 209–218.
16. ———. *Time-domain adaptive beamforming of HF backscatter radar signals.* IEEE Trans., vol. AP-24 (1976), no. 5, pp. 707–720.

17. Reed, T.S., Mallett, J.D., and Brennan, L.E. *Rapid convergence rate in adaptive arrays.* IEEE Trans., vol. AES-10 (1974), no. 6, pp. 853–863.
18. Washburn, T.W. and Sweeney, L.E. *An on-line adaptive beamforming capability for HF backscatter radar.* IEEE Trans., vol. AP-24 (1976), no. 5, pp. 721–731.

Chapter 9
Over-the-Horizon Radar Design Principles

9.1 INTRODUCTION

Over-the-horizon radars operate in particular conditions which are very different from those of ordinary radars operating at higher frequencies. One of the principal differences is the need for continuous optimization of the operating frequency [10, 18] in order to minimize the propagation loss.

Another important difference is the constant presence of intense and numerous sources of unintentional interference from HF radio transmitters.

The sensitivity of radars operating in the higher frequency bands is usually determined by the internal receiver noises; external noises generally do not affect these radars. In the HF band, on the other hand, external interference (atmospheric, cosmic, industrial, and radio transmissions) significantly exceed the internal receiver noises. For normal operation, it is necessary for the radar to analyze the interference conditions in the chosen optimum sub-band, find those channels in which the interference levels are lowest [10, 18], and tune the receiver to these frequencies. This is also necessary to solve the problem of minimizing the interference of the OTH radar on other radio systems [10].

Another feature of these systems is the fact that large areas of the earth's surface are illuminated during operation. The result is that the target signals are, as a rule, observed in powerful clutter, formed by the backscatter from land or sea. The clutter level may exceed the signal level by several orders of magnitude [9, 10, 17].

The useful signal is extracted from clutter in OTH radars using Doppler techniques, inasmuch as the spectral components of the clutter are concentrated close to the frequencies in the transmitted signal spectrum, while the target signal exhibits a Doppler shift corresponding to the radial velocity of the target [9, 10, 17]. Operation in the HF band allows OTH radars to be used to obtain information on the Aurora Borealis, meteors, and sea state [1, 2, 4, 6, 10, 11, 12, 14, 18].

In this last problem, the radar makes use of signals reflected from the sea surface (when locating relatively quickly moving objects, such as aircraft, sea surface reflections constitute clutter). The spectrum of these signals contains substantial information about the nature of the reflecting object (the height of sea waves, and their direction and speed).

9.2 OVER-THE-HORIZON RADAR DESIGN PRINCIPLES

Basic principles for over-the-horizon radars may be formulated based on data presented in the literature on foreign HF radars [1, 6, 7, 10, 11, 12, 18]. It should be noted that OTH radars are Doppler radars, in that they extract the desired signal from clutter on the basis of the Doppler shift of the signals reflected by the targets [10]. Backscatter systems, owing to the significant difficulties in isolating the highly sensitive receiver from the very powerful transmitter, are in most cases built with the transmitter and receiver separated by some distance (from tens to one or two hundred kilometers) [9, 17]. We will consider the design of the different components of OTH radars—the antenna-feed system, the transmitter and the receiver—by describing the characteristics of foreign radars (TEPEE, MADRE, WARF) and their design.

9.2.1 Antenna-Feed Systems

The specific operating conditions of OTH radars, formulated in a number of works by US authors [6, 8, 10, 17, 18], determine the basic requirements for the antenna system. The antenna should have a high gain (20-30 dB) [10, 17], cover a wide-band of frequencies (the frequency overlap coefficient of the radar as a whole should be 5-6 dB [10, 17]), and provide rapid scanning in a wide azimuthal sector [10]. In addition, the transmitting antenna should be able to radiate signals at high power (the average power is on the order of several hundred kilowatts [10]). These requirements indicate that the antenna should be designed as a phased array. Examples of OTH radars with phased array antennas are the US MADRE system (the antenna of which was briefly described in Ch. 1), and the WARF (Wide Aperture Research Facility) complex [17, 18].

To obtain the necessary signal power in the HF band, non-Soviet radars use several transmitters, operating at the radiating elements forming the antenna array [14]. Special phasing elements, connected to the array elements, are used to provide wide-angle scanning by varying the relative phase at the element.

Wide-angle coverage in azimuth is obtained in the receiving antenna with special beam-forming systems, attached to the elements of the receiver array. By switching delay lines of various lengths in the beamformer, it is possible to scan the beam, or to use a splitting phase system in the beamformer to create multiple (fanned) beams. The receivers are connected to the output of the beamformer.

In the elevation plane, the beam must be steered as close as possible to the horizon, due to the propagation properties of HF signals. When using horizontal polarization, the need to place the beam at the horizon leads to a requirement for very high antenna structures. When using vertical polarization, to place the beam near the horizon and to reduce losses in the ground, it is necessary to employ a metallic ground plane in front of the antenna. This fence is usually a net (wire) screen, placed in the ground [14], or, to avoid additional losses in snow cover, placed about 1.5–2 m above the surface.

The main requirement on the radiating elements composing the transmitting antenna is that their input impedance be constant over the operating band and in a given scanning sector. Meeting this requirement is a complicated engineering problem, due to the coupling of the array elements. A shunted wide-band dipole is often used as the basic array element in HF systems.

Log-periodic antennas are also used in a number of foreign OTH radars [5, 9]. Various modifications of log-periodic antennas are widely used: self-supporting and those with stays, horizontal and vertical polarization, symmetrical and asymmetrical inputs.

Examples of log-periodic antennas are the antennas of the experimental radars in the "Polar Cap" program, located in Caribou, Maine; New Kent, Virginia; and Hall Beach in Melville, Canada [4, 14, 17]. These antennas consist of horizontally situated antenna fields, the elements of which are vertical log-periodic dipole antennas, placed parallel to one another. The dipole antennas are suspended in curtains, fastened to terminal towers.

The antenna array in New Kent consists of four log-periodic antennas. In Caribou, the antenna array contains 32 antennas, four of which compose the transmitting section, the remaining 28 making up the receiver array. The transmitting section is designed for power levels up to 800 kW. The receiving antenna provides sidelobe suppression in the receiving pattern to −26 dB. The transmitter array in Hall Beach contains 32 parallel antennas, separated by 12 m, for a total array length of 320 m. The length of each log-periodic antenna is 60 m. The antenna array radiates at a low angle to the horizon. The normal antenna beam is oriented to the northwest, with a coverage of 60°. The radiated pulse power reachs 3 MW. The receiving antenna array has an analogous form, and comprises 28 antennas, separated by 15 m. These three sites operate in a pulsed-Doppler mode. The band of operating frequencies is 6–30 MHz. An example of an OTH radar with a phased-array antenna is the US MADRE system.

9.2.2 The Transmitter

The transmitter comprises two main parts: the transmitter apparatus, and the antenna system. The basic requirements for an OTH radar transmitter operating in the HF band, in accordance with what has been said above, are: large overall width of overlapping operating frequencies; high power output; maximum fidelity of the spectral composition for the given transmitter signal waveforms.

According to [10], if it is required that the system be able to cover the area from 1000–4000 k simultaneously, then it is necessary to be able to choose the operating frequency within the approximate limits of ±25 percent of the nominal operating frequency. The variations in the maximum usable frequency (MUF) relative to its median value, and the requirement for the extent of the covered range, together lead to a requirement for a fully tunable band with a frequency overlap coefficient equal to 2–3 MHZ. For radars with a large azimuthal sector to cover, the necessary band may be 4–32 MHz [10].

With propagation to the object being detected, scattering from that object, and propagation back to the receiver, the overall propagation losses may be extremely high. To maintain a signal level at the receiver input which is sufficient for processing, the average radiated signal power should be from hundreds of kilowatts to several megawatts [10].

This requirement for such a high signal power leads to the design of the transmitting system in the form of a multichannel amplifier and phased array antennas [14]. The summation of the signals from the individual elements, combined at the corresponding channels of the power amplifier, occurs in the far zone of the antenna. The effective signal power is therefore additionally increased in proportion to the antenna gain. The need to control the directivity of the radiation to cover the given azimuthal sector leads to an additional requirement for the formation of the necessary signal phase distribution in the antenna elements, and for rapid control of the phase distribution to cover the azimuthal sector.

The requirement for spectral fidelity in the transmitted signal is connected with the fact that moving targets in clutter are detected on the basis of their Doppler shift. The presence of spurious noises and discrete spectral components in the transmitted signal complicate the extraction of the target signal from clutter, the level of which, as a rule, is several orders of magnitude higher than the signal level [10].

Both CW and pulsed waveforms are used, and frequency-modulated signals or signals with various forms of coding. The pulse length in OTH radars is from hundreds of microseconds to several milliseconds, and the repetition frequency is from several to tens of hertz [10, 13, 15]. The operating (carrier) frequency must be highly accurate and stable in order to perform coherent processing and Doppler filtering of the received signal.

9.2.3 The Transmitting Complex

The transmitting complex should provide all of the equipment necessary to perform the functions discussed above. In the transmitter complex, information about the signal modulation parameters, the chosen operating frequency, and the required phase distribution in the power amplifier channels, which is obtained from the receiver, is transformed to control commands which are sent to the controlling elements. The receiver also provides signals which are used to synchronize the operation of the transmitting and receiving systems. The formation of a signal with the desired waveform and transmission through the corresponding power amplifier channels is performed in the transmitter complex. Phasing and amplification to the required level is performed in each power amplifier channel, after which the signal is passed to the input of the feed, such that the output of each channel is connected with the corresponding antenna array element. The functional control equipment assesses the operating condition of the various transmitter components, and checks the signal parameters against the corresponding command values.

9.2.4 Waveform Generation System

In one of the main signal generation systems, all of the signals are formed from one or another primary reference signal, obtained from a special highly stable oscillator. The required waveform is generated at a relatively low-power level.

9.2.5 Power Amplification Channel

The functions of each power amplification channel are: giving the signal the necessary phase to produce the desired phase distribution in the phased-array antenna aperture; amplification of the signal to the required level with minimal distortion of the amplitude and phase structure of the signal.

Each power amplification channel may be included in a feedback loop which effects automatic gain and phase control. The automatic control system is designed to compensate for amplitude and phase fluctuations and to maintain the necessary spectral fidelity in the transmitted signal.

9.2.6 Control and Synchronization Equipment

This apparatus provides the connection between the transmitter complex and the rest of the radar system, and forms the necessary control signals in accordance with the established operating scheme, using information from the computer. The control signals are formed with allowance for the operability of the system components, data on which are provided by the control systems.

9.2.7 Functional Control System

This equipment obtains information about the operating condition of the transmitter components and the main transmission parameters.

9.2.8 The Receiving System

The following components are present in one of the possible versions of an OTH radar receiver [10, 18]:

- antenna-feed system; receiver channels for detection, determination of the optimum operating sub-band, and operating channel selection;
- computer system, consisting of special-purpose processors and a general purpose computer, performing primary signal processing, detection, determination of the optimum sub-band and choice of operating channel, on the basis of information from the corresponding receiver channels;
- synchronization system, containing a highly stable reference frequency oscillator and blocks for forming the set of frequencies necessary to synchronize and control the operation of all the receiving equipment;
- communications for the synchronization and control signals, and also technical operating information.

In recent years, thanks to advances in electronic techniques, it has become possible to make use of digital signal processing, which possesses a number of important advantages in comparison with analog processing.

This allows the wide application of adaptive information processing, which improves the basic performance of the radar [9, 17, 18].

We will examine some questions relating to the receiver in more detail.

9.2.9 The Detection Channel

This is the primary radar channel, and performs detection of objects, hidden far beyond the horizon. The structure of the channel, its processing algorithms and its hardware design are determined by the application and characteristics of the radar system. In any version, however, it is possible to identify certain main features intrinsic to the detection channels of OTH radars [6, 9, 10, 17, 18]:

- the operation of the detection channel at several frequencies simultaneously, which reduces the information loss connected with the rather strong dependence of propagation loss on frequency;
- the simultaneous or quasisimultaneous coverage of zones by several partial beams, which leads to a multichannel design for the detection block;

Over-the-horizon radars, as a rule, operate with complex waveforms, with linear frequency modulated (LFM) or phase-shift keyed (PSK) signals. The repetition frequency in a pulsed mode is determined by the range boundaries of the area to be covered. The spectral width of the transmitted signal is limited by the capabilities of the HF band, and also by the need to reduce the interference of transmitters operating in neighboring channels, and is in the range from 100 Hz to tens of kilohertz. In accordance with this spectral width, the range resolution is no better than several kilometers [10].

In OTH radars, signal detection is performed in intense and extremely nonstationary interference and clutter. It is advantageous to use adaptive processing for signal detection in these conditions. In particular, it is necessary to incorporate spatial adaptivity, so as to minimize the effects of signals which are received not in the main beam, but in the sidelobes, and also frequency adaptivity, which allows the system to respond to changes in the spectral characteristics of the clutter (the Doppler shift of the spectral lines and changes in their width due to propagation effects), thus maximizing its cancelation [9, 17, 18].

The theoretically optimum processors are usually extremely complex and may not be realized, due to the large amount of computer power required. As a rule, quasioptimal systems are used, in which the processing is divided into a number of separate stages, thus greatly simplifying the processing. A consequence of this simplification, however, is an unavoidable drop in the effectiveness of the processing, and a corresponding reduction in target detection capability. It is therefore necessary to observe the appropriate safeguards when designing the processor stages, and to make intelligent compromises. The correct solution to the problem depends on the conditions in which the radar must operate, and may be different for different concrete cases.

9.2.10 Spatial Processing Systems

Adaptive formation of the receiver antenna pattern in the presence of localized interference sources is one of the most important methods for increasing the signal-to-interference ratio in the detection channel. Spatial processing consists of weighted summation of the signals received coherently through the various antenna system channels [9]. The signal y_j at the output of the spatial processing system, corresponding to the signal received from the jth azimuth direction, is determined by the scalar product of the vectors $\mathbf{X}$ and $\mathbf{W}_j$:

$$y_j = \sum_{n=1}^{N} x_n w_{nj} = \mathbf{X}^T \mathbf{W}_j = \mathbf{W}_j^T \mathbf{X} \tag{9.1}$$

where n is the number of the receiver channel; $\mathbf{X}$ is the column vector of the samples, taken from the outputs of the receiver at the time of processing; $\mathbf{W}$ is the column vector of interchannel weights; and T indicates transposition.

The vector of weights $\mathbf{W}_{j\,\mathrm{opt}}$, maximizing the signal-to-interference ratio for signals arriving from the jth azimuth direction, is determined for a Wiener filter through the equation:

$$\mathbf{W}_{j\,\mathrm{opt}} = \mathbf{R}^{-1}\mathbf{g}_j^*$$

where $\mathbf{R}$ is the interchannel covariance matrix of the interference samples, taken from the output of the receiver channels at an arbitrary moment; $\mathbf{g}_j^*$ is the complex conjugate of the vector of antenna gain coefficients for the receiver channels in the jth azimuth direction.

We note that when forming an adaptive antenna array with N elements, inverting the covariance matrix $\mathbf{R}$ requires N^3 arithmetic operations. Therefore, for real-time operation with large N, extremely fast processors must be used.

An effective means of simplifying the processing is to divide the antenna into subarrays with some number of elements L, each of the subarrays forming a nonadaptive beam [17,18]. To form an adaptive beam, the subarrays are treated as individual elements, so that the number of adaptive processing channels, and the dimensions of the matrix $\mathbf{R}$, are reduced by a factor of L. It should be noted, however, that the number of interference sources which may be canceled by the system is also reduced by the factor of L. According to data in the literature [17,18], the use of adaptive spatial processing increases the signal-to-interference ratio by an average of 10–15 dB.

9.2.11 Spectral-Time Processing

The signal at the input of the spectral-time processor is an additive mixture of the desired signal, clutter and interference. The characteristics of the useful signal are determined by the features of the target. The characteristics of clutter and interference, along with the question of detecting fluctuating signals with unknown parameters in nonstationary interference, were treated in Chs. 5, 7, and 8.

When detecting a signal with unknown parameters, the receiver should be split into multiple channels in both time and frequency, effecting the detection algorithm for each resolution element within a given search area. When detecting objects such as aircraft, for which the reflected signals exhibit a narrow fluctuation spectrum (significantly narrower than the repetition frequency), the practical realization of the time-frequency processor is significantly simplified. Due to the narrow signal bandwidth, its spectrum is concentrated within a single detection channel (when the pulse-to-pulse processor is designed as a spectrum analyzer). However, due to the unknown Doppler of the useful signal, several channels should be included, covering the interval of Doppler shifts from zero to the repetition frequency.

With narrow-band signals, there arises the problem of "blind speeds," which are those speeds for which the Doppler shift is a multiple of the pulse repetition frequency. There are well-known methods for solving this problem, one example being "wobbulation" of the pulse repetition frequency.

9.2.12 Determining the Optimum Sub-Band of Operating Frequencies

In order for an OTH radar to operate effectively, it is important to obtain real-time information on the characteristics of the propagation path, along with data on the usage of the frequency band. It is also important to match the radar parameters to the parameters of the external medium [10, 18].

One of the possible methods for performing this adjustment of the radar parameters (such as the operating frequency) to match the propagation characteristics (amplitude-frequency and range-frequency), is to include in the radar a block devoted to determining the optimum sub-band of operating frequencies [18]. The goal of this block should be to choose the sub-band, in which the propagation losses are minimized, in order to optimize the operation of the detection channel.

The primary information which is used for this function is that derived from the amplitude of the backscattered signals and their time delays as functions of frequency, which are determined by the amplitude-frequency and range-frequency characteristics of the propagation medium. In order to obtain this data, the frequency selection system must continuously scan a wide band of frequencies. This block may be realized as an independent radar contained within the main radar, with its own oscillators and receivers. The power amplifier, transmitter complex, and transmitting and receiving antennas may be combined with those of the detection channel [18]. This scanning system should transmit at frequencies different from the main operating frequency, during the pauses between the primary pulses.

The optimum operating frequency is chosen through analysis of the signals' amplitude-frequency and range-frequency responses. The most important signals for this purpose are backscattered signals and round-the-world signals.

9.2.13 Operating Channel Selection

In addition to selecting the optimum sub-band, it is necessary to analyze this entire range of frequencies to determine its usage by other radio systems. As was noted in [10, 18], the object is to select the precise operating frequency and bandwidth such that the interference level is minimized, along with effects on other radio systems. The problem is to choose the particular nominal operating frequencies for the detection channel, within the band of optimal frequencies. The optimum frequency is chosen so as to minimize the interference level, taking account of the detection channel bandwidth. This equipment consists of a receiver which analyzes the interference level as a function of frequency.

9.2.14 Computer

The computer should be very fast, and have large amounts of program and data memory, inasmuch as it is responsible for performing the secondary processing of data from the radar's main receiver channels, and also tests, controls and records the operating parameters. Other components which are found in OTH receiver systems do not have specific features, and therefore are not discussed.

9.3 THE WARF OVER-THE-HORIZON RADAR

The WARF (Wide Aperture Research Facility) system [18] is constructed as a polygon in the state of California, and is intended for the study of OTH radar design methods and the associated components. This system may be used to detect aircraft and ships, to observe the state of the sea surface, and to study the ionosphere [18].

9.3.1 Features of the WARF System

One of the main features of the system is its gigantic receiving array, which is about 2.5 km long. The antenna is formed of two rows of 256 asymmetric vertical monopoles, each about 5.5 m long, spaced uniformly. The rows of monopoles are 4.7 m apart. The antenna is divided into eight sections.

The elements are connected to the central processor with cables and switching devices. The antenna array may be electronically steered $\pm 32°$ in azimuth in both the east and west directions, and may be tuned over the band 6–30 MHz, with an azimuthal beamwidth of 0.5° in the middle of this band (15 MHz). The pattern may be trimmed an additional 0.25°. The gain of the receiving antenna is about 30 dB. This radar is distinguished by its fine azimuth resolution (0.5°) and range resolution (1.5 km). The transmitting antenna has a beamwidth of 6°.

The target signal is extracted from interference and clutter using correlation and filter processing, along with Doppler processing.

Adaptive beam-forming techniques are used to cancel interference received through the sidelobes.

9.3.2 Block Diagram of the Radar

The radar makes use of CW LFM transmission (Fig. 9.1). The transmitting system includes those devices usually found in FM systems. Internal synchronization is maintained by transmitting signals between points in

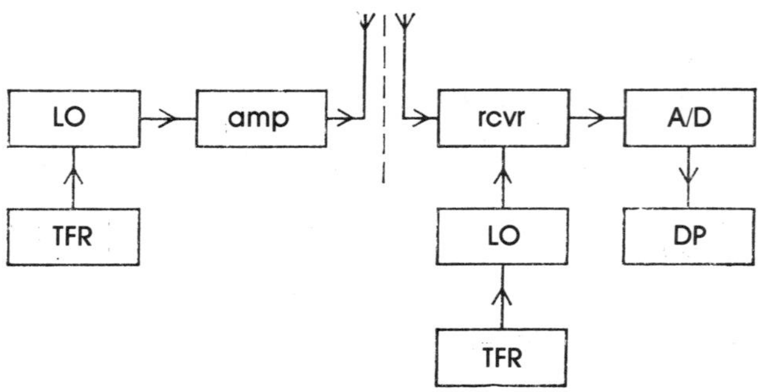

Fig. 9.1 Simplified block diagram of the WARF (Wide Aperture Research Facility) radar. LO = local oscillator, amp = power amplifier, TFR = time-frequency reference, DP = digital processor.

the receiver and transmitter through underground cables.

The receiving antenna array, which has a total length of 2.5 km, is divided into eight 32-element subarrays, each of which is 320 m long. Each of the subarrays has its own receiver, so that the receiving system has eight channels, balanced in phase and gain.

One of these receiver channels is shown in the diagram in Fig. 9.1. The receiver is stabilized in phase (to 0.5°) and gain (to 0.5 dB), and operates in the band 3–30 MHz. Automatic gain control systems are employed in the receivers, providing control within 100 dB. The internal noise level does not exceed 3 dB.

The target signal taken from the output of the receiver is passed to an analog-to-digital converter (11 bits plus sign). A minicomputer is used to perform the spectral analysis of this signal. The transmitted signal is kept coherent from period to period, and the use of coherent integration over several periods in the spectral analysis allows the system to determine the Doppler frequency for each range cell.

9.3.3 Methods Used to Improve the Anti-Interference Performance

Experience has shown that in order to cover a given zone, the operating frequency must be changed five or more times a day. Due to this heavy use of the HF band, it is necessary to employ special methods, both to improve the anti-interference performance of the radar, and to reduce the interfering effects of the radar on other radio systems. A brief description of the methods used in the WARF system are given below.

Receiver channel selection

Seeking the receiver channel which is "free" from interference is performed in a frequency scanning system. The search is usually performed in a band of 1-5 MHz. At the middle frequency of 14 MHz, it is possible to find a "clean" channel approximately 10-100 kHz wide. As the operating frequency is lowered, the width of the free channel decreases. At 6-10 MHz, especially at night, it is not possible to find a free channel 100 kHz wide. It is therefore necessary to use interference cancelation techniques.

Time blanking

Interference signals which are discrete in either frequency or time may be eliminated in the receiver using various "notching" methods. One of the methods used in the WARF system makes it possible to "notch" interference which occupies only a small part of the scanned frequency band by blanking a portion of the received signal.

Removing a bias in the Doppler filters

Coherent processing of radar signals is performed "from scan to scan", i.e., during one period. Interference signals which are coherent during a scan period are distributed in range and have a constant Doppler frequency. This interference introduces a constant Doppler bias in the Doppler spectrum of the target and in the Rayleigh distribution of the interference and noises. The existence of interference is detected by calculating the ratios of the average bias obtained over all range cells to the shift in each individual Doppler filter, and comparing it with a calculated threshold. A special algorithm is employed for this operation.

This algorithm for eliminating the bias and suppressing the interference is called the BRISA *(Bias Removal Interference Suppression Algorithm)*. Powerful interference, appearing at some Doppler frequency, may be eliminated almost entirely, without reducing the detection threshold at other Doppler frequencies. When using this algorithm in the WARF system, it was possible to suppress the interference 30 dB, this level being determined by the coherency of the interference source.

Canceling interference received through the sidelobes

Adaptive beam-forming algorithms were studied in the WARF system, for reducing the level of interference received through the sidelobes. The signals received in each of the eight receiver channels passed through corresponding matched tapped delay lines. The resulting signal, obtained from all the channels, was passed through an A/D converter to the computer, where it was processed.

The coefficients determining the antenna pattern (the "shaper coefficients") were calculated by several methods: the least squares method; an optimization method making use of the inverse of the signal covariance matrix, with the criterion being the maximum signal-to-interference ratio; and general sidelobe cancelation techniques. The characteristics of these methods were given in Ch. 6. The primary data on the WARF radar are presented in Table 9.1.

In experiments with adaptive beam-forming, the active interference signal was attenuated on the average by 15 dB. Adaptive beam-forming, in addition to reducing the level of interference penetrating the sidelobes, also reduced the clutter from local surface features. The most typical values for the reduction in the strength of interference from point sources are in the range 10-15 dB; in some cases the suppression reached 20dB.

In the opinion of Washburn *et al.* [18], the methods employed in the WARF system greatly increased the signal-to-interference and signal-to-noise ratios.

Table 9.1

Characteristic	*Detection Channel*	*Sub-band Determination*
	TRANSMITTER	
Power	20 kW, average	10 kW, average
Type of antenna	18-element array in eastern direction, 18-element array of log-periodic elements in western direction Beamwidth: 6° at 15 MHz	Revolving horizontal log-periodic antenna
Frequency band	9–26 MHz, east 6–30 MHz, west	—
Antenna gain	20 dB at 15 MHz	13 dB
Azimuth scan	±32° (in both east and west directions)	
	RECEIVER	
Type of antenna	Linear array of 256 vertical monopoles with a total length of 2.5 km. Beam width = 0.5° at 15 MHz	—
Antenna gain	27 dB	—
Azimuth scan	±32° n 0.25° increments, in both east and west directions	—

REFERENCES

1. *Zagorizontnoy radiolokator sistemy rannevo preduprezhdeniya (Early warning over-the-horizon radar systems)*. Elektronika, vol. 50 (1977), no. 4, pp. 4–5.
2. Barrik, D.E. and Snider, J.B. *The statistics of HF sea-echo doppler spectra.* IEEE Trans., vol. AP-25 (1977), no. 1, pp. 19–28.
3. Crombie, D.D. *Doppler spectrum of sea-echo at 13.56 mc/s.* Nature, vol. 175 (1955), no. 4459, pp. 681–682.
4. *Cyprus radars.* Aviation Week and Space Technology. vol. 101 (1974), no. 5, p. 11.
5. Dong, J.H. *World's largest log-periodic antenna.* J. Struct. Div. Proc. of the Amer. Soc. of Civil Engineers, vol. 97 (1971), no. 9, pp. 2371–2381.
6. Fenster, W. *The applications, design and performance of the over-the-horizon radar.* (Radar-77. Internat. Conf., London, 1977), pp. 36–40.
7. Frank J. and Ruze, J. *Beam steering increments for a phased array.* IEEE Trans., vol. AP-15 (1967), no. 6, pp. 820–821.
8. *General Electric to develop over-the-horizon radar.* Interavia. Air Letter, 1975, no. 8229, p. 7–8.
9. Griffiths, L.J. *Time-domain adaptive beam-forming of HF back-scatter radar signals.* IEEE Trans., vol. AP-24 (1976), no. 5, pp. 707–720.
10. Headrik, J.M. and Skolnik, M.I. *Over-the-horizon radar in the HF Band.* Proc. IEEE, (1974), no. 6, pp. 664–672.
11. Klass, P.J. *HF radar detects Soviet ICBMs.* Aviation Week and Space Technology, vol. 95 (1971), no. 23, pp. 38–40.
12. Mason, J.E. *Backscatter radar on 2 coasts to detect planes over horizon.* Electronic Design, vol. 20 (1972), no. 14, pp. 30–32.
13. *OTH-B radar installation begins.* Interavia. Air Letter, 1976, no. 8602, p. 4.
14. Pitt, W.A. *A 7—25 MHz high-power ten-element electronically scanned array for ionospheric backscatter measurement.* IEEE Trans., vol. AP-19 (1971), no. 5, pp. 584–593.
15. *Project "Tepee" may cost less, outperform BMEWS.* Missiles and Rockets, 1959, vol. 5, August 17, p. 45.

16. Wait, J.R. *Theory of HF ground wave backscatter from sea waves* J. Geophys. Res., vol. 71, (1966), pp. 4832–4839.
17. Washburn, T.W. and Sweeney, L.E. *An on line adaptive beamforming capability for HF backscatter radar.* IEEE Trans., vol. AP-24 (1976), no. 5, pp. 721–731.
18. Washburn, T.W., Sweeney, L.E., Barnum, J.R., and Zavoli, W.B. *Development of HF skywave radar for remote sensing applications.* Special Topics in HF Propagation/AGARD Conf. Proc. no. 263, 28.05 1.06.1979. Lisbon, 32/1 32/17, New York, 1979.

Index

Absorption,
coefficient, 34, 70
of high frequency waves along path, equation, 33
Accuracy,
influence of transmitted waveform, 171–173
of signal parameter estimation, 227–228
Active interference, 103, 112–113, 118–119, 195–196, 214–215, 244–246
net (combined), 122–123, 136–137
Adaptive,
antenna systems, 144–153, 153–164
band-rejection filters, 169–170
beam-forming system, 235–236, 276
beam steering, 140
filtration, 169–170
frequency selection, 136
effectiveness, 136
spacial filtering, 140–143
Adiabatic,
approximation, 37
invariants, 38, 46
Aircraft, radar cross section, 98–99
Algorithms,
for adaptive spatial filtering, 153–164
detection, 177–178, 187, 208, 210, 236, 244, 254, 274
track processing, 229
Altitude factor, 10
Ambiguity function, 171
Antenna,
area, 21
feed systems, 264–266, 268
gain, 20, 22
systems, adaptive, 144–152, 153–164
using correlation loops, 144–152
Anti-interference,
devices, 164–168, 274–276
performance, influence of transmitted waveform, 171–173
techniques, 135–140
Anti-waveguide propagation, 47
Approximation, ray optics, 51–56
Ascending rockets, radar cross section, 93–97
Aspect sensitivity, 90
Atmospheric interference, 104–108, 122, 124
Attenuation estimation, using semiphenomenological model, 68–69
Autocorrelation function, 171

Backscatter cross section, 90
Band-rejection filters, 169
Bandwidth, noise level dependence on, 124–125
Bayesian decision rules, 181
Beam,
diffusion, 47
equations, 35
steering, adaptive, 140
Bias removal interference suppression algorithm (BRISA), 274
Bistatic,
reflectivity pattern, 88, 98
system, 1, 88
Blind speeds, 271
Bohr,
approximation, 42
expansion, 42
BRISA, 274

Cancelers, 144–153, 169–170
Canceling interference, 274
Caustic zone, 10

Channel,
  discrimination function, 208
  selection, 136–140, 272
Characteristic,
  equation, 35
  function, 35
Clutter, 125–130, 177
  cross section, 128
  hopped signals, 126
  model, 200–205
  parameter estimation, 205–217
  reflectivity, 126–127
  signal detection in, 201–205, 221–227, 248–250
  types, 125
Combined (net) interference, 122–125, 136–140
Control and synchronization equipment, 268–269
Correlation loops, in antenna systems, 144–153
Cosmic noise, 104, 109–112
Cramer-Rao lower bound, 207–227
Critical elevation angle, 10

Debye method, 35
Decision rules, 181, 244, 248, 254 (*see also* Minimax decision rule and Maximum likelihood decision rule)
Detection,
  algorithms, 177–178
  channel, 269–270
  characteristics, 17
  probability of, (*see* Probability of detection)
  of signals, 181–186, 217–228, 254–260
    in background of active interference, 195–200, 248–250
    in background of stationary interference, 186–195, 236–243
    in clutter, 200–205
Dielectric permittivity of medium, equation, 28, 30–31
Diffusion coefficient, 69, 74
Doppler filters, removing bias, 274
Ducts,
  excitation of, 47
  ionosphere, 31–35
    wave capture in, 35
    wave propagation in, 47
  qualitative characteristics, 30–31
  seepage through walls, 47

Effective receiving antenna area, 21
Electromagnetic field, total loss in, equation, 10
Energy characteristics of ionospheric propagation, 42
Excitation of ducts, 47

False alarm, probability of, (*see* Probability of false alarm)
Faraday effect, and polarization losses, 23
Feed losses, 22
Filtration, adaptive, 169
Fluctuating signals, (*see* Rapidly fluctuating signal, detection and Slowly fluctuating signal, detection)
Free channels, 116
  probability of, 136
Frequency,
  notation, 57
  selection, 272
    adaptive, 136
    effectiveness, 136
Functional control system, 268

General radar equation, 10–16
Geometric optics, method of, 35

Hamming filter, 171
Helmholtz equation, 28, 35
High frequency,
  anti-interference techniques, 135–173
  interference, 103–130
Hopped signals, 125, 126, ~~129~~, 129–130

Incoherent integration, losses due to, 25
Industrial interference, 110–112
Intensity distribution by mode spectrum, equation, 69–73
Interference,
  active, 103, 119, 120, 136–140, 177, 195–200
  atmospheric, 104–108, 122, 124
  canceling, 274
  channel selection for minimum, 136
  classification, 179
  combined (net), 122–123
  free channels, 116
    probability of, 136
  high frequency radar, 103–130
  industrial, 110–112
  level, prediction, 120
  from lightning, 104–108
  models, 179–181, 186
  net (combined) active, 122–125

out-of-band radio transmission, 120–122
parameter detection in, 187–205, 217–219, 221–227
passive, 103
protection device, 164–168
from radio stations, 112–120, 124, 136, 195–200
characteristics, 112–113
distribution, 114–118
sources, 113–114
from receiver imperfections, 130
spurious, 120–122
suppression coefficient, 144
Interference suppression coefficient, 144
Internal receiver noises, 104
Ionospheric,
ducts, 31-35
wave capture in, 35, 47
wave propagation in, 47
electron density, radar cross section of magnetically
oriented disturbances in, 89–91
irregularities and inhomogeneities, 125
in beam diffusion, 47
in reflected signals, 129
in refraction, 47
propagation, energy characteristics, 42

Jamming, 171

Least mean square error criterion (*see* LMSE)
LFM, 269
Lightning, causing interference, 104–108
Linear frequency modulated (LFM) signals, 269
LMSE, 140, 153
LOF, 57
Log-periodic antenna, 264
Long range propagation of high frequency radio waves, theoretical study methods, 28–50
Loss coefficient, 22, 23
Losses,
due to mismatched passband, 25
due to use of incoherent integration, 25
feed, 22
in free space, 9
polarization, 23
signal processing, 24
system, 22–25

Lowest observed frequency (LOF), 57
LOF-RTW and MUF relationship, 57

MADRE, over-the-horizon radar, 88, 135, 171, 264
Maximum likelihood decision rule, 181, 198, 203, 207, 208, 221, 229, 236, 244, 248, 250, 254
Maximum observed frequency, (*see* MOF)
Maximum usable frequency, (*see* MUF)
Maxwell's equations, 28–30
Measurement of signal parameters, 208
Meteor trails, 129
radar cross section, 91–93
Method,
of adiabatic invariance, 35
of geometric optics, 35
of multiple-scattering, 42
of small perturbations, 42
Minimax decision rule, 181, 187, 195, 203, 207, 221, 236, 244, 248, 254
Models,
of clutter, 200–201
of radio frequency interference, 195
signal, 205–206
of signals and interference, 179–181, 186, 236
MOF, 57
MOF-RTW AND MUF relationship, 57
Monostatic system, 1, 88–89
Monte Carlo method, 221
MUF, 57
and LOF-RTW relationship, 57
and MOF-RTW relationship, 57
Multiple-scatter method, 45
Multiple target detection problem, 187, 193, 198, 201, 203, 236, 244
Multi-scatter approximation, 42

Narrow-band interference,
protection against, 164–168
suppressing, 168–170
Net (combined) interference, 122–125
Neyman-Pearson rule, 181, 254
Neyman series, 42
Noise,
bandwidth, 19
level, 103–104 (*see also* Interference)
dependence on receiver bandwidth and receiving antenna, 124–125
from receiver imperfections, 130 (*see also* Interference)

'Notching' methods, 274

Obliquely propagating signals, spectral characteristics, 129
Operating,
  band, 118
  frequency, determination, 272
Optical scattering, 88
OTH, (*see* Over-the-horizon radar)
Out-of-band transmissions, interference from 120–122
Over-the-horizon radar (OTH), 57
  characteristics, 1
  design principles, 264–273
  MADRE, 88, 135, 171, 264
  object scattering characterisitcs, 88–89
  problems in, 1
WARF, 1, 4, 135, 153, 264, 273

Parameter signal, 205–217
  accuracy, 227–228
  in active interference, 214–216
  estimation, 177–232
  in interference with unknown angular distribution, 235–260
  measurement, 208
Passive interference, 103
Path loss, calculation methods, 27–80
Permittivity of medium, 20, 30–31
Phase-shift keyed (PSK) signals, 269
Pilot signal, 144, 153
Polarization losses, 23
Potential wells, 31
Power, 10
  amplification channel, 267–268
  density, equation, 10
Preliminary spatial filtering, 144
Probability of,
  detection, 17–19, 25, 177–178, 181–186, 187–195, 195–198, 217–219, 221–227, 254–260
  false alarm, 17–19, 177–178, 181–186, 187–198, 201–202, 203–205, 217–219, 221–227, 229–232, 236–243
  free channel, 136–140
Processing,
  spatial, 270–271
  spectral-time, 271
Propagation of radio waves, (*see* Wave propagation)
Protection, against narrow-band interference, 164–168

PSK, 269

Quasiwhite noise, 164

Radar,
  cross section (RCS), 88–89
    of aircraft, 98–99
    of ascending rockets, 93–97
    of magnetically oriented disturbances, 89–91
    of meteor trails, 91–93
  detection, 177–186, 186–195
  frequency response, optimal form, 25
  loss in free space, 9
  range equation, 9
  range equation, in determination of reflected signal level, 128
  systems, backward scattering, 1
    forward scattering, 1
    losses, 22–25
    performance, 18–21
    problems in over-the-horizon, 4–6
Radiant intensity, 42
Radiated energy, 19
Radio stations,
  characteristics, 112–113, 136
  distribution, 113–114, 115–116, 116–118
  free channels, probability of, 116–118
  interference, 112–119, 124–125, 195–200
  sources, 113–114
Rayleigh scattering, 88
Rapidly fluctuating signal, detection, 193–195, 198–200, 203–204, 209, 213, 216, 236, 244, 248, 250, 254
Ray optics approximation, 51–56
RCS, (*see* Radar cross section)
Receiver,
  bandwidth, 124–125
  channel, 274
  noises, internal, 104
Receiving,
  antenna, effect on noise level, 124–125, 130
  system, 268–269, 274
Reflected signal level, determination, 128
Reflectivity pattern, 88, 98
Resonant scattering, 88
Rockets, radar cross section, 93–97
Round-the-world signals (RTW)
  clutter, 130
  frequency, dependencies, 57
    –energy characteristics, modeled, 76

responses, 57
LOF-RTW and MUF relationship, 57
MOF-RTW and MUF relationship, 57
spectral characteristics, 57–68
RTW, (*see* Round-the-world signals)

Scattered wave, average intensity, 42
Scattering,
optical, 88
Rayleigh, 88
resonant, 88
Second order method, 35
Seepage of a field through a duct wall, 47
Semiphenomenological model,
for attenuation estimation, 68–80
design, 68–69
functional dependencies, 74
Shift coefficients, 74
Signal,
detection, 177–232, 154–260
in active interference, 195–200, 221–227, 236–244
algorithms, 177–178
in clutter, 200–205, 221–227
in interference with unknown angular distribution, 235–260
optimal, 181–186
of rapidly fluctuating signals, 193–195, 198–200
of slowly fluctuating signals, 187–198
in stationary interference, 217–219, 236–243
and interference model, 236
models, 179–181, 186, 205–206
parameter, accuracy, 227–228
estimation, 205–217
measurement, 208–209
of rapidly fluctuating signal, 210–213, 216
of slowly fluctuating signal, 210–213, 216
in stationary interference, 209
with unknown angular distribution, 250–254
processing losses, 24
Signal-to-interference ratio, 16–18, 25, 144, 164, 181, 187, 195, 203, 210, 216, 217, 221, 227, 236, 254, 270
Signal-to-noise ratio, 153–171
Skip propagation, 47
Sliding propagation, 47
Slowly fluctuating signal, detection, 187–193, 195–198, 201–203, 209, 210–213, 216, 236, 244, 248, 250, 254
Small perturbations, method of, 42
Spatial,
field energy, characteristics, 51–56
filtering, adaptive, 140–143
preliminary, 144
processing, 270–271
Spectral,
density, 19
-time processing, 271
Specular points, 88
Spherical spreading factor, 10
Spurious,
interference, 120–122
radiation, 121
Standard radar equation, 9–10
Suppressing narrow-band interference, 168–170
Synchronization and control equipment, 268

Target,
cross section, 87–99
reflectivity pattern, 88, 98
Time,
blanking, 274
division, defined, 179
Track processing, 229–232
Transfer equations, 35–69
Transmitted waveform, influence on anti-interference and accuracy of radar, 171–173
Transmitter, 266–267
Transmitting complex, 267
Transport equation, 42
Two-dimensional autocorrelation function, 171

WARF, (*see* Wide aperture research facility)
Wave capture in ionospheric ducts, 35
Wave equation, 28–30
Waveform generation system, 264
Wave propagation, 47
anti-waveguide mechanism, 47
characteristics, 57, 129
skip, 47
sliding mechanism, 47

through ionospheric ducts, 47
Weighting coefficients, 140–143
limiting, 144, 153
Whitening filter, 164
White noise, 177
Wide aperture research facility (WARF) over-the-horizon radar, 1, 4, 135, 153, 264, 273–277
features, 273
Wideband active interference, 177
Wobbulation, 271